Mathematical Sciences Research Institute
Publications

9

Editors

S.S. Chern
I. Kaplansky
C.C. Moore
I.M. Singer

Calvin C. Moore   Claude Schochet

Global Analysis on Foliated Spaces

With 16 Illustrations

Springer-Verlag
New York Berlin Heidelberg London Paris Tokyo

Calvin C. Moore
Department of Mathematics
University of California
Berkeley, CA 94720, USA

Claude Schochet
Department of Mathematics
Wayne State University
Detroit, MI 48202, USA

Mathematical Sciences Research Institute
1000 Centennial Drive
Berkeley, CA 94720, USA

AMS Classification: 19K56, 28D15, 46L05, 46L50, 57R30, 58F11, 58G10, 58G11, 58G15

Library of Congress Cataloging-in-Publication Data
Moore, C. C. (Calvin C.)
  Global analysis on foliated spaces.
  (Mathematical Sciences Research Institute
publications ; 9)
  Bibliography: p.
  1. Global analysis.   2. Foliations (Mathematics)
I. Schochet, Claude.   II. Title.
III. Series.
QA614.M65   1988   514'.74   87-28631

© 1988 by Springer-Verlag New York Inc.
All rights reserved. This work may not be translated or copied in whole or in part without the written permission of the publisher (Springer-Verlag, 175 Fifth Avenue, New York, NY 10010, USA), except for brief excerpts in connection with reviews or scholarly analysis. Use in connection with any form of information storage and retrieval, electronic adaptation, computer software, or by similar or dissimilar methodology now known or hereafter developed is forbidden.
The use of general descriptive names, trade names, trademarks, etc. in this publication, even if the former are not especially identified, is not to be taken as a sign that such names, as understood by the Trade Marks and Merchandise Marks Act, may accordingly be used freely by anyone.

Camera-ready copy provided by the authors.
Printed and bound by R.R. Donnelley & Sons, Harrisonburg, Virginia.
Printed in the United States of America.

9 8 7 6 5 4 3 2 1

ISBN 0-387-96664-1 Springer-Verlag New York Berlin Heidelberg
ISBN 3-540-96664-1 Springer-Verlag Berlin Heidelberg New York

CB-SłK
SCIMON

PREFACE

This book grew out of lectures and the lecture notes generated therefrom by the first named author at UC Berkeley in 1980 and by the second named author at UCLA, also in 1980. We were motivated to develop these notes more fully by the urgings of our colleagues and friends and by the desire to make the general subject and the work of Alain Connes in particular more readily accessible to the mathematical public. The book develops a variety of aspects of analysis and geometry on foliated spaces which should be useful in many contexts. These strands are then brought together to provide a context and to expose Connes' index theorem for foliated spaces [Co3], a theorem which asserts the equality of the analytic and the topological index (two real numbers) which are associated to a tangentially elliptic operator. The exposition, we believe, serves an additional purpose of preparing the way towards the more general index theorem of Connes and Skandalis [CS]. This index theorem describes the abstract index class in $K_0(C_r^*(G(M)))$, the index group of the C^*-algebra of the foliated space, and is necessarily substantially more abstract, while the tools used here are relatively elementary and straightforward, and are based on the heat equation method.

We must thank several people who have aided us in the preparation of this book. The origins of this book are embedded in lectures and seminars at Berkeley and UCLA (respectively) and we wish to acknowledge the patience and assistance of our colleagues there, particularly Bill Arveson, Ed Effros, Marc Rieffel and Masamichi Takesaki. More recently, we have benefitted from conversations and help from Ron Douglas, Peter Gilkey, Jane Hawkins, Steve Hurder, Jerry Kaminker, John Roe, Jon Rosenberg, Bert Schreiber, George Skandalis, Michael Taylor, and Bob Zimmer.

We owe a profound debt to Alain Connes, whose work on the index theorem aroused our own interest in the subject. This work would not exist had we not been so stimulated by his results to try to understand them better.

Table of Contents

INTRODUCTION

Global analysis has as its primary focus the interplay between the local analysis and the global geometry and topology of a manifold. This is seen classically in the Gauss-Bonnet theorem and its generalizations, which culminate in the Atiyah-Singer Index Theorem [AS1] which places constraints on the solutions of elliptic systems of partial differential equations in terms of the Fredholm index of the associated elliptic operator and characteristic differential forms which are related to global topological properties of the manifold.

The Atiyah-Singer Index Theorem has been generalized in several directions, notably by Atiyah-Singer to an index theorem for families [AS4]. The typical setting here is given by a family of elliptic operators $P = \{P_b\}$ on the total space of a fibre bundle $F \longrightarrow M \longrightarrow B$, where P_b is defined on the Hilbert space $L^2(p^{-1}(b), dvol(F))$. In this case there is an abstract index class $ind(P) \in K^0(B)$. Once the problem is properly formulated it turns out that no further deep analytic information is needed in order to identify the class. These theorems and their equivariant counterparts have been enormously useful in topology, geometry, physics, and in representation theory.

A smooth manifold M^n with an integrable p-dimensional subbundle F of its tangent bundle TM may be partitioned into p-dimensional manifolds called *leaves* such that the restriction of F to the leaf is just the tangent bundle of the leaf. This structure is called a *foliation* of M. Locally a foliation has the form $\mathbb{R}^p \times N$, with leaves of the form $\mathbb{R}^p \times \{n\}$. Locally, then, a foliation is a fibre bundle. However the same leaf may pass through a given coordinate patch infinitely often. So globally the situation is much more complicated.

Foliations arise in the study of flows and dynamics, in group representations, automorphic forms, groups acting on spaces (continuously or even measurably), and in situations not easily modeled in classical algebraic topology. For instance, a diffeomorphism acting ergodically on a manifold M yields a 1-dimensional foliation on $M \times_{\mathbb{Z}} \mathbb{R}$ with each leaf dense. The space of leaves of a foliation in these cases is not decent topologically (every point is dense in the example above) or even measure-theoretically (the space may not be a standard Borel space). Foliations carry interesting differential operators, such as signature operators along the leaves. Following the Atiyah-Singer pattern, one might hope that there would be an index class of the type

$$\text{ind}(P) = \text{Average } \text{ind}(P_x).$$

There are two difficulties. First of all, leaves of compact foliations need not be compact, so an elliptic operator on a leaf may well have infinite dimensional kernel or cokernel, and thus "$\text{ind}(P_x)$" makes no sense. This problem aside, the fact that the space of leaves may not be even a standard Borel space suggests strongly that there is no way to average over it. There was thus no analytic index to try to compute for foliations.

Alain Connes saw his way through these difficulties. He realized that the "space of leaves" of a foliation should be a non-commutative space -- that is, a C^*-algebra $C_r^*(G(M))$. In the case of a foliated fibre bundle this algebra is stably isomorphic to the algebra of continuous functions on the base space. This suggests $K_0(C_r^*(G(M)))$ as a home for an abstract index $\text{ind}(P)$ for tangentially elliptic operators. [Subsequently Connes and Skandalis proved [CS] an abstract index theorem which identifies this class.]

Next Connes realized that in the fibre bundle case there is an invariant transverse measure ν which corresponds to the volume measure on B. So we must assume given some invariant transverse measure in general. [These may not exist. If one exists it may not be unique up to scale.] An invariant transverse measure ν gives rise to a trace ϕ_ν on $C_r^*(G(M))$ and thus a real number

$$\text{ind}_\nu(P) = \phi_\nu(\text{ind}(P)) \in \mathbb{R}$$

which Connes declared to be the *analytic index*. [Actually we are cheating here; the most basic definition of the analytic index is in terms of locally traceable operators as we shall explain below and in

Chapters I and IV.] With an analytic index to compute, Connes computed it.

Connes Index Theorem. Let M be a compact smooth manifold with an oriented foliation and let ν be an invariant transverse measure with associated Ruelle-Sullivan current C_ν. Let P be a tangentially elliptic pseudodifferential operator. Then

$$\text{ind}_\nu(P) = <\text{ch}(P)\text{Td}(M), [C_\nu]>.$$

Connes' theorem is very satisfying and its proof involves a tour of many areas of modern mathematics. The authors decided to expose this theorem and to use it as a centerpiece to discuss this region of mathematics. Along the way we realized that the setting of *foliated spaces* (local picture $\mathbb{R}^p \times N$ with N not necessarily Euclidean) was at once simpler pedagogically and yielded a somewhat more general theorem, since foliated spaces which are not manifolds occur with some frequency.

The local picture of a foliated space is simply a space of the form $\mathbb{R}^p \times N$, where we regard sets of the form $\mathbb{R}^p \times \{n\}$ as leaves and N is a transversal.

N

To such a space is canonically associated a p-plane vector bundle

$F|_{\mathbb{R}^p \times N}$ with $F_{(t,n)} \cong T(\mathbb{R}^p)$. The global picture of a foliated space X is somewhat more complex. We stipulate that X be a separable metrizable space with coordinate patches $U_x \cong \mathbb{R}^p \times N_x$ and continuous change of coordinate maps of the form

$$t' = \varphi(t,n)$$

$$n' = \psi(n) \;.$$

which are smooth along the leaves, in the sense that a set in U_x of the form $\mathbb{R}^p \times n$ is sent to a set of the form $\mathbb{R}^p \times \psi(n)$ by a smooth map. This guarantees that the leaves in each coordinate patch coalesce to form leaves ℓ in X which are smooth p-manifolds. The bundles $F\big|_{U_i}$ coalesce to form a p-plane bundle F over X such that $F\big|_{\ell} \cong T(\ell)$ for each leaf ℓ.

Any foliated manifold is a foliated space. There are interesting examples of foliated spaces which are not foliated manifolds. For instance, a solenoid is a foliated space with leaves of dimension 1 and with N_i homeomorphic to Cantor sets. If M^n is a manifold which is foliated by leaves of dimension p and if N is a transversal of M^n then any subset of N determines a foliated subspace of M simply by taking those leaves of M^n which meet the subset. This includes the laminations of much current interest in low dimensional topology. Finally, X may well be infinite dimensional: take $\Pi_1^\infty S^1$ foliated by lines corresponding to algebraically independent irrational rotations. Then $\{1\} \times \Pi_2^\infty S^1$ is a transversal!

If X is a foliated space then $C_\tau^\infty(X)$ is the ring of continuous functions on X which are smooth in the leaf directions. If $E \xrightarrow{\pi} X$ is a foliated bundle (i.e., E is also foliated, π takes leaves to leaves, and π is smooth on each leaf) then $\Gamma_\tau(E) \equiv \Gamma_\tau(X,E)$ denotes continuous tangentially smooth sections of E. We let $\Omega_\tau^k(X) = \Gamma_\tau(\Lambda^k F^*)$ and define the *tangential cohomology* groups of a foliated space by

$$H_\tau^k(X) = H^k(\Omega_\tau^*(X))$$

where d: $\Omega_\tau^k(X) \longrightarrow \Omega_\tau^{k+1}(X)$ is the analogue of the de Rham differential obtained by differentiating in the leaf directions. Similar (but not the same) groups have been studied by many authors. Tangential cohomology groups are based upon forms which are *continuous* transversely (even if X is a foliated manifold.) It turns out that this small point has some major consequences. The groups may be described as

$$H_\tau^k(X) = H^k(X; \mathcal{Q}_\tau)$$

where $\mathcal{Q}_\tau$ is the sheaf of germs of continuous functions [which are constant along leaves. The tangential cohomology groups are functors from foliated spaces and leaf-preserving tangentially smooth maps to graded commutative ℝ-algebras. They vanish for $k > p$. There is the usual apparatus of long exact sequences, suspension isomorphisms, and a Thom isomorphism for oriented k-plane bundles.

The groups $H_\tau^*(X)$ have a natural topology and are not necessarily Hausdorff; we let $\bar{H}_\tau^k(X) = H_\tau^k(X)/[\bar{0}]$ denote the maximal Hausdorff quotient. For example, if X is the irrational flow on the torus then $H_\tau^1(X)$ has infinite dimension but $\bar{H}_\tau^1(X) \cong ℝ$. The parallel between de Rham theory and tangential cohomology theory extends to the existence of characteristic classes. Given a tangentially smooth vector bundle $E \longrightarrow X$ we construct tangential connections, curvature forms, and Chern classes. This leads to a tangential Chern character, a tangential Todd genus and hence a topological index

$$\iota_D^{top} = \pm\Phi^{-1}ch_\tau(D)Td_\tau(X) \in H_\tau^p(X)$$

where Φ denotes the tangential Thom isomorphism.

Next we recall the construction of the groupoid of a foliated space; the idea is due to Ehresmann, Thom and Reeb and was elaborated upon by Winkelnkemper. If X is a foliated space then there is a natural equivalence relation: x ⁓ y if and only if x and y are on the same leaf. The resulting space $\mathcal{R}(X) \subset X \times X$ is not a well-behaved topological space. The holonomy groupoid G(X) of a foliated space is designed to by-pass this difficulty. It contains holonomy data not given by $\mathcal{R}(X)$; holonomy is essential for diffeomorphism and structural questions about the foliated space. The *holonomy groupoid* G(X) consists of triples (x,y,[α]) where x and y lie on the same leaf ℓ of X. α is a path from x to y in ℓ, and $[\alpha]$ denotes the holonomy class of the path α. The map $G(X) \longrightarrow \mathcal{R}(X)$ is simply $(x,y,[\alpha]) \longrightarrow (x,y)$. The preimages of (x,y) correspond to holonomy classes of maps from x to y. The space $G(X)$ is a (possibly non-Hausdorff) foliated space. If N is a complete transversal (i.e. N is Borel and for each leaf ℓ, $1 \leq \#(N \cap \ell) \leq \aleph_0$) then G_N^N is the subgroupoid of $G(X)$ consisting of triples $(x,y,[\alpha])$ with $x,y \in N$. In a sense which we make quite precise, G_N^N is a good discrete model for $G(X)$.

Next we turn to a study of differential and pseudodifferential operators on X. Suppose that E_0 and E_1 are foliated bundles over X and D: $\Gamma_\tau(E_0) \longrightarrow \Gamma_\tau(E_1)$. D is said to be *tangential* if D restricts to D_ℓ: $\Gamma(E_0|_\ell) \longrightarrow \Gamma(E_1|_\ell)$ for each ℓ, and D is *tangentially elliptic* if each operator D_ℓ is an elliptic operator. If D is a tangential, tangentially elliptic operator then Ker D_ℓ and Ker D_ℓ^* consist of smooth functions on ℓ. However these spaces may well be infinite-dimensional, and hence expressions such as

$$\dim \text{Ker } D_\ell - \dim \text{Ker } D_\ell^*$$

make no sense. However there is some additional structure at our disposal, for Ker D_ℓ and Ker D_ℓ^* are $C^\infty(\ell)$-modules. We shall show that these spaces are for each ℓ locally finite dimensional in a sense that we now describe.

Let Y be a locally compact space endowed with a measure (in the application to index theory $Y = \ell$ is a leaf and the measure is a volume measure) and suppose that T is a positive operator on $L^2(Y,E)$ for some bundle E over Y. Then

$$\text{Trace}(f^{1/2}Tf^{1/2}) = \text{Trace } (T^{1/2}fT^{1/2})$$

for every bounded positive function f. We define a measure μ_T by

$$\text{Trace } (f^{1/2}Tf^{1/2}) = \int_Y f d\mu_T$$

and declare T to be *locally traceable* with local trace μ_T provided that $\mu_T(Y_i) < \infty$ where the Y_i are compact sets with union Y. If $T = \Sigma \ \lambda_i T_i$ with each T_i locally traceable then T is locally traceable with local trace $\mu_T = \Sigma \lambda_i \mu_{T_i}$. We identify a closed subspace V with the orthogonal projection onto it and say that the subspace is *locally finite dimensional* if the projection is locally traceable. Any closed subspace of C^∞-functions is locally finite dimensional.

If Y is a C^∞ manifold and D is an elliptic pseudodifferential operator on Y then DD^* and D^*D are locally traceable so Ker D and Ker D^* are locally finite dimensional. The *local index* of D is defined to be

$$\ell_D = \mu_{\ker D} - \mu_{\text{Ker } D^*} \ .$$

If Y is a compact manifold then $\int_Y \ell_D = \text{ind}(D)$, the classical Fredholm index.

The notion of locally traceable operator makes it possible to discuss the index of an elliptic operator on a non-compact manifold. As we observed previously, if D is a tangential, tangentially elliptic operator on a compact foliated space X then D_ℓ is an elliptic operator on the leaf ℓ and its local index

$$\ell_{D_\ell} = \mu_{\text{Ker } D_\ell} - \mu_{\text{Ker } D_\ell^*}$$

does make sense as a (signed) Radon measure on ℓ. Write $\ell_D^x = \ell_{D_\ell}$ for each $x \in \ell$. Then $\ell_D = \{\ell_D^x\}$ is a *tangential measure*; that is, a family of Radon measures supported on leaves of X with suitable invariance properties (cf. 4.11). We regard ℓ_D as the index of D. If the foliation bundle F is oriented then a tangential measure determines a class in $\bar{H}_\tau^p(X)$. The task of an index theorem is to identify that class.

To proceed further along these lines and because they are of substantial independent interest, we introduce transverse measures. For this we move temporarily to a measure-theoretic context. Suppose that $(X, \mathcal{R})$ is a standard Borel equivalence relation. We assume that there is a complete Borel transversal (which holds easily in the setting of foliated spaces) and that we are given a one-cocycle $\Delta \in Z^1(\mathcal{R}, \mathbb{R}^*)$. A *transverse measure* of modulus Δ is a measure ν on the σ-ring of all Borel transversals which is σ-finite on each transversal and such that $\nu|_T$ is quasi-invariant with modulus $\Delta|_T$ for the countable equivalence relation $\mathcal{R} \cap (T \times T)$ for each transversal T. If $\Delta \equiv 1$ then ν is an *invariant* transverse measure. For example, if X is the total space of a fibration $\ell \longrightarrow X \longrightarrow B$ foliated with fibres as leaves then an invariant transverse measure on X is precisely a σ-finite measure on B.

Recall that a tangential measure λ is an assignment $\ell \rightarrow \lambda_\ell$ of a measure to each leaf (or class of $\mathcal{R}$) which satisfies suitable Borel smoothness properties (cf. 4.11). For example, if D is a tangential, tangentially elliptic operator on X then the local index ι_D is a tangential measure. If we choose a coherent family of volume measures for each leaf ℓ then these coalesce to a tangential measure.

Given a tangential measure λ and an invariant transverse measure ν, we wish to describe an integration process which produces a measure $\lambda d\nu$ on X and then a number $\int \lambda d\nu$ obtained by taking the total mass of the measure. Choose a complete transversal N. There is a Borel map $\sigma: X \longrightarrow N$ with $\sigma(x) \sim x$. Then $\sigma^{-1}(n)$ is contained in the leaf containing n. Regard X as fibring measure-theoretically over N. Let λ_n be the restriction of λ_ℓ to $\sigma^{-1}(n)$. Then

$$\int_N \lambda_n d\nu(n) = \lambda d\nu$$ is a measure on X. This integration process is related to the pairing of currents with foliation cycles in Sullivan [Su].

How many invariant transverse measures are there? Let MT(X) be the vector space of Radon invariant transverse measures. The construction above provides a pairing

$$MT(X) \times \Omega_\tau^p(X) \longrightarrow \mathbb{R}$$

8

and hence a Ruelle–Sullivan map

$$MT(X) \longrightarrow \mathrm{Hom}_{\mathrm{cont}}(H^p_\tau(X), \ \mathbb{R}) \cong H^\tau_p(X).$$

We prove a Riesz representation theorem: this map is an isomorphism. For example, if X is foliated by points then $H^0_\tau(X) = C(X)$ and an invariant transverse measure is just a measure, so our result reduces to the usual Riesz representation theorem. We see also that X has <u>no</u> invariant transverse measure if and only if $\bar{H}^p_\tau(X) = 0$.

    With this machinery in hand we can state and prove the remarkable index theorem of A. Connes. Let D be a tangential, tangentially elliptic pseudodifferential operator on a compact oriented foliated space of leaf dimension p. As described above, we obtain the analytic index of D as a tangential measure ι_D. For any invariant transverse measure ν the real number $\int_X \iota_D d\nu$ is the analytic ν-index $\mathrm{ind}_\nu(D)$ defined by Connes. The Connes index theorem states that for any invariant transverse measure ν,

$$\int \iota_D d\nu = \int \iota_D^{top} d\nu$$

where $\iota_D^{top} = \pm\Phi_\tau^{-1}\mathrm{ch}_\tau(D)\mathrm{Td}_\tau(X)$ is the topological index of the symbol of D. Using the Riesz representation theorem we reformulate Connes' theorem to read

$$[\iota_D] = [\iota_D^{top}] \in \bar{H}^p_\tau(X)$$

which, as is evident, does not involve invariant transverse measures. Of course if X has no invariant transverse measures then $\bar{H}^p_\tau(X) = 0$ and $\iota_D \in \{\bar{0}\}$.

    There is a stronger form of the index theorem for foliated manifolds which is due to Connes and Skandalis. To state it we need to introduce the reduced C^*-algebra of the foliated space. The compactly supported tangentially smooth functions on G(X) form a $*$-algebra under convolution. (If G(X) is not Hausdorff then a modification is required.) For each leaf G^X of $G(X)$ with its natural volume measure there is a natural regular representation of this *-algebra on $\mathfrak{B}(L^2(G^X))$. Complete the *-algebra with respect to these representations and one obtains $C^*_r(G(X))$. This algebra enters into index theory because there is a natural pseudodifferential operator extension

$$0 \longrightarrow C^*_r(G(X)) \longrightarrow \bar{\wp}^0 \xrightarrow{\ \sigma\ } \Gamma(S^*F.\mathrm{End}(E)) \longrightarrow 0$$

and hence the tangential principal symbol of D yields an element of $K_0(C^*_r(G(X)))$. Connes and Skandalis [CS2] identify this element and thereby obtain a sharper form of the index theorem which is useful in the Type III situation. Even in the presence of an invariant transverse measure, if the symbol of an operator D has finite order in $K_0(C^*_r(X))$ then $[\iota_D] = 0$ in $H^p_\tau(X)$.

We conclude this introduction with a brief summary of the contents of each chapter.

I.   LOCALLY TRACEABLE OPERATORS

Given an operator T on $L^2(Y,E)$ for a locally compact space Y, we explain the concept of local traceability and we construct the local trace μ_T of T. The local index ι_D of an elliptic operator on a noncompact manifold is one motivating example. We also discuss several situations outside the realm of foliations where locally traceable operators shed some light. In particular, we interpret the formal degree of a representation of a unimodular locally compact group in these terms.

II.   FOLIATED SPACES

Here we set forth the topological foundations of our study. We give many examples of foliated spaces and construct tangentially smooth partitions of unity. Then follow smoothing results which enable us, for instance, to assume freely that bundles over our spaces are tangentially smooth. It is perhaps worth noting that $K^0(X)$ coincides with the subgroup generated by tangentially smooth bundles. Next we explain holonomy and, following Winkelnkemper, introduce the holonomy groupoid of a foliated space. We consider the relationship between $G(X)$ and its discrete model G_N^N and determine the structure of G_N^N in several examples.

III. TANGENTIAL COHOMOLOGY

In this chapter we define the tangential cohomology groups $H_\tau^*(X)$ as the cohomology of the de Rham complex $\Gamma_\tau(\Lambda^*F^*)$ and equivalently as the cohomology of X with coefficients in the sheaf of germs of continuous functions on X which are constant along leaves. There is an analogous compactly supported theory $H_{\tau c}^*(X)$ and an analogous tangential vertical theory $H_{\tau v}^*(E)$ on bundles. We develop the properties parallel to the expected properties from de Rham theory. There is a Mayer-Vietoris sequence (for open subsets) and a Künneth isomorphism

$$H_\tau^*(X) \otimes H^*(M) \xrightarrow{\cong} H_\tau^*(X \times M)$$

provided that M is a manifold foliated as one leaf and $X \times M$ is foliated with leaves $\ell \times M$. We establish a Thom isomorphism theorem (3.30) of the type

$$\clubsuit: H_\tau^k(X) \xrightarrow{\cong} H_{\tau v}^{n+k}(E)$$

for an oriented tangentially smooth n-plane bundle $E \longrightarrow X$. Finally we indicate the definition of tangential homology theory. In an appendix we rephrase these constructions in terms of Lie algebra cohomology.

IV. TRANSVERSE MEASURES

We develop here the general theory of groupoids, both in the measurable and topological contexts, in order to give a proper home to transverse measures. The prime examples are $G(X)$ and G_N^N, of course. We introduce transverse measures and their elementary properties. The proper integrands for transverse measures are **tangential measures**, as we have previously explained in the foliation context. We carefully explain the integration process

$$(\lambda,\nu) \rightsquigarrow \lambda d\nu \rightsquigarrow \int \lambda d\nu$$

and indicate the necessary boundedness results. Specializing to topological groupoids and continuous Radon tangential measures, we recount the Ruelle–Sullivan construction of the current $C_\nu \in \Omega_p^{\mathcal{T}}(X)$ associated to the transverse measure ν. The current is a cycle if and only if ν **is** invariant. We relate invariant transverse measures on X to invariant measures on a complete transversal N. Finally we establish the Riesz representation theorem: finite invariant transverse measures are exactly the group $\mathrm{Hom}_{\mathrm{cont}}(H_{\mathcal{T}}^p(X),\mathbb{R})$. One useful consequence of this result is that a linear functional F on $MT(X)$ is representable as $F(\nu) = \int \omega d\nu$ for some $\omega \in H_{\mathcal{T}}^p(X)$ if and only if the functional is continuous in the weak topology on $MT(X)$.

V.   CHARACTERISTIC CLASSES

This chapter contains the Chern–Weil development of tangential characteristic classes. This comes down to carefully generalizing the usual constructions of connections, curvature, and their classes. This results in tangential Chern classes $c_n^{\mathcal{T}} \in H_{\mathcal{T}}^{2n}(X)$, tangential Pontrjagin classes $p_n^{\mathcal{T}} \in H_{\mathcal{T}}^{4n}(X)$, and a tangential Euler class, as well as the now classical universal combinations of these. We construct these classes at the level of forms, so that, for a fixed tangential Riemannian connection, the topological index is a uniquely defined form. We verify the necessary properties of the tangential Chern character and the tangential Todd genus which relates the K–theory and tangential cohomology Thom isomorphisms.

VI.   OPERATOR ALGEBRAS

Each foliated space has associated to it the reduced C^*-algebra $C_r^*(G(X))$ introduced by A. Connes. In this chapter we present its basic properties. Central to our treatment is the Hilsum–Skandalis isomorphism

$$C_r^*(G(X)) \cong C_r^*(G_N^N) \otimes K$$

which shows that, at the level of C^*-algebras, the foliated space "fibres" over a complete transversal N. The C^*-algebra $C_r^*(G_N^N)$ is the C^*-algebra of the discrete model G_N^N of G(X). An invariant transverse measure ν induces a trace ϕ_ν on $C_r^*(G(X))$ and one then may construct the von Neumann algebra $W^*(G(X), \hat{\mu})$. The analogous splitting

$$W^*(G(X), \hat{\mu}) \cong W^*(G_N^N, \hat{\mu}) \otimes B(\mathcal{H})$$

at the von Neumann algebra level is expected, of course. In the ergodic setting this corresponds to the usual decomposition of a II_∞ factor into the tensor product of II_1 and I_∞ factors. We conclude with a brief introduction to K-theory and the construction of a partial Chern character c: $K_0(C_r^*(G)) \longrightarrow \bar{H}_\tau^p(X)$.

VII.   PSEUDODIFFERENTIAL OPERATORS

The usual theory of pseudodifferential operators takes place on a smooth manifold. In this chapter we "parametrize" the theory to the setting of foliated spaces. This involves constructing the pseudodifferential operator algebra and its closure, defining the tangential principal symbol, and showing that the analytic index class ι_D depends only upon the homotopy class of the principal symbol. We construct the pseudodifferential operator extension which has the form

$$0 \longrightarrow C_r^*(X) \longrightarrow \bar{\wp}^0 \longrightarrow \Gamma(S^*F, \mathrm{End}(E)) \longrightarrow 0.$$

Turning to tangential differential operators, we introduce bounded geometry and finite propagation techniques to demonstrate that ι_D is well-defined. We establish the McKean-Singer formula: for t > 0,

$$\text{ind}_\nu(D) = \phi_\nu([e^{-tD^*D}] - [e^{-tDD^*}]) = \phi_\nu^s(e^{-t\widehat{D}})$$

where $\widehat{D}$ is an associated self-adjoint superoperator and ϕ_ν^s is the supertrace. Next we prove that as t $\longrightarrow$ 0 there is an asymptotic expansion

$$\phi_\nu^s(e^{-t\widehat{D}}) \sim \sum_{j \geq -p} t^{j/2p} \int_X \lambda_j(\widehat{D}) d\nu$$

where each $\lambda_j(\widehat{D})$ is a signed tangential measure independent of t. As $\text{ind}_\nu(D)$ is independent of t, it is immediate that

$$\text{ind}_\nu(D) = \int \omega_D(g,E) d\nu$$

where ω_D is a tangentially smooth p-form which depends on the bundle E of D and upon the tangential Riemannian metric.

VIII. THE INDEX THEOREM

If D is a tangential, tangentially elliptic pseudodifferential operator on a compact foliated space with oriented foliation bundle of dimension p, then we have defined the analytic index ι_D and the topological index ι_D^{top} as tangential measures. We establish the Connes index theorem which asserts that for any invariant transverse measure ν,

$$\int \iota_D d\nu = \int \iota_D^{top} d\nu.$$

We reformulate this result, in light of the Riesz representation theorem, as

$$[\iota_D] = [\iota_D^{top}] \in \bar{H}_\tau^p(X).$$

Chapter VIII is devoted to the proof of the index theorem. We verify the theorem for tangential twisted signature operators and then argue on topological grounds that this suffices.

There are three appendices to the book; each applies the index theorem in concrete situations and so demonstrates some possible uses of the theorem. The first appendix, by Steven Hurder, develops some interesting examples and applications of the theorem to the case when the leaves of the foliation have a complex structure. The second appendix, by the authors and Robert J. Zimmer, explores the use of the index theorem to demonstrate the existence of square-integrable harmonic forms on certain non-compact manifolds. The third appendix, by Robert J. Zimmer, discusses the application of some of the Gromov-Lawson ideas regarding the existence of a tangential metric which has positive scalar curvature along the leaves. These provide a complement to the general development.

CHAPTER I: LOCALLY TRACEABLE OPERATORS

Our object in this chapter is to develop the notion of what we call locally traceable operators -- or, more or less equivalently, the notion of locally finite dimensional subspaces relative to an abelian von Neumann algebra $\mathfrak{A}$. The underlying idea here is that certain operators, although not of trace class in the usual sense, are of trace class when suitably localized relative to $\mathfrak{A}$. The trace, or perhaps better, the *local* trace of such an operator is not any longer a number, but is rather a measure on a measurable space X associated to the situation with $\mathfrak{A} = L^{\infty}(X)$. This measure is in general infinite but σ-finite, and it will be finite precisely when the operator in question is of trace class in the usual sense, and then its total mass will be the usual trace of the operator. Heuristically, the local trace, as a measure, will tell us how the total trace – infinite in amount – is distributed over the space X. Once we have the notion of a locally traceable operator, and hence the notion of locally finite dimensional subspaces, one can define then the local index of certain operators. This will be the difference of local dimensions of the kernel and cokernel, and will therefore be, as the difference of two σ-finite measure, a σ-finite signed measure on X. One has to be slightly careful about expressions such as $\infty - \infty$ that arise, but this is a minor matter and can be avoided easily by restricting consideration to sets of finite measure. These ideas are developed to some extent in Atiyah [At3] for a very similar purpose to what we have in mind here, and we are pleased to acknowledge our gratitude to him.

To be more formal and more exact about this notion, we consider a separable Hilbert space H with an abelian von Neumann algebra $\mathfrak{A}$ inside of $\mathfrak{B}(H)$, the algebra of all bounded operators on H. (We could dispense in part with this separability hypothesis, but it would make life unnecessarily difficult; all the examples and applications we have in mind are separable.) For example, suppose that X is a standard Borel space (cf. [Ar], [Z4, Appendix A] for definitions and properties of such spaces). It is a fact that X is isomorphic to either the unit interval [0,1] with the usual σ-field of Borel sets or is a countable set with every subset a Borel set; cf. [Ar] for details. Now let μ be a σ-finite measure on X and let H_n be a fixed n-dimensional Hilbert space where n = 1,2,....,∞. Then $H = L^2(X,\mu,H_n)$, the set of equivalence classes of square integrable H_n-valued functions on X, is a separable Hilbert space. The algebra $L^\infty(X,\mu)$ of equivalence classes of bounded measurable functions acts as a von Neumann algebra on H. We recall that an H_n-valued function f on X is measurable if $(f(\cdot),\xi)$ is measurable for each fixed ξ on H_n and square integrability means that $|f(\cdot)|^2$ is integrable.

This example is almost the most general such example of an abelian von Neumann algebra acting on a separable Hilbert space. Indeed, let us choose standard measure spaces (X_n,μ_n), one for each n = 1,2,....,∞, with the understanding that some X_n's may be the void set and so will not contribute anything; then form $H^{(n)} = L^2(X_n,\mu_n,H_n)$ as we did before and finally form the direct sum $H = \Sigma H^{(n)}$. The measure spaces (X_n,μ_n) may be assembled by disjoint union into a standard measure space (X,μ) and then $L^\infty(X,\mu)$, which is essentially the product of the spaces $L^\infty(X_n,\mu_n)$, acts as a von Neumann algebra on H by $(f\cdot\phi)_n = f|_{X_n}\cdot\phi_n$ where

$f \in L^\infty(X,\mu)$, $\phi = (\phi_n) \in H$. It is a standard theorem that if $\mathfrak{m}$ is any abelian von Neumann algebra acting on a separable Hilbert space K, then there are (X_n,μ_n) as above and a unitary equivalence U of K with $H = \Sigma L^2(X_n,\mu_n,H_n)$ such that $U\mathfrak{m}U^{-1} \cong L^\infty(X,\mu)$, (cf. Dixmier [Di1] p. 117.)

Thus whenever we have an abelian subalgebra $\mathfrak{A}$ of $\mathfrak{B}(H)$, H may be regarded by this result as a space of functions f on $X = \cup X_n$ with $f(x) \in H_n$ for $x \in X_n$. It is often convenient to introduce the notation $H_x = H_n$ for $x \in X_n$ so that $\{H_x\}$ may be thought of as a "field" of Hilbert spaces or a Hilbert bundle; the functions f satisfy $f(x) \in H_x$ and can be thought of as (square integrable) sections. The notion of measurability of such a function is clear: it should be measurable on each set X_n as a function into $H_n = H_x$. What we have in fact described is the *direct integral* construction defined by the abelian subalgebra $\mathfrak{A}$, and one writes

$$H = \int_X H_x d\mu(x)$$

as the direct integral of the spaces H_x. In the sequel we will freely think of elements f of H in this situation as vector-valued functions.

A more specific kind of example that we have in mind is described as follows: X is a connected C^∞ manifold, μ is a σ-finite measure absolutely continuous with respect to Euclidean measure on X, and E $\longrightarrow$ X is a Hermitian vector bundle on X -- that is, a complex vector bundle with each fibre given an Hermitian inner product which varies continuously from fibre to fibre. Denoting the fibre of E over $x \in X$ by H_x, we obtain a field of Hilbert spaces $\{H_x\}$ of constant (finite) dimension. It is easy to find a Borel trivialization of E, that is, a field of unitary isomorphisms φ_x of H_x with a fixed Hilbert space H_n so that these maps define a Borel isomorphism of the total space E of the bundle with $X \times H_n$. With H the set of square integrable measurable sections of E, (equivalently $H = \int_X H_x d\mu(x)$ or $H = L^2(X, H_n)$), and with $\alpha = L^\infty(X, \mu)$ acting by multiplication on H, we obtain exactly the kind of abstract structure described above.

Given such a pair H,α, we want to define what it means for an operator T on H to be locally traceable relative to α. To motivate this, consider a one dimensional subspace V of H and choose a unit vector φ in V. Viewing H as a direct integral of a field H_x

$$H = \int_X H_x d\mu(x)$$

we can think of φ as a function $\varphi(x)$ with $\varphi(x) \in H_x$ and then form $|\varphi(x)|^2$. This is an integrable function of norm one, or equivalently the measure $|\varphi(x)|^2 d\mu(x)$ is a probability measure which we denote $\mu_{P(V)}$. Its measure class is intrinsic to V and in particular does not depend on the choice of the measure μ used to write $\alpha = L^\infty(X, \mu)$. (Recall that μ could be replaced by any measure equivalent to μ in the sense of absolute continuity.) This measure $\mu_{P(V)}$ has μ-total mass one -- the dimension of V -- and can be thought of as describing how the dimension of V is "spread out over" the space X or also how the dimension of V "localizes." More generally, if V is any finite dimensional subspace of H, let us choose an orthonormal basis $\varphi_1,...,\varphi_n$ for V. Then it is an elementary and well known calculation that $\sum_{i=1}^{n} |\varphi_i(x)|^2$ is independent of the choice of the orthonormal basis and consequently that the measure $\mu_{P(V)}$ defined by

$$d\mu_{P(V)} = \sum_{i=1}^{n} |\varphi_i(x)|^2 \, d\mu(x)$$

is independent of all choices. Its total mass is n, the dimension of V, and again $\mu_{P(V)}$ can be thought of as describing how the total dimension of V is distributed or localized over the space X.

In the same way we argue that if T is any finite rank operator, and if $\varphi_1,...,\varphi_n$ is any orthonormal basis for the range of T (or for the orthogonal complement of the kernel of T), then the measure μ_T defined by

$$d\mu_T(x) = \sum_{i=1}^{n} ((T\varphi_i)(x), \varphi_i(x)) \, d\mu(x),$$

where the inner product is taken pointwise in H_x, is a signed measure of total mass equal to the trace of T and which again describes how this total trace is distributed over the space X. If T = P(V) is the orthogonal projection onto a finite dimensional subspace V, then this clearly coincides with the previous definition as the notation itself suggests.

With these simple examples in mind, the path of development is fairly clear and leads us to consider operators T for which a suitably defined μ_T is a σ-finite measure; or, if as in many examples X is naturally a locally compact space, then operators T for which μ_T is a Radon measure (finite on compact sets). We begin with the trace function which we view as defined on all nonnegative operators on a Hilbert space H with values in the extended positive real numbers. Denote this cone of nonnegative operators by $\mathcal{B}(H)^+$ and for $T \in \mathcal{B}(H)^+$ define $Tr(T) = \Sigma(T\xi_i, \xi_i)$ where ξ_i is any orthonormal basis for H and where we define $Tr(T)$ to be $+\infty$ if the series (of nonnegative terms) diverges. It is elementary, using the positive square root $S = T^{1/2}$ of T, to see that the sum is independent of the choice of basis. As a map from $\mathcal{B}(H)^+$ to $\bar{\mathbb{R}}^+$, Tr satisfies

(1) $Tr(T_1 + T_2) = Tr(T_1) + Tr(T_2)$

(2) $Tr(\lambda T) = \lambda Tr(T)$, $\lambda \geq 0$

(3) $Tr(A^*A) = Tr(AA^*)$, $A \in \mathcal{B}(H)$

(4) For any increasing net T_α in $\mathcal{B}(H)^+$ with $T = \text{lub } T_\alpha$
     in the sense of the order on $\mathcal{B}(H)^+$, $Tr(T) = \text{lub } Tr(T_\alpha)$

(cf. Dixmier [Di2] p. 93 and p. 81). Such mappings defined on the positive cone in any von Neumann algebra are called *normal traces*. Condition (3) is equivalent to the condition

(3') $Tr(UTU^{-1}) = Tr(T)$ for T in $\mathcal{B}(H)^+$ and U unitary.

If one drops (3) altogether such functions are called *normal weights*; in this connection see Haagerup [Haa] for a discussion of the continuity condition (4).

Suppose now that $\mathcal{O}$ is an abelian von Neumann algebra on H; then $\mathcal{O} \simeq L^\infty(X,\mu)$ and for convenience we use the same symbol for a function and the corresponding operator. (We note parenthetically that for most of this $\mathcal{O}$ could be any von Neumann algebra, but as we do not have any significant applications in mind except for abelian $\mathcal{O}$ we shall not pursue this level of generality). Our first observation is the following.

Proposition 1.1. Let $f \in \mathcal{O} \cong L^\infty(X;\mu)$ and nonnegative, and let $T \in \mathcal{B}(H)^+$. Then

$$Tr(\ f^{1/2}Tf^{1/2}\ ) = Tr(T^{1/2}fT^{1/2})$$

where $f^{1/2}$ and $T^{1/2}$ are the nonnegative square roots of f and T.

20

Proof. Let $S = f^{1/2}T^{1/2}$; then $S^* = T^{1/2}f^{1/2}$ and the formula of the statement results immediately from the fact that $\text{Tr}(AA^*) = \text{Tr}(A^*A)$. $\square$

This shows first of all that for fixed nonnegative T, the left hand side above is linear in f for f nonnegative. The continuity and additivity properties of the trace and the fact that $g \longrightarrow T^{1/2}gT^{1/2}$ is order preserving and weak operator continuous show that if we defined for any measurable subset E of X,

$$\mu_T(E) = \text{Tr}(f_E Tf_E)$$

where f_E is the characteristic function of E, then μ_T is a positive countably additive measure on X, absolutely continuous with respect to μ in that $\mu(E) = 0$ implies $\mu_T(E) = 0$. The same reasoning and an approximation argument shows that for any $f \geqslant 0$

$$\text{Tr}(f^{1/2}Tf^{1/2}) = \int_X fd\mu_T.$$

The crucial problem, and this will lead us to the definition, is that μ_T may and often does fail to be σ-finite in the sense that $X = \bigcup_{i=1}^{\infty} X_i$ where X_i is an increasing sequence of sets of finite μ_T measure. At this point one has a choice of two closely related definitions of local traceability of T. On the one hand one could say that T is locally traceable if μ_T is σ-finite, and this is perfectly satisfactory, but for applications we want something a bit different which reflects extra structure on X. Namely suppose we are given in X an increasing family of subsets X_i which exhaust X. The idea is that μ_T should be not just σ-finite relative to any exhaustion of X, but that $\mu_T(X_i) < \infty$ for this particular choice of X_i. We have in mind the example of X a locally compact second countable space with X_i a countable fundamental family of compact sets. The condition above just means that μ_T is a Radon measure.

Definition 1.2. If (X_i) is an exhaustion of X by increasing Borel sets, one says that a positive operator T on H is *locally traceable* (relative to this exhaustion) if $\mu_T(X_i) < \infty$ for all i. The measure μ_T is called the *local trace* of T.

Agreeing to call a Borel subset of X *bounded* if it is contained in some X_i, we can rephrase slightly the definition of local traceability as follows: a positive operator T is locally traceable iff fTf is trace class for every nonnegative f in $\mathcal{O}\!\!\mathcal{I} = L^\infty(X,\mu)$ of bounded support.

It is evident from Proposition 1.1 and the continuity properties of the trace that we have the following properties for local traces which we state without proof.

Proposition 1.3.

(1) $\mu_{T+S} = \mu_T + \mu_S$.

(2) $\mu_{\lambda T} = \lambda \mu_T$.

(3)   if $0 \leqslant S \leqslant T$ and T is locally traceable then so is S.

(4)   if $T(\alpha)$ is a net converging upward to T then

$$\mu_T(E) = \lim \mu_{T(\alpha)}(E)$$

for every measurable set E.

For non-positive operators one extends the notion of local traceability by linearity.

Definition 1.4. If T is any operator on H, T is *locally traceable* (relative to a given exhaustion of X) if we can write $T = \sum_{i=1}^{n} \lambda_i P_i$ where P_i are nonnegative locally traceable operators and λ_i complex numbers. The *local trace* of such a T is by definition $\sum \lambda_i \mu_{P_i}$.

This last statement requires a little explanation. First, the local trace is indeed well defined, for if T can be written in two different ways as a linear combination of positive locally traceable operators, it is easy to see using the additivity properties that μ_T comes out to be the same. Second, the measure μ_T is not quite a standard kind of object, for as a "measure" defined on all Borel subsets of X, it is all too likely to involve inadmissible expressions like $\infty - \infty$. What we have is a complex valued measure defined on the σ-ring of all Borel sets of X which are contained in some X_i (i.e. the bounded Borel sets) for the given exhaustion and which is countably additive on the (relative) σ-field of Borel subsets of each X_i.

If an operator T is locally traceable, with local trace μ_T, then for every positive f in $\mathfrak{A}$ of bounded support, $f^{1/2}Tf^{1/2}$ is a trace class operator and we have

$$\text{Tr}(f^{1/2}Tf^{1/2}) = \int f d\mu_T$$

where the integral on the right is well defined since f has bounded support.

We record some elementary consequences of these definitions which extend the integral formula above. For part two below note that the set of complex valued measures defined above is a (two-sided) module over $\mathfrak{A} = L^\infty(X,\mu)$ by multiplication of measures by functions with the left and right actions being the same.

Proposition 1.5.

(1)   If $P = (P_1-P_2) + i (P_3-P_4)$ is the canonical representation of an operator P in terms of positive operators (i.e. P_1 is the positive part of the real part of P, etc) then P is locally traceable iff each P_i is.

(2)   The class of locally traceable operators is closed under adjoints and is a two-sided module over $\mathfrak{A}$. Moreover the local trace is a two-sided module map.

Proof. (1) If each P_i is locally traceable, then by definition P is

locally traceable. Conversely, if P is locally traceable so P =
$\sum_{i=1}^{n} \lambda_i T_i$, T_i positive and locally traceable, then P^* has a similar
expression with $\bar{\lambda}_i$ instead and so is locally traceable. Consequently
the real and imaginary parts of P are locally traceable, and so it will
suffice to show that if P is self adjoint and locally traceable, then
$P^{\pm}$, its positive and negative parts, are also. We may assume
$P = \Sigma \lambda_i T_i$ with T_i positive locally traceable, λ_i real; then by
combining terms, $P = T_1 - T_2$, T_i positive locally traceable. Then P +
$T_2 = T_1$ is a positive operator greater than P and hence $P + T_2 = T_1$
$\geqslant P^+$. By (3) of Proposition 1.3 it follows that P^+ is locally
traceable, and then that P^- is also.

(2)   We have already seen that the locally traceable
operators are closed under adjoints. To see that this class is a
two-sided module over $\mathfrak{A}$, it suffices, by taking linear combinations, to
show that gP is locally traceable when P is nonnegative locally
traceable, and $g \in \mathfrak{A}$. To do this we show that the self adjoint
operators $gP + Pg^*$ and $i(gP - Pg^*)$ are locally traceable. Writing $P =
Q^2$ and observing that

$$R = (Q + Qg^*)^*(Q+Qg^*) = Q^2 + gQ^2 + Q^2 g^* + gQ^2 g^*,$$

we see that $gP + Pg^*$ is a linear combination of the positive operators
P, gPg^*, and R. The first is given as locally traceable. To see that
the second is also, let f be an element of bounded support in $\mathfrak{A}$ and
observe that $fgPg^*f = g(fPf)g^*$ is of trace class since fPf is. Hence
gPg^* is locally traceable. For the third, the definition of R shows
that $gP + Pg^* \leqslant P + gPg^*$ and hence that $R \leqslant 2(P + gPg^*)$. By
monotonicity, R is locally traceable and it follows that $gP + Pg^*$ is
locally traceable. A similar argument can be used for the imaginary
part of gP, establishing that gP is locally traceable.

To see that the local trace is a bimodule map, consider an
operator S = hTk with T locally traceable, and h,k positive elements
in $\mathfrak{A}$. The local trace μ_S satisfies

$$(1.6) \qquad \mathrm{Tr}(f^{1/2}Sf^{1/2}) = \int f d\mu_S$$

for every positive f in $\mathfrak{A}$ of bounded support, and this property characterizes μ_S since μ_S is uniquely determined by the integrals above. But now

$$\mathrm{Tr}(f^{1/2}hTkf^{1/2}) = \mathrm{Tr}(h^{1/2}h^{1/2}f^{1/2}Tf^{1/2}k^{1/2}k^{1/2})$$

$$= \mathrm{Tr}(k^{1/2}h^{1/2}f^{1/2}Tf^{1/2}k^{1/2}h^{1/2})$$

using the fact that $f^{1/2}Tf^{1/2}$ is trace class and the commutativity properties of the trace. By the definition of μ_T the last expression can be written as the integral of the nonnegative function khf, which is of bounded support, against the measure μ_T. Combining these equalities we see that

$$\mathrm{Tr}(f^{1/2}Sf^{1/2}) = \int f d\mu_S = \int fhk d\mu_T.$$

By unicity we find

$$\mu_{hTk} = kh(\mu_T)$$

at least for h and k positive. By linearity this holds for all h and k and so the local trace is a bimodule map. □

We isolate as a separate statement a useful formula implicit in the above proof.

Corollary 1.7. If T is locally traceable and h, k $\in$ $\mathfrak{A}$ are of bounded support, then hTk is traceable and

$$\mathrm{Tr}(hTk) = \int hk d\mu_T. \qquad □$$

The local trace has a further rather straightforward invariance property. Suppose that u is a unitary operator in the normalizer of the abelian algebra $\mathfrak{A}$; that is, $u\mathfrak{A}u^{-1} = \mathfrak{A}$. Then conjugation by u defines a *-isomorphism of $\alpha = L^\infty(X,\mu)$ and by point realization theorems, cf. Mackey [Ma2], there is a Borel automorphism θ of X with $\theta_*\mu \sim \mu$ so that $(ufu^{-1})(x) = f(\theta^{-1}(x))$ for $f \in L^\infty(X,\mu) = \alpha$. Recall that $\theta_*\mu(E) = \mu(\theta^{-1}(E))$ for any Borel set E. Now if (X_i) is a given exhaustion of X as introduced earlier in this section, we know what bounded sets are and we want θ to map bounded sets to bounded sets. Then the expected fact concerning this situation is true, and we omit the short proof.

Proposition 1.8.   If u and θ are as above, and if T is a locally traceable operator with local trace μ_T, then uTu^{-1} is locally traceable with local trace $\theta_*(\mu_T)$.   □

Many of the most common examples of locally traceable operators are self adjoint projections. If $V \subset H$ is a closed subspace and P(V) the orthogonal projection onto it, then we say that V is *locally finite dimensional* if P(V) is locally traceable. The local trace $\mu_{P(V)}$ is called the *local dimension* and for brevity we will write it simply as μ_V.

Now let us suppose that X is a locally compact space denumerable at ∞, and let the exhaustion X_i of X consist of a fundamental sequence of compact sets (every compact set K is eventually in some X_i). Further suppose that ξ is a finite dimensional Hermitian vector bundle over X and that the Hilbert space H is the space of (equivalence classes of) L^2 sections of ξ relative to some Radon measure μ on X, which without loss of generality we take to have support equal to all of X. Then it makes sense to talk about the continuous sections in H; this is the (dense) subspace C of H consisting of those equivalence classes (mod null sections) which contain a continuous section of ξ. If such a continuous section exists in a given class, it is of course unique. We make the following definition.

Definition 1.9.   An operator S from H to H is *smoothing of order zero* if $S(H) \subset C$, the continuous sections.

The following result will provide large classes of interesting and important locally traceable operators -- the exhaustion here is understood to be by compact subsets of X.

Theorem 1.10.   Let S_i, i = 1,....n be operators on H which are smoothing of order zero. Then $T = \sum_{i=1}^{n} S_i S_i^*$ is locally traceable.

Proof.   It clearly suffices to consider one such S. If $v \in H$, then the element S(v) of H lies in C and is represented by a unique continuous section S(v)($\cdot$). Then for fixed $x \in X$ and for a fixed vector ϕ in the dual space E_x^* of the fibre E_x of ξ at x, we can define $\phi(S(v)(x))$. By a standard argument in functional analysis using the closed graph theorem, this is a continuous linear functional $b(x,\phi)$ of v.   Moreover, if $\phi(x)$ is a continuous section of the dual bundle ξ^* of ξ, it is clear that $b(x,\phi(x))$ is a continuous function of x.   From all of this it follows that we can find for each $x \in X$, a measurable section K(x,$\cdot$) of the bundle End(ξ) with $|K(x,\cdot)|$ square integrable for each x such that

$$S(v)(x) = \int K(x,y)v(y)d\mu(y).$$

It is an easy matter to choose this function K to be jointly measurable in its two variables by the von Neumann selection theorem (cf. [Z4, p. 196]), and, by continuity in x, the L^2 norm of $|K(x,\cdot)|$ is bounded as x runs over compact sets.   Since the Hilbert–Schmidt norm of K(x,y) is at most a constant multiple of its operator norm because the fibre is finite dimensional, the same statement holds for this norm.   Thus if f is a bounded Borel function of compact support viewed both as a function and as the corresponding multiplication operator, the operator fS can be written as

$$(fS)(v)(x) = \int f(x)K(x,y)v(y)d\mu(y).$$

The kernel f(x)K(x,y) has compact support in x and it follows from our remarks above and the Fubini theorem that the Hilbert–Schmidt norm $|f(x)K(x,y)|_{HS}$ is an L^2 function on X $\times$ X. This implies that fS is a

Hilbert Schmidt operator, and hence that $(fS)(fS)^* = fSS^*\bar{f}$ is a traceable operator. This means by definition that SS^* is locally traceable, and we are done. □

As an example of this theorem, consider a closed subspace V of H which consists of continuous functions. Then it follows immediately that the projection P(V) onto V is locally traceable and that V is locally finite dimensional.

By far the most important example of this for us is the following: X a C^∞ manifold which is not necessarily compact, ξ an Hermitian vector bundle over X, and D' a differential operator from ξ to ξ which we assume to be elliptic, (cf. Taylor [Tay]). We form the space H of square-integrable sections of ξ and form the corresponding unbounded operator D on H. This is of course somewhat inexact, for one could form many such operators with different domains. The smallest such would be the closure of the operator D' acting on the space of compactly supported sections. The largest would be the Hilbert space adjoint of the formal adjoint $(D')^*$ defined on the compactly supported sections. For our purposes here D can be any closed operator between these two. (As a remark for future chapters, we note that in the specific cases to be treated later these two extreme operators defined by D' coincide [cf. (7.24)] so there is no ambiguity about the unbounded operator D on H). With such a D we form its kernel V = Ker(D). The elements v of V will be by definition weak solutions of the differential equation D'v = 0 and hence by ellipticity actually C^∞ sections. By Theorem 1.10 and the comments following it, Ker(D) is locally finite dimensional; its local dimension, which we write μ_D, is a Radon measure on X. If D^* is the Hilbert space adjoint of D, the same considerations apply and we can form the local dimension μ_{D^*} of the kernel of D^*.

Definition 1.11. The *local index* ι_D of D is the difference $\mu_D - \mu_{D^*}$, a signed Radon measure on X.

If X is compact, then of course these are all finite measures and the total mass of ι_D, necessarily an integer, is the usual index of D. The classical Atiyah-Singer index theorem provides a formula for this in terms of topological invariants. The object of the Connes index theorem for foliations is to provide a similar formula in the following context: X a compact foliated space, D a differential operator assumed to be tangentially elliptic (see Chapter VII) so that for any leaf ℓ of the foliated space the differential operator $D\big|_\ell \equiv D_\ell$ will be elliptic in the usual sense and will define a local index on each leaf ℓ. The leaves ℓ are not necessarily compact and hence $\text{Ker}(D_\ell)$ is not necessarily finite dimensional. The theorem then provides a formula for the average of these local indices, the average being taken over all leaves. This averaging process is by no means straightforward and requires a whole subsequent chapter, Chapter IV, to explain.

The framework of locally traceable operators provides a convenient bridge to the work of Atiyah [At3] on the index theorem for covering spaces. Let X be a manifold (not necessarily compact) and let $\tilde{X} \longrightarrow X$ be a covering space with fundamental domain U and covering group Γ. Then

$$L^2(\tilde{X}) = L^2(U) \otimes L^2(\Gamma)$$

where X is given volume measure, $\tilde{X}$ is given the pullback measure, and Γ acts by the left regular representation. With respect to this decomposition the commutant of Γ is the von Neumann algebra

$$\tilde{A} = \mathcal{B}(L^2(U)) \otimes \mathcal{R},$$

where $\mathcal{R}$ is the algebra corresponding to the right regular representation. There is a natural trace τ on $\tilde{A}$ corresponding to the usual trace on $\mathcal{B}$ tensor with the canonical trace on $\mathcal{R}$. Suppose that $\tilde{T}$ is a bounded operator on $L^2(\tilde{X})$ which commutes with the action of Γ. Then $\tilde{T}$ has the form

$$\tilde{T} = \Sigma_\gamma \ T_\gamma \otimes R_\gamma$$

with respect to the decomposition above. Atiyah defines

$$\text{ind}_\Gamma(\tilde{T}) = \tau(f\tilde{T}f)$$

where f is the characteristic function of U. We may simplify this to read

$$\text{ind}_\Gamma(\tilde{T}) = \text{Trace}(fTf)$$

where $T = T_{id} \otimes I$.

Let $A = L^\infty(X)$ acting upon $L^2(\tilde{X})$ by multiplication, and, by restriction, A acts on $L^2(U)$. Then A is isomorphic to $L^\infty_c(U)$ acting upon $L^2(U)$ by multiplication. Write $T = T_{id}$ acting upon $L^2(U)$. Then it is clear that $\text{ind}_\Gamma(\tilde{T})$ is precisely the integral of the local trace:

$$\text{ind}_\Gamma(\tilde{T}) = \int_X d\mu_T \ .$$

Then Atiyah's theorem may be understood simply as relating μ_T to the lift of μ_T to $\tilde{X}$.

The content of Theorem 1.10 can be rephrased somewhat with no reference to topology; namely if $H = L^2(X,\mu)$ (or finite dimensional vector valued functions) on a measure space, let T be a bounded linear transformation on H to itself such that $T(H) \subset L^\infty(X,\mu)$. That is, the image of T consists of bounded functions. Then we claim that TT^* is locally traceable and that there is a very simple formula for the local trace. Actually the same idea would work if $T(H)$ were contained in a suitably defined space of locally bounded functions too, but for simplicity let us stick to globally bounded functions.

First we observe that an application of the closed graph theorem shows that T is bounded as a map of $L^2(X)$ into $L^\infty(X)$. Further it is an easily shown fact, (cf. Dunford–Schwartz [DS] p. 499) that whenever T is a bounded linear map from a separable Banach M into $L^\infty(X)$, there is a measurable bounded function k(x) from X into the dual M^* of M such that $(Tm)x = k(x)(m)$ for $m \in M$. Application of this yields a bounded map $x \longrightarrow k(x)$ from X to $L^2(X)$ which serves as a "kernel" for T. The following is proved in exactly the same way that Theorem 1.10 is.

Proposition 1.12. If T maps $L^2(X,\mu)$ into $L^\infty(X,\mu)$ then TT^* is locally traceable (relative to any exhaustion by sets of finite μ-measure) and its local trace is the measure $|k(x)|^2 d\mu(x)$ where k is as above.

It is a standard fact that the L^2 valued measure function can be written as $k(x)(y) = K(x,y)$ for a jointly measurable function. Then

$$(Tf)(x) = \int K(x,y)f(y)d\mu(y)$$

is, as we observed already, an integral kernel operator.

Because the issue will come up in the construction of operator algebras associated with groupoids and foliations, we recall briefly some sufficient conditions for a kernel $K(x,y)$ to define a bounded operator.

Definition 1.13. A kernel $K(x,y)$ on $X \times X$ is *integrable* (with respect to a measure μ on X) if

$$\text{ess}\sup_{x} \int |K(x,y)| d\mu(y) < \infty$$

$$\text{ess}\sup_{y} \int |K(x,y)| d\mu(x) < \infty .$$

One may define an operator $T = T_K$ from functions on X to functions on X formally by

$$(Tf)(x) = \int K(x,y)f(y)d\mu(y) .$$

If $f \in L^1 \cap L^\infty$ then the integral at least makes sense and the two conditions in the definition above show immediately that $|Tf|_1$ is bounded by a constant times $|f|_1$ and that $|Tf|_\infty$ is bounded by a constant times $|f|_\infty$. It is an easy and standard interpolation result using the Riesz convexity theorem (cf. Dunford–Schwartz [DS], p. 525) that T defines a bounded operator on each L^p to L^p for each p with a norm no worse than the larger of the two bounds in the definition.

Proposition 1.14. If the kernel K is integrable, $T = T_K$ defines a bounded operator on $L^2(X)$. If in addition $k(x) = \left[\int |K(x,y)|^2 dy\right]^{1/2}$ is essentially bounded in x, then T maps $L^2(X)$ to $L^\infty(X)$ and the local trace of TT^* is $k^2(x)du(x)$.

The ideas developed above find other interesting applications and it is our purpose in the balance of this chapter to look at some of these. Specifically, let G be a locally compact second countable abelian group. Let $H = L^2(G,\mu_G)$ with $\alpha = L^\infty(G,\mu_G)$ acting by multiplication, where μ_G is Haar measure. If E is any Borel subset of the dual group $\hat{G}$, we construct the subspace V(E) of H consisting of functions $\varphi \in H$ whose Fourier transform $\hat{\varphi}$ vanishes outside of E. We recall that if μ_G is any Haar measure on G, then there is a uniquely determined choice of Haar measure $\mu_{\hat{G}}$ on $\hat{G}$ with the property that the Fourier inversion formula holds exactly, not just up to a scalar, when μ_G and $\mu_{\hat{G}}$ are used. Specifically if

$$\hat{\varphi}(\alpha) = \int \overline{(\alpha, x)}\varphi(x) \, d\mu_G(x)$$

and if

$$\hat{\psi}(x) = \int (\alpha,x) \, \psi(\alpha) \, d\mu_{\hat{G}}(\alpha),$$

then $(\hat{\varphi})^\wedge = \varphi$ for suitable functions φ where $(\cdot,\cdot)$ is the duality pairing of $\hat{G} \times G$ to the circle group.

Let us assume that the subset E of $\hat{G}$ has finite dual Haar measure. Then by the Fourier inversion theorem, the elements of V(E) are back transforms of elements of $L^2(E) \subset L^2(\hat{G})$. But since E has finite measure, $L^2(E) \subset L^1(E)$, and consequently the elements of V(E) are back transforms of integrable functions on $\hat{G}$. It follows that V(E) consists of continuous functions and so by Theorem 1.10, V(E) is locally finite dimensional. Let μ_E be the local dimension of V(E). The unitary operator u_g induced by left translation leaves V(E) invariant and normalizes α. Proposition 1.8 tells us then that μ_E is invariant under left translation by elements of G. Thus μ_E is a Haar measure; the only question is which one. This is not difficult to answer.

Proposition 1.15. Let μ_G be a Haar measure on G, let $\mu_{\hat{G}}$ be its dual Haar measure on $\hat{G}$, and let E be a subset of finite measure in $\hat{G}$. Then the local dimension of V(E) is given by

$$\mu_E = \mu_{\hat{G}}(E)\mu_G.$$

Proof. First we observe that the answer written above does not depend on the original choice of μ_G in view of the way $\mu_{\hat{G}}$ changes when we change μ_G. To obtain the result we note that the projection operator P_E onto V(E) is given as a convolution operator with the kernel $K(x,y) = \int_E (xy^{-1},\alpha)\, d\mu_{\hat{G}}(\alpha)$. Now for f positive, bounded and of compact support, the operator $P_E f^{1/2}$ is given by convolution with the L^2 kernel $K(x,y)f^{1/2}(y)$. Since $P_E^2 = P_E$, we have

$$f^{1/2}P_E f^{1/2} = (P_E f^{1/2})^*(P_E f^{1/2})$$

and is given as a convolution operator with kernel

$$f^{1/2}(x)\, K(x,y)f^{1/2}(y)$$

which is the convolution of $K(x,y)f^{1/2}(y)$ with its adjoint. Consequently we can calculate the trace of $f^{1/2}P_E f^{1/2}$ by integrating the kernel on the diagonal x = y. So

$$\mathrm{Tr}(f^{1/2}P_E f^{1/2}) = \int (f^{1/2}(x))^2\, K(x,x)\, d\mu_G(x)$$

$$= \int f(x) \int_E (1,\alpha)d\mu_{\hat{G}}(\alpha)\, d\mu_G(x)$$

$$= \int f(x)\, \mu_{\hat{G}}(E)\, d\mu_G(x).$$

Thus $\mu_E = \mu_{\hat{G}}(E)\mu_G$ as desired. □

Let us continue this discussion a little further; suppose that G is a unimodular locally compact second countable group, and let π be a square integrable irreducible representation. This means that π occurs as a summand of the left regular representation on $L^2(G)$, or that one (equivalently each) of its matrix coefficients is square integrable. Associated to such a representation is a number d_π called the *formal degree* of π (cf. Dixmier [Di2, 14.4]) which can be defined by the equation

$$\int (\pi(g)x,y)\ \overline{(\pi(g)u,v)}\ d\mu_G(g) = d_\pi^{-1}(x,u)\overline{(y,v)}.$$

Of course d_π depends on the choice of Haar measure μ_G, but it is clear that the product $d_\pi\mu_G$ is intrinsic. This suggests, as we shall show in a moment, that the formal degree is not properly a number, but rather a Haar measure.

Proposition 1.16. Let G be unimodular, π a square integrable irreducible representation, and let $V(\pi)$ be any irreducible subspace of the left regular representation equivalent to π. Then $V(\pi)$ has locally finite dimension; the local dimension is a multiple of Haar measure given by $d_\pi\mu_G$ where d_π is usual formal degree.

Proof. It follows from the usual discussion of square integrable representations that any subspace $V(\pi)$ can always be realized as the set of matrix coefficients $\{(\pi(g)^{-1}y,x_0)\}$ where x_0 is fixed and y varies over $H(\pi)$, the Hilbert space upon which π is realized. This demonstrates immediately that $V(\pi)$ consists of continuous functions and hence by Theorem 1.10 is locally finite dimensional. The same argument as in the abelian case shows that the local dimension is a multiple of Haar measure. In order to compute the multiple, we realize $V(\pi)$ as the set of matrix coefficients $\{c_x : x \in H(\pi)\}$ where $c_x = (\pi(g^{-1})x,x_0)$. By the orthogonality relations the square norm of c_x is $d_\pi^{-1}(x_0,x_0)(x,x)$. Normalizing x_0 by $(x_0,x_0) = d_\pi$, we see that $x \longrightarrow c_x$ is an isometry. Now let $\{e_i\}$ be an orthonormal basis in $H(\pi)$ and let c_i be the corresponding vectors in $V(\pi)$. Further let V_n be the span of $(c_1,...,c_n)$. By the introductory comments in the chapter, the local trace μ_n of V_n is given by

$$d\mu_n = \sum_{i=1}^{n} \left| c_i(g) \right|^2 d\mu_G(g).$$

As n tends to ∞, the projection onto V_n increases monotonically to the projection onto V. By (4) of Proposition 1.3, $\mu_n(E)$ increases upward

to $\mu_\pi(E)$ where μ_π is the local dimension of $V(\pi)$. But $\sum_{i=1}^{n} \left| c_i(g) \right|^2$ increases monotonically to the infinite sum

$$\sum_{i=1}^{\infty} \left| c_i(g) \right|^2 = \sum_{i=1}^{\infty} \left| (\pi(g^{-1})e_i, x_o) \right|^2 = \left| \pi(g)x_o \right|^2 = \left| x_o \right|^2 = d_\pi.$$

It follows that $d\mu_\pi = d_\pi d\mu_G$ as desired.   □

    If the group G is non-unimodular the situation becomes more complicated as one might guess from Duflo-Moore [DM], Pukanszky [Puk].   Suppose that π is an irreducible square integrable representation of G. This means that π occurs as summand of the left regular representation, but now some, but not all matrix coefficients are square integrable. Let $P(\pi)$ be the closed linear span of all irreducible summands of $L^2(G)$ equivalent to π.   Then $P(\pi)$ is also invariant under right translation and as a $G \times G$ module is isomorphic to $\pi \times \tilde{\pi}$ where $\tilde{\pi}$ is the contragredient of π (cf. Mackey [Ma6]).   As $\tilde{\pi}$ is also square integrable, and as $P(\pi)$ is primary for the left and the right actions, there are, once we fix a left Haar measure on G, two canonically defined formal degree operators on $P(\pi)$, D_π for the left action and $\tilde{D}_\pi$ for the right action [Ma6]. Each is an unbounded positive operator affiliated to the von Neumann algebras associated to the left and right actions respectively, and semi-invariant under these actions. If we change Haar measure by a scalar factor c, then D_π and $\tilde{D}_\pi$ change by c^{-1} so that symbolically the products $D_\pi d\mu_G$ and $\tilde{D}_\pi d\mu_G$ are intrinsic. We recall that both the left and right von Neumann algebras are isomorphic to $\mathcal{B}(H)$, the algebra of all bounded operators, and so have canonically defined traces.
    Now suppose that $V \subset P(\pi)$ is a subspace of $P(\pi)$ invariant

under the right action. We would like to know when V is locally finite dimensional and in those cases we want a formula. As before, the local dimension, if it exists, is a multiple of left Haar measure. The subspace V, being left invariant, defines a projection P_V in the right von Neumann algebra on $P(\pi)$ as these two algebras are commutants of each other. We now try to make sense out of the expression $(\tilde{D}_\pi)^{1/2} P_V (\tilde{D}_\pi)^{1/2}$ as a bounded positive operator. In fact it will be a well–defined bounded operator precisely when the range of P_V is included in the domain of $(\tilde{D}_\pi)^{1/2}$. When this happens and when in addition this bounded positive operator has a trace, we see that P_V or V itself is finite relative to $\tilde{D}_\pi$. Another way to say this very much in the spirit of Pedersen–Takesaki [PT] is to observe that $\tilde{D}_\pi$ defines a weight ψ on the von Neumann algebra of the right action given by $T \longrightarrow \mathrm{Tr}(D_\pi^{1/2} T D_\pi^{1/2})$ (cf. Moore [Mr1]) and the condition on P_V is that ψ is finite on this element. Our result is the following.

Proposition 1.17. The subspace V of $P(\pi)$ has locally finite dimension if and only if P_V is finite relative to $\tilde{D}_\pi$. The local dimension is then $\mathrm{Tr}(\tilde{D}_\pi^{1/2} P_V \tilde{D}_\pi^{1/2}) \, \mu_G$. □

We omit the proof of this fact and simply remark that if G is unimodular, then D_π and $\tilde{D}_\pi$ become scalar multiples of the identity, namely $d_\pi \cdot 1$ where d_π is the usual (scalar) formal degree. Then the statement above is exactly the same as in the unimodular case. It is interesting that, contrary to the unimodular case, not all irreducible summands of $P(\pi)$ have finite local dimension, and moreover that there are irreducible subspaces of $P(\pi)$ with arbitrarily small local dimension.

These special cases suggest the form of the general result which is as follows: let μ_G be left Haar measure on G. Then there are semi-finite normal weights semi-invariant for the modular functions ψ on $\mathcal{L}$, the von Neumann algebra of the left regular representation, and $\tilde{\psi}$ on $\mathcal{R}$, the von Neumann algebra of the right regular representation. Normalize these so that Fourier transform becomes an isometry. Then if V is an invariant subspace for the left regular representation, the projection P_V onto it is in the algebra $\mathcal{R}$ of the right regular representation.

Proposition 1.18. The subspace V has finite local dimension if and only if $\tilde{\psi}(P_V) < \infty$; in this case the local dimension is $\tilde{\psi}(P_V)\mu_G$.
□

We again omit the proof of this result.

CHAPTER II: FOLIATED SPACES

In this chapter we introduce the basic definitions and elementary properties of foliated spaces.

Definition 2.1. A *foliated space* X of dimension p is a separable metrizable space X together with a collection of open sets $\{U_x | x \in X\}$ with $x \in U_x$ and homeomorphisms

$$\varphi_x : U_x \longrightarrow L_x \times N_x$$

with L_x open in $\mathbb{R}^p$ which satisfy the following conditions:

1) Writing $\varphi_x = (t,n)$, then coordinate changes are given by
$$t' = \varphi(t,n)$$
$$n' = \psi(n) \quad \text{for some local homeomorphism } \psi.$$

2) If $U_x \cap U_y$ is nonempty then the composite
$$n \rightarrow \varphi_y \varphi_x^{-1}(\cdot, n)$$
gives a continuous map $N_x \longrightarrow C^\infty(L_x, L_y)$.

Further, the collection $\{U_x\}$ is assumed maximal among such collections.

Since coordinate changes smoothly transform the level surface n = constant to n' = constant, the level surfaces coalesce to form maximal connected sets called *leaves*, and the space X is foliated by these leaves. Each leaf is a smooth manifold of dimension p.

The main examples of foliated spaces are, of course, foliated manifolds (cf. Lawson [L]), of class C^∞, or of class $C^{\infty,0}$ as in Connes [Co3]. We pause to exhibit some simple examples of foliated manifolds. These are quite standard; our reference is Lawson [L] upon whom we have relied heavily.

The simplest example of a foliated manifold is just M = $L^p \times N^q$ where L and N are smooth manifolds and M is foliated with leaves of the form $L \times (n)$. The projection map f: M $\longrightarrow$ N is a *submersion* (i.e., df_x: $TM_x \longrightarrow TN_x$ is surjective for all x). More generally, if $f: M^{p+q} \longrightarrow N^q$ is any submersion then M has a p-dimensional foliation with leaves corresponding to connected components of some $f^{-1}(n)$. For instance, suppose that

$$F^p \longrightarrow M \longrightarrow B^q$$

is a fibre bundle in the category of smooth manifolds with F connected. Then M^{p+q} is foliated by the inverse images $F_b \equiv f^{-1}(b) \cong$ F. The Hopf fibration

$$S^1 \longrightarrow S^3 \longrightarrow S^2$$

and a closed connected subgroup H of a Lie group G

$$H \longrightarrow G \longrightarrow G/H$$

yield foliations of S^3 and of G respectively.

A different sort of example arises by taking a connected Lie group G acting smoothly on a manifold M. Assume that the isotropy group at x, $\{g \in G \mid gx = x\}$, has dimension independent of x. Then M is foliated by the orbits of G. (If H acts on G for H a closed connected subgroup then this coincides with the previous example.)

Foliations may also be described in terms of the foliation bundle FM. Let $M = T^2 = \mathbb{R}^2/\mathbb{Z}^2$ and fix a smooth one-form $\omega = a_1 dx_1 + a_2 dx_2$ with $a_1 a_2 \neq 0$. It is evident that $d\omega = 0$. Let $FM = \{v \in TM \mid \omega(v) = 0\}$. This is an involutive sub-bundle and hence foliates the torus. If $a_1/a_2 \in \mathbb{Q}$ then each leaf is a circle. If $a_1/a_2 \notin \mathbb{Q}$ then each leaf is dense, in fact a copy of $\mathbb{R}$ sitting densely in the torus, which corresponds to an irrational flow on the torus.

Next we construct bundles with discrete structural group. Let F be a space, let B^p be a manifold (connected for simplicity), and let $\tilde{B} \longrightarrow B$ denote the universal cover. Suppose given a homomorphism

$$\varphi: \pi_1(B) \longrightarrow \text{Homeo}(F).$$

Form the space

(2.2)
$$M = \tilde{B} \times_{\pi_1(B)} F$$

as a quotient of $\tilde{B} \times F$ by the action of $\pi_1(B)$ determined by deck transformations on $\tilde{B}$ and by φ on F. The action on $\tilde{B} \times F$ is free and properly discontinuous, hence M is a foliated space. It is foliated by leaves ℓ_x which are the images of $\tilde{B} \times (x)$ as $x \in F$. There is a natural map $M \longrightarrow B$ and the composite $\ell_x \longrightarrow M \longrightarrow B$ is a covering space. If F is a manifold and φ takes values in Diff(F) then M is a smooth manifold.

A very special case of the above construction is of considerable importance. Suppose given a single homeomorphism $\theta \in$ Homeo(F). Then $\pi_1(S^1) = \mathbb{Z}$ acts on Homeo(F) via θ and there results a bundle

(2.3)
$$M = \mathbb{R} \times_{\mathbb{Z}} F \longrightarrow S^1$$

called the *suspension* of θ. For instance, if $\theta \in$ Diff($\mathbb{R}$) is the map $\theta(y) = -y$ then $\mathbb{R} \times \mathbb{R}$ has a $\mathbb{Z}$-action given by $(x,y) \longrightarrow (x+1,-y)$ and $M = \mathbb{R} \times_{\mathbb{Z}} \mathbb{R}$ is the Mobius band

(2.4)

Each leaf ℓ_y is a circle wrapping around twice except for the core circle ℓ_o (corresponding to $\mathbb{R} \times (0)$) which wraps once.

Finally we describe the Reeb foliation of S^3. This is constructed in stages. First foliate the open strip $\mathbb{R} \times [-1,1]$ as shown:

Then spin the strip about the x-axis to obtain a solid infinite cylinder (thought of as a collection of snakes, each eating the tail of the next):

Next identify (x,y,z) with $(x+1,y,z)$ to obtain a solid torus foliated by copies of $\mathbb{R}^2$ and the boundary leaf which is of course the torus.

(2.5)

(This is to be thought of as a collection of snakes, each eating its own tail.) Finally, observe that S^3 may be obtained by gluing two copies of a solid torus along the boundary torus. Taking two copies of the solid torus above, one obtains S^3 foliated by leaves of dimension two. Along the boundary of the two solid tori there is a closed leaf diffeomorphic to T^2. All other leaves are copies of $\mathbb{R}^2$. No leaf is dense: the closure of a typical copy of $\mathbb{R}^2$ is $\mathbb{R}^2$ together with the closed leaf T^2. Note that each point p on the closed leaf is a sort of saddle point in the sense that curves in leaves nearby (above and below) have the following saddle property:

(2.6)

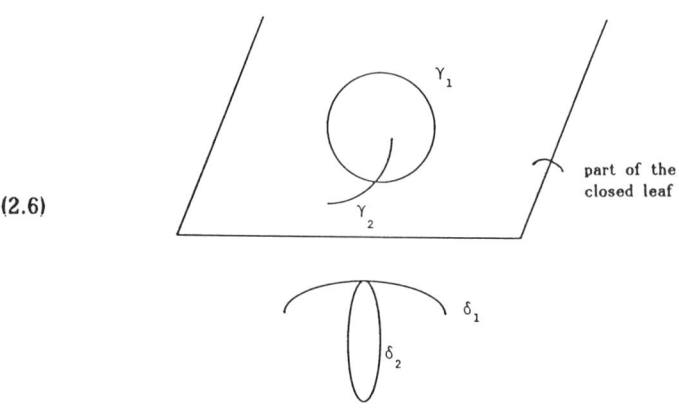

Curves γ_1, δ_1 are in the xz plane; curves γ_2, δ_2 are in the yz plane. Curves γ_1, δ_2 lie in the same leaf; curves γ_2, δ_1 lie in the same leaf. Schematically the snake below the closed leaf is moving left to right whereas the snake above the closed leaf is moving towards the reader. This is important for the resulting holonomy property as we shall see.

The notion of foliated space is strictly more general than that of a foliated manifold. A solenoid is a foliated space (p = 1) with each N_x homeomorphic to a subspace of a Cantor set. The infinite torus $T^\infty = \prod\limits_{j=1}^{\infty} T^1$ has a flow given by

$$r \rightarrow \lambda_o + \prod e^{ir\theta_j}$$

for fixed algebraically independent numbers $\{\theta_j\}$ and hence is a foliated space of dimension 1. Each N_x is homeomorphic to a subspace of T^∞, thought of as $1 \times T^\infty \subset T^1 \times T^\infty \equiv T^\infty$.

A continuous function f: $X \longrightarrow Y$ between foliated spaces of

possibly different dimensions which takes leaves to leaves is
tangentially smooth if in local coordinates the function

$$T_x \xrightarrow{\quad} T_x \times N_x \xrightarrow{\psi_{fx} f \varphi_x^{-1}} T'_{f(x)} \times N'_{f(x)} \xrightarrow{\quad} T'_{f(x)}$$

$$t \rightsquigarrow (t,n_o) \qquad\qquad (t',n') \xmapsto{\qquad} t'$$

is smooth for every n_o.

Each local patch U_x has a natural tangent bundle which is
induced by φ_x from the bundle $T_x \times T_x \times N_x \longrightarrow T_x \times N_x$. The transition
functions $\{\varphi_x\}$ preserve smoothness in the leaf direction and hence
these coalesce to form a p–plane bundle over X, called the *tangent
bundle* or foliation bundle of the foliated space and denoted p:
$FX \longrightarrow X$. We frequently write $F = FX$ and also write $FX_x \equiv F_x = p^{-1}(x)$ for the fibre over $x \in X$.

Proposition 2.7.   a)   A tangentially smooth map f: $X \longrightarrow Y$ induces a
bundle map df: $FX \longrightarrow FY$ which over leaves corresponds to the usual
differential.

b)   Let X^δ denote the disjoint union of the leaves of X
(each leaf having its smooth manifold topology).   Then X^δ is a
(usually non–separable) smooth manifold of dimension p, the identity
map i: $X^\delta \longrightarrow X$ is tangentially smooth, and $i^*FX = T(X^\delta)$, the
tangent bundle of X^δ.   □

A vector bundle p: $E \longrightarrow X$ of (real) dimension k over a
foliated space X of dimension p is *tangentially smooth* if E has
the structure of a foliated space of dimension p+k which is compatible
with the local product structure of the bundle and if p:$E \longrightarrow X$ is
tangentially smooth.   The tangent bundle is tangentially smooth.   We
let $C_\tau^\infty(X)$ denote the ring of (real–valued or complex–valued)
tangentially smooth functions on X and $\Gamma_\tau(E)$ or $\Gamma_\tau(X,E)$ denote
the $C_\tau^\infty(X)$–module of tangentially smooth sections of the bundle $E \longrightarrow X$.

The following series of propositions (2.8 - 2.15) serves to let us assume freely that all bundles which arise in our study are tangentially smooth. Transverse continuity is essential here; Proposition (2.8) is false if one assumes only transverse measurability.

Proposition 2.8.   Let X be a foliated space.   Then every open cover of X has a subordinate tangentially smooth partition of unity.

Proof: (Compare Hirsch [Hir, 2.2.1])   Let $\mathcal{U} = \{U_i\}_{i \in I}$ be an open cover of X.   There is a locally finite atlas on X, $\{\varphi_\alpha, V_\alpha\}$, such that $\{\bar{V}_\alpha\}$ refines $\mathcal{U}$; and we may assume that each $\varphi_\alpha(V_\alpha) \subset T_\alpha \times N_\alpha \subset \mathbb{R}^p \times N_\alpha$ is bounded and each $\bar{V}_\alpha \subset \bar{X}$ is compact.   There is a shrinking $\{W_\alpha\}_{\alpha \in J}$ of $V = \{V_\alpha\}_{\alpha \in J}$, and each $\bar{W}_\alpha \subset V_\alpha$ is compact.   It suffices to find a tangentially smooth partition of unity subordinate to V.

For each α, cover the compact set $\varphi_\alpha(\bar{W}_\alpha) \subset \mathbb{R}^p \times N_\alpha$ by a finite number of closed balls

$$B(\alpha,1),...,B(\alpha,k(\alpha))$$

contained in $\varphi_\alpha(V_\alpha)$.   Choose maps

$$h_{\alpha,j} \colon \mathbb{R}^p \times N_\alpha \longrightarrow [0,1] \quad j = 1,...,k(\alpha)$$

which are tangentially smooth (i.e., maps $N_\alpha \longrightarrow C^\infty(\mathbb{R}^p,[0,1])$) such that

$$h_{\alpha,j}(x) > 0 \quad \text{if and only if} \quad x \in \text{Int } B(\alpha,j).$$

Let

$$h_\alpha = \sum_{j=1}^{k(\alpha)} h_{\alpha,j} : \mathbb{R}^p \times N_\alpha \longrightarrow [0,\infty).$$

Then

$$h_\alpha(x) > 0 \quad \text{if } x \in \varphi_\alpha(\bar{W}_\alpha)$$

$$h_\alpha(cx) = 0 \quad \text{if } x \in \mathbb{R}^p \times N_\alpha - \cup_j B(\alpha, j).$$

Let $m_\alpha \colon M \longrightarrow [0, \infty)$ be defined by

$$m_\alpha(x) = \begin{cases} h_\alpha \varphi_\alpha(x) & \text{if } x \in V_\alpha \\ 0 & \text{if } x \in X - V_\alpha \end{cases}$$

Then m_α is tangentially smooth, $m_\alpha > 0$ on $\bar{W}_\alpha$, and supp $m_\alpha \subset V_\alpha$. Define $r_\alpha = m_\alpha / \sum_\alpha m_\alpha$. Then $\{r_\alpha\}$ is a tangentially smooth partition of unity subordinate to V. □

For foliated spaces X and Y, let $C^0_\ell(X,Y)$ denote the continuous functions from X to Y which take leaves to leaves and let $C^\infty_\tau(X,Y)$ denote the subset of tangentially smooth maps. We topologize $C^0_\ell(X,Y)$ by the strong topology. Let $\Phi = \{\varphi_i, U_i\}_{i \in \Lambda}$ and $\Psi = \{\psi_i, V_i\}_{i \in I}$ be locally finite sets of charts on X and Y respectively. Let $K = \{K_i\}_{i \in I}$ be a family of compact subsets of X, with $K_i \subset U_i$, let $\mathcal{E} = \{\epsilon_i\}_{i \in I}$ a family of positive numbers, and let $f \in C^0_\ell(X,Y)$ with $f(K_i) \subset V_i$. A *strong basic neighborhood* $N^0(f; \Phi, \Psi, K, \mathcal{E})$ is the set of maps $g \in C^0_\ell(X,Y)$ such that $g(K_i) \subset V_i$ for all $i \in I$ and $\|(\psi_i f \varphi_i^{-1})(x) - (\psi_i g \varphi_i^{-1})(x)\| < \epsilon_i$ for all $x \in \varphi_i(K_i)$. The strong topology has all possible sets of this form for a base. If X is compact then this topology coincides with the weak (= compact-open) topology on $C^0_\ell(X,Y)$. We refer the reader to Hirsch [Hir] from which we have freely borrowed.

Proposition 2.9. Let $X = T \times N$ and $X' = T' \times N'$ be trivial foliated spaces (with $T \subset R^p$, $T' \subset R^{p'}$). Then $C^\infty_\tau(X,X')$ is dense in $C^0_\ell(X,X')$.

Proof: Since all functions preserve leaves we may assume that $X' = T' = \mathbb{R}^n$, regarded as a foliated space with one leaf. We must show that $C^\infty_\tau(X,\mathbb{R}^n)$ is dense in $C^0(X,\mathbb{R}^n)$ in the strong topology.

Let $\{V_\alpha\}$ be a locally finite open cover of X and for each α let $\epsilon_\alpha > 0$. Let $f \colon X \longrightarrow \mathbb{R}^n$ be continuous, and suppose we want a

C^∞_τ map g to satisfy $|f-g| < \epsilon_\alpha$ on V_α for all α. For each $x \in$ X, let $W_x \subset X$ be a neighborhood of x meeting only finitely many V_α. Set

$$\delta_x = \min\{\epsilon_\alpha : x \in V_\alpha\} > 0.$$

Let $U_x \subset W_x$ be an open neighborhood of x so small that $|f(y)-f(x)|$ $< \delta_x$ for all $y \in U_x$. Define constant maps $g_x : U_x \longrightarrow \mathbb{R}^n$ by $g_x(y)$ $= f(x)$. Relabeling the cover $\{U_x\}$ and the maps $\{g_x\}$, we have shown: there is an open cover $\{U_i\}_{i \in I} = \mathcal{U}$ of X and C^∞_τ maps $g_i : X \longrightarrow \mathbb{R}^n$ such that whenever $y \in U_i \cap V_\alpha$ then

$$|g_i(y) - f(y)| < \epsilon_\alpha.$$

Let $\{r_i\}_{i \in I}$ be a C^∞_τ partition of unity subordinate to $\mathcal{U}$. Define g: $X \longrightarrow \mathbb{R}^n$ by

$$g(y) = \Sigma_i\, r_i(y)g_i(y).$$

Then $g \in C^\infty_\tau(X,\mathbb{R}^n)$, and

$$|g(y) - f(y)| = |\Sigma\, r_i(y)g_i(y) - \Sigma\, r_i(y)f(y)|$$

$$\leqslant \Sigma\, r_i(y)\,|g_i(y) - f(y)|.$$

Hence if $y \in V_\alpha$ then

$$|g(y) - f(y)| < \Sigma\, r_i(y)\epsilon_\alpha = \epsilon_\alpha. \qquad \square$$

The following relative approximation lemma allows us to globalize the preceeding proposition.

Proposition 2.10. Let $U = L \times N \subset \mathbb{R}^p \times N$ and $V = L' \times N' \subset \mathbb{R}^{p'} \times N'$ be open sets, $K \subset U$ a closed set, $W \subset U$ an open set, and $f \in C^0_\ell(U,V)$ such that f is tangentially smooth on a neighborhood of

46

K-W. Then every neighborhood N of f in $C^{o}_{\ell}(U,V)$ contains a map h: $U \longrightarrow V$ which is tangentially smooth on a neighborhood of K and agrees with f on U - W.

Proof: Since all maps are to send leaves to leaves and since $C^{o}(U,L')$ is open in $C^{o}(U,\mathbb{R}^{p})$ we may assume that $V = L' = \mathbb{R}^{n}$. Let $A \subset U$ be an open set containing the closed set K-W such that $f|A$ is C^{∞}_{τ}. Let $W_{o} \subset U$ be open with

$$K-A \subset W_{o} \subset \overline{W}_{o} \subset W.$$

Let $\{r_{o},r_{1}\}$ be a C^{∞}_{τ} partition of unity for the open cover $\{W,U-\overline{W}_{o}\}$ of U. Define

$$G: C^{o}(U,\mathbb{R}^{n}) \longrightarrow C^{o}(U,\mathbb{R}^{n})$$

by

$$G(g)(x) = r_{o}(x)g(x) + r_{1}(x)f(x).$$

Then

$$G(g)\big|_{W_{o}} = g\big|_{W_{o}}$$

and

$$G(g)\big|_{U-W} = f\big|_{U-W}.$$

Further, $G(g)$ is C^{∞}_{τ} on every open set on which both f and g are C^{∞}_{τ}, and G is clearly continuous.

Since $G(f) = f$, there is an open set $N_{o} \subset C^{o}(U,\mathbb{R}^{n})$ containing f such that $G(N_{o}) \subset N$. By Proposition 2.9 there is a C^{∞}_{τ} map $g \in N_{o}$ (since $C^{\infty}_{\tau}(U,\mathbb{R}^{n})$ is dense in $C^{o}(U,\mathbb{R}^{n})$). Then $h = G(g)$ has the required properties. □

We now prove the basic approximation theorem.

Theorem 2.11.　　Let X and Y be foliated spaces.　　Then $C_{\mathcal{T}}^\infty(X,Y)$ is dense in $C_{\ell}^0(X,Y)$ with the strong topology.

Corollary 2.12.　　Let X be a foliated space and let M be a C^∞-manifold, regarded as a foliated space with one leaf.　　Then $C_{\mathcal{T}}^\infty(X,M)$ is dense in $C^0(X,M)$.

Proof:　　Let f: X $\longrightarrow$ Y be in C_{ℓ}^0.　Let $\Phi = \{\varphi_i, U_i\}_{i \in I}$ be a locally finite atlas for X and let $\Psi = \{\psi_i, V_i\}_{i \in I}$ be a family of charts for Y such that for all i $\in$ I, $f(U_i) \subset V_i$.　Let $\mathcal{C} = \{C_i\}_{i \in I}$ be a closed cover of X, $C_i \subset U_i$.　Let $\mathcal{E} = \{\varepsilon_i\}_{c \in \Lambda}$ be a family of positive numbers and put $N = N(f;\ \Phi, \psi, \mathcal{C}, \mathcal{E}) \subset C_{\ell}^0(X,Y)$.　We look for a g $\in$ N which is $C_{\mathcal{T}}^\infty$.　The set I is countable; we therefore assume that I = {1,2,3,...} or, if X is compact, I = {1,2,...,s}.

Let $\{W_i\}_{i \in I}$ be a family of open sets in X such that $C_i \subset W_i \subset \bar{W}_i \subset U_i$.　We shall define by induction a family of $C_{\mathcal{T}}^\infty$ maps $g_k \in$ N, having the following properties: $g_o = f$ and for k $\geqslant$ 1,

$$g_k = g_{k-1} \text{ on } X - W_k$$

$$g_k \text{ is } C_{\mathcal{T}}^\infty \text{ on a neighborhood of } \bigcup_{0 \leqslant j \leqslant k} C_j$$

Assuming for the moment that the g_k exist, define g: X $\longrightarrow$ Y by $g(x) = g_{\kappa(x)}(x)$, where $\kappa(x) = \max\ \{k \,|\, x \in \bar{U}_k\}$.　Each x has a neighborhood on which $g = g_{\kappa(x)}$.　This shows that g $\in C_{\mathcal{T}}^\infty$ and g $\in$ N, and the theorem is proved.

It remains to construct the g_k.　Put $g_o = f$; then the hypotheses are true vacuously.　Suppose that 0 < m and we have maps $g_k \in$ N, 0 $\leqslant$ k < m satisfying the inductive hypothesis.　Define a space of maps

$$\mathcal{Y} = \{h \in C_{\ell}^0(U_m, V_m) \,\big|\, h = g_{m-1} \text{ on } U_m - W_m\}.$$

Define T: $\mathcal{Y} \longrightarrow C_{\ell}^0(X,Y)$ by

$$T(h) = \begin{cases} h & \text{on } U_m \\ g_{m-1} & \text{on } X - U_m \end{cases}$$

It is evident that T is continuous, $T(g_{m-1}\vert U_m) = g_{m-1}$, and hence $T^{-1}(N) \neq \varnothing$.

Let $K = \cup_{k \leq m} C_k \cap U_m$. Then K is a closed subset of U_m and $g_{m-1}: U_m \longrightarrow V_m$ is C_τ^∞ on a neighborhood of $K - W_m$. Since U_m and V_m are trivially foliated spaces we can apply the previous proposition to $C_\ell^0(U_m, V_m)$. We conclude that the maps in $\mathcal{B}$ which are C_τ^∞ in a neighborhood of K are dense in $\mathcal{B}$. Therefore $T^{-1}(N)$ contains such a map h. Define $g_m = T(h)$; then $g_m \in N$ satisfies the inductive hypothesis at stage m, completing the proof. $\square$

Relative Approximation Theorem 2.13. Let $f \in C_\ell^0(X,Y)$ and suppose that f is tangentially smooth on some neighborhood of a (possibly empty) closed set $A \subset X$. Then every neighborhood N of f in $C_\ell^0(X,Y)$ contains a map $h \in C_\tau^\infty(X,Y)$ with $h = f$ on some neighborhood of A.

Proof: If X and Y are product foliations then this follows from the relative approximation lemma 2.10. The local-global process is essentially the same as in the proof of Theorem 2.11 where we show that $C_\tau^\infty(X,Y)$ is dense in $C_\ell^0(X,Y)$. In the construction of the maps $\{g_k\}$, add the additional condition that $g_k = f$ on A. In the induction assume that every map in $\mathcal{B}$ agrees with f on some neighborhood of A. The relative approximation lemma 2.10 allows the same argument to proceed. $\square$

Lemma 2.14. Let $f \in C_\ell^0(X, \mathbb{R}^p \times N)$, let A be a closed subset of X, and suppose that f is tangentially smooth on some neighborhood of A. Then there is a homotopy $H \in C_\ell^0(X \times \mathbb{R}, \mathbb{R}^p \times N)$ such that

$$1) \quad H(x,t) = \begin{cases} f(x) & \text{for } t \leq 0 \\ H(x,1) & \text{for } t \geq 1 \\ f(x) & \text{for } x \in A \end{cases}$$

2) $H(x,1) \in C_\tau^\infty(X, \mathbb{R}^p \times N)$

3)   For each t, $H(-,t)$ is arbitrarily close to f on compact subsets.

Proof: Write $f(x) = (f_1(x), f_2(x))$. By the previous theorem, there is a map $g \in C_\tau^\infty(X, \mathbb{R}^p \times N)$ with $g = f$ on some neighborhood of A and g arbitrarily close to f. We may assume that $g = (g_1, f_2)$. Let $\delta \in C^\infty(\mathbb{R}, \mathbb{R})$ be a monotone function with $\delta(t) = 0$ for $t \leq 0$ and $\delta(t) = 1$ for $t \geq 1$. Define H: $X \times \mathbb{R} \longrightarrow \mathbb{R}^p \times N$ by

$$H(x,t) = (f_1(x)(1-\delta(t)) + g(x)\delta(t), \; f_2(x)).$$

Then H has the required properties.   □

Theorem 2.15.   Let $f \in C_\ell^o(X,Y)$ and suppose that f is tangentially smooth on some neighborhood of a closed subset A. Then there is a homotopy $H \in C_\ell^o(X \times \mathbb{R}, \; Y)$ such that

1) $H(x,t) = \begin{cases} f(x) & \text{for } t \leq 0 \\ H(x,1) & \text{for } t \geq 1 \\ f(x) & \text{for } x \in A \end{cases}$

2) $g = H(x,1) \in C_\tau^\infty(X,Y)$

3) $H(-,t)$ is arbitrarily close to f on compact subsets.

The function $g \in C_\tau^\infty(X,Y)$ is unique up to tangentially smooth homotopy (rel A) and hence defines a unique map $f^*: H_\tau^*(Y) \longrightarrow H_\tau^*(X)$. Here H_τ^* is tangential cohomology which will be formally introduced in Chapter III.

Proof: Let $\Psi = \{V_i\}$ be a family of coordinate patches for Y and let $\Phi = \{U_i\}$ be a family of coordinate patches for X with $f^{-1}(V_i) \subset U_i$. Let

$C = \{C_i\}$ be a closed cover of X with $C_i \subset U_i$. Let $\mathcal{E} = \{\mathcal{E}_i\}$ be a family of positive numbers, and let

$$N = (f\pi, \Psi \times \mathbb{R}, \Psi \times \mathbb{R}, \mathcal{C}, \mathcal{E}) \subset C^o_\ell(X \times \mathbb{R}, Y)$$

where $\pi: X \times \mathbb{R} \longrightarrow X$ is the projection and $\Phi \times \mathbb{R}$ is the pullback along π of Φ. Choose open sets W_i with $C_i \subset W_i \subset \bar{W}_i \subset U_i$.

We shall define by induction a family of maps $g_k \in C^o_\ell(X \times \mathbb{R}, Y)$ with the following properties:

1)
$$g_k(x,t) = \begin{cases} f(x) & \text{for } t \leq 0 \\ g_k(x,1) & \text{for } t \geq 1 \\ f(x) & \text{for } x \in A \end{cases}$$

2) $g_k(x,1)$ is tangentially smooth on a neighborhood of the set $(L_1 \cup ... \cup L_k) \times [1, \infty)$

3) $g_k(-,t)$ is close to f on compact subsets

4) $g_o(x,t) = f(x)$

5) $g_k = g_{k-1}$ on $(X \times \mathbb{R}) - (W_k \times \mathbb{R})$.

Suppose for the moment that the g_k exist. Define H: $X \times \mathbb{R} \longrightarrow Y$ by

$$H(x,t) = g_{\kappa(x)}(x,t)$$

where $\kappa(x) = \max\{k \mid x \in \bar{U}_k\}$. Each point (x,t) has a neighborhood on which $H(x,t) = g_{\kappa(x)}(x,t)$. Thus $H(x,1) \in C^\infty_\tau(X,Y)$. The other conditions on H are evident, so H has been constructed as required.

Here is the construction of the g_k. Set $g_o(x,t) = f(x)$. Suppose that $m > 0$ and we have maps $g_k \in N$ with $0 \leq k < m$ satisfying the inductive hypotheses. Define a space of maps $\mathfrak{V}$ by

$$\mathfrak{d} = \{h \in C^o_\ell(U_m \times \mathbb{R}, V_m) \mid h = g_{m-1} \text{ on } (U_m - W_m) \times \mathbb{R} \text{ and}$$

$$h = f \text{ on a neighborhood of } (U_m \cap A) \times \mathbb{R}\}.$$

Define $T: \mathfrak{d} \longrightarrow C^o_\ell(X \times \mathbb{R}, Y)$ by

$$T(h) = \begin{cases} h & \text{on } U_m \times \mathbb{R} \\ g_{m-1} & \text{on } (X - U_m) \times \mathbb{R}. \end{cases}$$

It is evident that T is continuous and $T(g_{m-1}|_{U_m}) = g_{m-1}$, so $T^{-1}(N)$ is non-empty.

Let $K = \bigcup_{k \leq m} (C_k \cap U_m) \times \mathbb{R}$. Then K is a closed subset of

$U_m \times \mathbb{R}$ and $g_{m-1} \in C^o_\ell(U_m \times \mathbb{R}, V_m)$ is tangentially smooth on a neighborhod of $K \times (W_m \times \mathbb{R})$. Since $U_m \times \mathbb{R}$ and V_m are product foliated spaces, we may apply the previous proposition to $C^o_\ell(U_m \times \mathbb{R}, V_m)$. We conclude that the maps in $\mathfrak{d}$ which are tangentially smooth on some neighborhood of K are dense in $\mathfrak{d}$. Therefore $T^{-1}(N)$ contains such a map h. Define $g_m = T(h)$. Then $g_m \in N$ satisfies the inductive hypotheses at stage m. This completes the proof of the existence of the homotopy H.

It remains to demonstrate that $g = H(-,1)$ is unique up to tangentially smooth homotopy which fixes A. Suppose that g and $\bar{g}$ are both constructed by the above procedure with $g = H(-,1)$ and $\bar{g} = \bar{H}(-,1)$ and $g = \bar{g}$ on A. An obvious construction yields a homotopy $M \in C^o_\ell(X \times \mathbb{R}, Y)$ with

$$M(x,t) = \begin{cases} g(x) & t \leq 0 \\ \bar{g}(x) & t \geq 1 \\ g(x) & x \in A \end{cases}$$

and M is close to g as usual. Let $\tilde{A} = X \times [(-\infty, 0] \cup [1, \infty)] \cup (A \times \mathbb{R})$. Apply the first part of the theorem with X replaced by $X \times \mathbb{R}$, f replaced by M, and A replaced by $\tilde{A}$. We obtain a function $\tilde{M} \in C^\infty_T(X \times \mathbb{R}, Y)$ with

$$\tilde{M}(x,t) = \begin{cases} g(x) & t \leqslant 0 \\ \bar{g}(x) & t \geqslant 1 \\ g(x) & x \in A \end{cases}$$

and $\tilde{M}$ is close to g as usual. Thus g is homotopic to $\bar{g}$ via a tangentially smooth homotopy fixing A. □

We consider next the consequences of Theorem 2.15 for vector bundles.

Proposition 2.16: Every continuous (real or complex) vector bundle E over a compact foliated space X has a compatible C_τ^∞ bundle structure; and such a structure is unique up to C_τ^∞ isomorphism.

Proof: Let g: $X \longrightarrow G_n$ be a classifying map for $E \longrightarrow X$, where $G_n = G_n(\mathbb{R}^{n+k})$ or $G_n(\mathbb{C}^{n+k})$ denotes a suitable compact Grassmann manifold with canonical smooth bundle $E^n \longrightarrow G_n$. Then g can be approximated by, and so is homotopic to, a C_τ^∞ map h by Theorem 2.15. Then E is equivalent to $h^* E^n$, and $h^* E^n$ is a C_τ^∞-bundle, since E^n is a smooth bundle and h is of class C_τ^∞.

If E_0 and E_1 are C_τ^∞ bundles that are isomorphic as $C^\bullet$-bundles then there is a continuous map H: $X \times I \longrightarrow G_n$ such that $H_i^*(E^n) = E_i$ i = 0,1. Approximate H by a map $\hat{H}$ in C_τ^∞ fixing H_0 and H_1; then $\hat{H}_\tau^*(E^n)$ is a C_τ^∞ equivalence between E_0 and E_1. □

The proposition implies that tangentially smooth K-theory (i.e., K-theory defined via tangentially smooth bundles) on locally compact foliated spaces coincides with the usual K-theory. In the next chapter we shall introduce tangential de Rham cohomology. This does not agree with ordinary de Rham cohomology, as will become evident.

Let X be a foliated space of dimension p. The next order of business is the construction of the holonomy groupoid or graph of X, denoted G or G(X). Our construction follows that of Winkelnkemper [Wi] as closely as possible.

Recall that a *plaque* is a component of U ∩ ℓ, where ℓ is some leaf and U is some coordinate patch. Every point of X has a neighborhood which consists of a union of plaques with respect to some U and, with respect to the same U, two different plaques $\hbar$, $\hbar'$, can be on the same leaf. A *regular covering* is a covering of X by open coordinate patches U_i such that each plaque in U_i meets at most one plaque in U_j. We henceforth assume (without any loss of generality) that our covers are always regular.

We recall the definition and elementary properties of the concept of *holonomy*. Let ℓ be a leaf of X and α an arc in ℓ starting at a and ending at b. Subdivide the arc α into small enough subarcs by means of points $a = a_0, a_1, ..., a_k = b$ so that each point a_i has a neighborhood U_i consisting entirely of plaques, so that if we choose a plaque $\hbar_0$ of U_0 then there is a unique plaque $\hbar_1 \subset U_1$ which intersects $\hbar_0$, a unique plaque $\hbar_2 \subset U_2$ which intersects $\hbar_1$, etc., and finally a unique plaque $\hbar_k \subset U_k$.

Let N_a and N_b be transversals X through a and b respectively. For points $n \in N_a$ which are sufficiently close to a, we define $H^{\alpha}_{ab}(n) \in N_b$ by the above procedure. That is, find the unique plaque $\hbar_0 \subset U_0$ which contains n, follow the plaque to plaque $\hbar_k \subset U_k$, and define $H^{\alpha}_{ab}(n)$ to be the unique element in $\hbar_k \cap N_b$. Then H^{α}_{ab} is a homeomorphism from a neighborhood of a in N_a to a neighborhood of b in N_b, and $H^{\alpha}_{ab}(n)$ lies on the same leaf as n. Choosing the partition $\{a_i\}$ and the neighborhoods $\{U_i\}$ differently changes H^{α}_{ab}, but the new and old maps will coincide on some smaller neighborhood. Thus the germ of H^{α}_{ab} does not depend on these choices. Altering α by a homotopy in ℓ which fixes endpoints preserves the germ of H^{α}_{ab}.

If $a = b$ and $N_a = N_b$ then composing the germs is a well-defined operation under which the holonomy germs form a group G^a_a. The natural map $\pi_1(\ell, a) \longrightarrow G^a_a$ given by $\alpha \to H^{\alpha}_{aa}$ is a surjective homomorphism, so if ℓ is simply connected then $G^a_a = \{0\}$ for each $a \in \ell$. The group G^a_a is the *holonomy group* of the leaf ℓ at the point a. (The notation comes from groupoids and will become apparent.) The set $\{x \in X \mid G^x_x = 0\}$ is a dense G_δ, by Epstein, Millett, and Tischler [EMT], so that in that sense at least trivial holonomy is generic.

The set $\{x \in G \mid G^x_x \neq 0\}$ may have positive measure. For example, let K be a Cantor subset of the unit circle of positive measure. Let ϕ be a homeomorphism of the circle which has K as its fixed point set. The associated foliation of the torus has closed leaves corresponding to each point of K and each of these leaves has non-trivial holonomy.

For example, if we foliate the annulus as shown

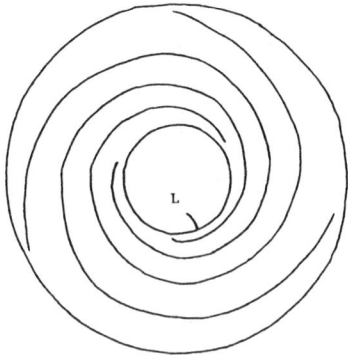

then $G_a^a = \mathbf{Z}$ for each $a \in \ell$, since with respect to the arc α which traverses the leaf once clockwise the holonomy map $H: I \longrightarrow I$ is monotone decreasing and hence of infinite order in G_a^a. Similarly, the other closed leaf has non-trivial holonomy. Each of the remaining leaves is homeomorphic to $\mathbb{R}$ (and thus simply connected) and hence has trivial holonomy.

Let $\tilde{X}$ be the foliated space shown:

(2.17)

This is noncompact, of course. Every leaf is simply connected, so all of the holonomy groups are trivial. The exponential map yields a tangentially smooth map $\tilde{X} \longrightarrow X$, and this is a covering space.

Here are some more examples to illustrate the concept.

Consider the torus, foliated as indicated:

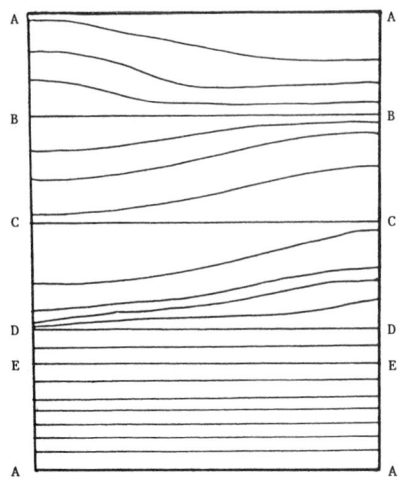

(2.18)

There are closed leaves through A,B,C, and a family of closed leaves intercepting the line segment DA (with ℓ_E as a typical closed leaf in this family). Each closed leaf ℓ is a circle, with $\pi_1(\ell) = \mathbf{Z}$. The leaf ℓ_E has trivial holonomy, since a small transverse disk meets only the adjacent family of closed leaves which are plaque paths. The leaves ℓ_A, ℓ_B, ℓ_C, and ℓ_D each have holonomy group $\mathbf{Z}$. Note that for the leaf ℓ_D a disk placed between D and E is acted upon trivially; the disk must overlap the C–D area to see the holonomy.

The Reeb foliation of $\mathbf{S}^3$ has a unique closed leaf ℓ_0 diffeomorphic to the torus T^2 with $\pi_1(\ell_0) = \mathbf{Z}^2$. The holonomy group G_x^x for $x \in \ell_0$ is also $\mathbf{Z}^2$, generated by the images of the paths δ_1 and δ_2 in figure 2.6.

The case of a bundle $M \longrightarrow B$ with discrete structural group (2.2) given by a homeomorphism $\varphi: \pi_1(B) \longrightarrow \mathrm{Homeo}(F)$ is particularly pleasing. For $x \in F$, let

$$\Gamma_x = \{g \in \pi_1(B) \mid \varphi(g)x = x\}$$

be the isotropy group. The leaf ℓ_x (which is the image of $\tilde{B} \times \{x\}$ in M) may be expressed as $\ell_x = \tilde{B}/\Gamma_x$ where Γ_x acts on $\tilde{B}$ by deck transformations. The holonomy group G_x^x is the image of the homeomorphism

$$\pi_1(\ell_x) \cong \Gamma_x \longrightarrow \text{Homeo}(F,x)$$

where Homeo(F,x) denotes the germs of homeomorphisms at x which fix x.

For instance, consider the Mobius strip

$$M = \mathbb{R} \times_{\mathbb{Z}} \mathbb{R}$$

foliated by circles corresponding to the images of $\mathbb{R} \times \{y\}$ for various values of $y \in \mathbb{R}$ (cf. 2.4). If $y \neq 0$ then $\pi_1(\ell_y) = \mathbb{Z}$ acts trivially upon Diff($\mathbb{R},y$), and hence $G_y^y = 0$. However, the holonomy group G_0^0 of the core circle is the group $\mathbb{Z}/2$, since the diffeomorphism $\theta(y) = -y$ which creates M does lie in Diff($\mathbb{R},0$), and $\theta^2 = 1$.

Holonomy is a critically important internal property of foliations. As evidence we cite a special case of the Reeb stability theorem and refer the reader to Lawson [L] for more information.

Theorem 2.19 (Reeb). Let M^{p+q} be a smooth foliated manifold with a compact leaf ℓ with trivial holonomy. Then there exists a neighborhood U of ℓ in M such that U is a union of leaves and a diffeomorphism

$$\ell \times D^q \xrightarrow{\ f\ } U$$

which preserves leaves.

Thus M has a family of compact leaves near ℓ. We see this theorem at work in example (2.18). The leaves $\ell_A, \ell_B, \ell_C, \ell_D,$ and ℓ_E are all compact. Only ℓ_E has trivial holonomy; it does have a family of closed leaves near it, of the form

$$\ell \times (E-\epsilon, \ E+\epsilon).$$

We next introduce the graph (or groupoid) of a foliation and verify its elementary properties. This is due to Ehresman [Eh],

Reeb, and Thom [Tho] and expounded by Winkelnkemper [Wi].
See also Phillips [Ph]. A more systematic discussion of groupoids
will be found in Chapter IV.

Definition 2.20. The *holonomy graph* or *groupoid* G(X) of the
foliated space X is defined to be the collection of all triples
$(x,y,[\alpha])$ where x and y lie on the same leaf ℓ, α is a
(piecewise-smooth) path from x to y in ℓ, and $[\alpha]$ is the
holonomy equivalence class of α: α is equivalent to β if $\alpha\beta^{-1} =$
1 or id in G_y^y.
There are canonical maps as follows:

1) Δ: X $\longrightarrow$ G(X) by $\Delta(x) = (x,x,[0])$, where 0 denotes
the constant arc at x.

2)   an involution i: G(X) $\longrightarrow$ G(X) given by $i(x,y,[\alpha]) =$
$(y,x,[\alpha^{-1}])$

3)   projections p_1,p_2: G(X) $\longrightarrow$ X defined by

$$p_1(x,y,[\alpha]) = x$$
$$p_2(x,y,[\alpha]) = y.$$

Frequently p_1 is written as r (= range) and p_2 is written as s
(= source). Note that if ℓ_x is the leaf through x in X, then $p_1^{-1}(x)$
will turn out to be $\tilde{\ell}_x$, the covering space of ℓ_x corresponding to
the holonomy kernel and $\tilde{\ell}_x/G_x^x = \ell_x$. Thus the construction
"'unwraps' all leaves of X simultaneously with respect to their correct
topology as well as their holonomy." (Winkelnkemper [Wi], 0.3)

4)   Let
$$G(X) \oplus G(X) = \{(u,v) \in G(X) \times G(X) \mid p_1(u) = p_1(v)\}.$$

Then we have m: G $\oplus$ G $\longrightarrow$ G defined by

$$m((x,y,[\alpha]),(x,z,[\beta])) = (y,z,[\beta\alpha^{-1}])$$

with $m \circ \text{diag} = \Delta p_2$, $m(u,v) = i \circ m(v,u)$, and

$$m(u, \Delta \circ p_1(u)) = i(u).$$

The foliation on X induces a 2p-dimensional foliation on G(X): the leaf in G(X) through the point $(x_o, y_o, [\alpha_o])$ does not depend on $[\alpha_o]$ and consists of all triples $(x, y, [\alpha])$ with $x, y \in \ell_{x_o} = \ell_{y_o}$ with $[\alpha]$ arbitrary. With the leaf topology it is diffeomorphic to $p_1^{-1}(x) \times p_2^{-1}(y) \cong \tilde{\ell}_x \times \tilde{\ell}_y$.

Next we define the topology on G(X). Let $z = (a, b, [\alpha])$ be a point in G(X). Choose a path α which represents $[\alpha]$, a family $\mathcal{U} = \{U_1, ..., U_k\}$ of coordinate patches which implements the holonomy map H_{ab}^{α} and an open transversal N upon which H_{ab}^{α} is defined. We may assume that the projection n: $U_1 \longrightarrow N$ is surjective. Suppose that x is an element of U_1. There is a path s_x in the plaque of x (unique up to holonomy) from x to n(x). By the setup above, there is a canonical (up to holonomy) path α_x from n(x) to $H_{ab}^{\alpha}(n(x))$. Let q(x) be the unique plaque in U_k which contains H_{ab}^{α}. Then for $y \in q(x)$ there is a path t_{yx} in q(x) (unique up to holonomy) from $H_{ab}^{\alpha}(n(x))$ to y. As a subbase for the topology of G(X) we take the subsets

$$V_{z,\alpha,\mathcal{U},N} = \{(x, y, [t_{yx}\alpha_x s_x]): x \in U_1, y \in q(x)\}.$$

Since each leaf, being a smooth p-manifold, has countably generated fundamental groups, it follows that this topology has a countable base.

Proposition 2.21. With the above topology G(X) is Hausdorff if and only if for all x and y, the holonomy maps along two arbitrary arcs α and β from x to y and with respect to the same transversals N_x, N_y are already the same if they coincide on an open subset of their domain whose closure contains x.

Proof: Since X is Hausdorff, it is enough to separate points $z = (x, y, [\alpha])$ and $z' = (x, y, [\alpha'])$ in order to show that G(X) is Hausdorff.

Suppose any two neighborhoods of z and z' had a point z" = (x",b",[a"]) in common.   Then

$$z" \in V_{z,\alpha} \cap V_{z',\alpha'} =$$

$$\{(\tilde{x},\tilde{y},\beta_n) \in G(X) \mid \tilde{x} \in \wedge_x(n), \ \tilde{y} \in H^{\alpha}_{xy}(\wedge_x(n)) \cap H^{\alpha'}_{xy}(\wedge_x(n))\}$$

where $\beta_n = s_{\tilde{x}} \cup \alpha_n \cup s_{\tilde{y}} \simeq s'_{\tilde{x}} \cup \alpha'_n \cup s_{\tilde{y}}$. Since the short arcs $s_{\tilde{x}}, s_{\tilde{y}}$ do not affect holonomy, the holonomy along both α and α' would have to coincide with the holonomy defined by $\alpha"$ on its domain.   The domain of the holonomy of $\alpha"$ contains x in its closure.

Conversely, if the holonomy along α and α' coincided on an open set, containing x in its closure, then from the definition of the sets $V_{z\alpha}$ above any neighborhood of z will intersect any neighborhood of z'.   □

<u>Corollary</u> 2.22.   If $G^x_x = 0$ for all $x \in X$ then G(X) is Hausdorff.

This is the case, for instance, if each leaf is simply-connected.

Consider the graph of Example (2.18).   Is it Hausdorff? Following Proposition 2.21, it suffices to examine the foliation at leaves with non-trivial holonomy, in this case leaves $\ell_A, \ell_B, \ell_C,$ and ℓ_D.   Intuitively the question is whether the holonomy is one-sided.   Leaves ℓ_B and ℓ_C cause no difficulty.   However, leaves ℓ_A and ℓ_D do indeed cause difficulty.   Take ℓ_D, for example.   Here is the picture:

Let α be the horizontal circle through D and Let β be the constant path at D.   Let N be the transversal (r,s) and let N' be the

60

transversal (D,s).   We have already determined that G_D^D = $\mathbf{Z}$
generated by $[\alpha]$, and thus

$$(D,D,[\alpha]) \neq (D,D,[\beta])$$

in the graph.   However the transversal N' (which contains D in its
closure) does not detect the presence of holonomy.   Proposition 2.21
implies   that   the   points $(D,D,[\alpha])$ and $(D,D,[\beta])$ cannot   be
separated by disjoint open sets, so the graph is not Hausdorff.

   The graph of the Reeb foliation of S^3 is also not Hausdorff,
though this is for more subtle reasons.   The point is that the
holonomy corresponding to the spreading out in the γ_2 direction is
seen by a closed path in the leaf parallel to δ_1, so that if one cuts
the foliation a cross-section appears just as figure 2.6 and the same
non-separation problem occurs.

Proposition 2.23.   Each point of G(X) has a neighborhood which is
tangentially diffeomorphic to an open neighborhood of $\mathbb{R}^{2p} \times N_a$.

Proof: Pick $(a,b,[\alpha]) \in$ G(X) and α representing $[\alpha]$.   Choose
neighborhoods $U_1,...,U_k$ and tangential coordinate patches (t_i,n_i): $U_i \longrightarrow$
$\mathbb{R}^p \times N$ corresponding to the path α.

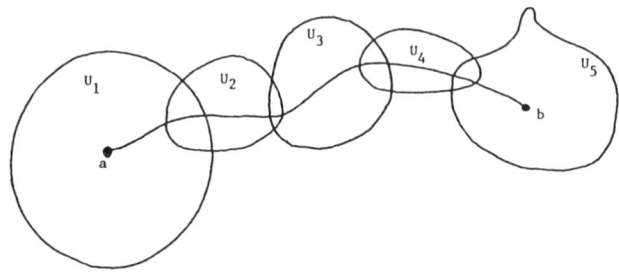

Let N_a = $n_1(U_1)$ and N_b = $n_k(U_k)$.   After a possible shrinking of U_1
and U_k there results the holonomy map H_{ab}^α: $N_a \longrightarrow N_b$.
   Given $x \in U_1$, there is a unique plaque path relating $n_1 x$ with

$H^{\alpha}_{a\,b}(n_1x)$. If $y \in U_k$ with $n_ky = H^{\alpha}_{a\,b}(n_1x)$ then the unique plaque path determines a path β from x to y in $\cup_i U_i$. The path β is not unique, but $[\beta]$ is unique, and of course $H^{\beta}_{x\,y}(n_1x) = n_ky$. Let

$$W = \{(x,y,[\beta]) \mid x \in U_1, y \in U_k,\ n_ky = H^{\alpha}_{a\,b}(n_1x),\ \beta$$
$$\text{determined as above}\}.$$

Note that if $(x,y,[\beta]) = (x,y,[\beta'])$ in W then $[\beta] = [\beta']$.

Define $\Phi: W \longrightarrow \mathbb{R}^p \times \mathbb{R}^p \times N$ by

$$\Phi(x,y,[\beta] = (t_1x, t_ky, n_1x).$$

We claim that Φ is a bijection. It is clear that Φ must be injective by our restriction on β. Suppose that $(r_1, r_2, r_3) \in \mathbb{R}^p \times \mathbb{R}^p \times N$, (or an open subset if the t_i and n_i are not surjective). Choose $x \in U_1$ with $t_1x = r_1$ and $n_1x = r_3$. Choose $y \in U_k$ with $t_ky = r_2$ and $n_ky = H^{\alpha}_{a\,b}(r_3)$. Since $n_ky = H^{\alpha}_{a\,b}(n_1x)$, there is a leaf path β in $\cup U_i$ from x to y. Then $\Phi(x,y,[\beta]) = (r_1, r_2, r_3)$, so Φ is surjective. It is clear that Φ is tangentially smooth. $\square$

Our final topic in this chapter is a close examination of the equivalence relation and (in anticipation of the Chapter IV discussion) the topological groupoid of a foliation in the case of a foliated bundle with discrete structural group and the case of the Reeb foliation.

Recall (2.2) that the initial data for a foliated bundle are a manifold B^p with universal cover $\widetilde{B}$, a space F and a homomorphism $\varphi: \pi_1(B) \longrightarrow \text{Homeo}(F)$. The resulting space $M = \widetilde{B} \times_{\pi_1(B)} F$ is a foliated space of dimension p, and the natural map $M \xrightarrow{\pi} B$ restricts to a covering space map $\widetilde{B} \times \{x\} \longrightarrow \ell \xrightarrow{\pi} B$.

Let Γ be the image of $\pi_1(B)$ in $\text{Homeo}(F)$, and for each $x \in F$ let

$$\Gamma_x = \{\gamma \in \Gamma \mid \gamma x = x\}$$

denote the isotropy group at x and let

(2.24)

$$\Gamma^x = \{\gamma \in \Gamma \mid \gamma y = y \text{ for all y in some neighborhood of x in F}\}$$

denote the *stable isotropy group* at x. The stable isotropy group Γ^x is a normal subgroup of the isotropy group Γ_x and our previous results imply that $\Gamma_x/\Gamma^x \cong G_x^x$, the holonomy group at x.

Let $b \in B$ be a basepoint, let $\tilde{b} \in \tilde{B}$ be some preimage of b, and let N be the image of $\tilde{b} \times F$ in M. The map $\tilde{b} \times F \longrightarrow N$ is a homeomorphism since $\pi_1(B)$ acts freely on $\tilde{B}$, so N is a copy of F sitting as a complete transversal to the foliated space.

Let G_N^N be the subgroupoid $\bigvee_{m,n \in N} G_m^n$, so that elements of G_N^N are triples $(n, m, [\alpha])$ with $n, m \in N$ and $[\alpha]$ some holonomy class of a path in the leaf ℓ_n of M from n to m. Regarding G as a category, then G_N^N is the full subcategory with objects N. Results of Hilsum-Skandalis [HS] (see Ch. VI) imply that the C^*-algebra of the foliation of M is determined by the C^*-algebra of the groupoid G_N^N. (In fact G_N^N is Morita equivalent to G(X); see A4.1.) As G_N^N is much simpler to understand than the full groupoid of the foliation, we explore its structure.

Theorem 2.25. If $M = \tilde{B} \times_{\pi_1(B)} F$ is a foliated bundle as above with complete transversal $N \cong F$ then there is a natural homeomorphism of topological groupoids

$$G_N^N \cong (F \times \Gamma)/\approx$$

where $(x, \gamma) \approx (y, \delta)$ if and only if $x = y$, and $\delta^{-1}\gamma$ lies in the stable isotropy group Γ^x. Thus G_N^N is completely determined by the action of $\Gamma = Im(\pi_1(B) \longrightarrow Homeo(F))$ on F.

We note some consequences of the result.

Corollary 2.26. If the holonomy groups G_x^x are trivial for all $x \in F$ then

$$G_N^N \cong (F \times \Gamma)/\sim$$

where $(x, \gamma) \sim (y, \delta)$ if and only if $x = y$ and $\delta^{-1}\gamma$ lies in the isotropy group Γ_x (i.e., if and only if $x = y$ and $\gamma x = \delta x$).

The corollary is immediate from the identification $G_x^x \cong \Gamma_x / \Gamma^x$.

Note that the stable isotopy groups Γ^x vanish for all x if and only if for each $\gamma \in \Gamma$ the fixed point set of γ has no interior. This condition is quite frequently satisfied in practice. For instance, if F is a Riemannian manifold F^q, Γ acts as isometries and each non-zero element γ moves some element of F, then the fixed point set of each $\gamma \in \Gamma$ is a manifold of dimension at most $(q-1)$ and hence has no interior. Indeed, any real analytic action satisfies this condition. For an example where the condition is violated, see (2.28).

Corollary 2.27. If for each $\gamma \in \Gamma$ the fixed point set of γ has no interior then there is a natural isomorphism of topological groupoids

$$G_N^N \cong F \times \Gamma.$$

Proof: Each stable isotropy group Γ^x vanishes and so the result follows from Theorem 2.25. □

Proof of Theorem 2.25. We shall show that $G_N^N \cong (N \times \Gamma)/\approx$, which suffices. Define a map $\sigma: (N \times \Gamma)/\approx \longrightarrow G_N^N$ as follows. Let $(n, \gamma) \in N \times \Gamma$. Represent γ by some based loop α in B. Lift α to a path $\hat{\alpha}$ in the leaf ℓ_n of $n \in M$ with $\hat{\alpha}(0) = n$. Then $\hat{\alpha}(1) \in N \cap \ell_n$ and $(n, \hat{\alpha}(1), [\hat{\alpha}])$ represents an element $\sigma(n, \gamma) \in G_N^N$. We argue that σ is well-defined as follows. Independence of choice of lifts $\hat{\alpha}$ of α is clear. Suppose that $(n, \gamma) \approx (n, \delta)$, so that $\delta^{-1}\gamma \in \Gamma^n$. Represent γ and δ by loops α and β respectively, and lift these loops to paths $\hat{\alpha}$ and $\hat{\beta}$ in ℓ_n with $\hat{\alpha}(0) = \hat{\beta}(0) = n$. Then $\hat{\beta}^{-1} \circ \hat{\alpha}$ is a loop in ℓ_n whose holonomy class is trivial, since $\delta^{-1}\gamma \in \Gamma^n$. Thus $\sigma(n, \gamma) = \sigma(n, \delta)$ and σ is well-defined.

The map σ is obviously continuous. If $\sigma(n, \gamma) = \sigma(n, \delta)$

then $\delta^{-1}\gamma$ must lift to a loop $\hat{\beta}^{-1}\hat{\alpha}$ with trivial holonomy in G_n^n. This implies that $\delta^{-1}\gamma \in \Gamma^n$ and hence σ is a monomorphism. If $(n,m,[\alpha]) \in G_N^N$ then the composite $[0,1] \xrightarrow{\alpha} M \xrightarrow{\pi} B$ is a loop (since $\pi n = \pi m$) and $\sigma(n,\pi\alpha) = (n,m,[\alpha])$. Thus σ is a homeomorphism.

The groupoid structure on $(N \times \Gamma)/\approx$ is obtained as follows. The unit space is N, of course, and $s(n,\gamma) = n$. The range map r is given by $r(n,\gamma) = \hat{\alpha}(1)$ where $\hat{\alpha}$ is a lift of a realization of the loop γ as earlier in this proof. Thus (n,γ) and (m,δ) may be multiplied when $\hat{\alpha}$ lifts γ and $\hat{\alpha}(1) = m$, and then

$$(n,\gamma) \cdot (m,\delta) = (n,\delta\gamma)$$

With this structure it is clear that σ is a homeomorphism of topological groupoids. $\square$

In practice Theorem 2.25 is rather easy to use. For instance, consider the Mobius strip $M = \mathbb{R} \times_{\mathbb{Z}} \mathbb{R}$ (cf. 2.4). The equivalence relation is simply the union of the $y = x$ and $y = -x$ lines in the plane, a figure "X". The group Γ is $\mathbb{Z}/2$ acting non-freely since Γ fixes $0 \in \mathbb{R}$. This is an isolated fixed point and certainly has no interior; thus $G_N^N \cong \mathbb{R} \times \mathbb{Z}/2$ with the obvious structure of a (Hausdorff) smooth manifold by Corollary 2.27. The map $G_N^N \longrightarrow \{(x,y) \mid y = \pm x\}$ is a homeomorphism except at the origin, where it is two-to-one in the obvious way, corresponding to the fact that $G_0^0 \cong \mathbb{Z}/2$, $G_n^n = 0$ for $n \neq 0$.

Next consider the manifold M constructed as the suspension of the action of $\mathbb{Z}$ on $\mathbb{R}$ given by θ where

$$\theta(n)(t) = \begin{cases} 2^n t & t > 0 \\ t & t \leq 0. \end{cases}$$

The element $\theta(1)$ fixes $(-\infty,0]$ and hence the condition of Corollary 2.27 is violated. The equivalence relation for G_N^N has the form

(2.28)

The groups Γ_t are given by $\Gamma_t = \mathbb{Z}$ for $t \leqslant 0$ and $\Gamma_t = 0$ for $t > 0$, and using Theorem 2.25 we have

$$G_N^N = (\mathbb{R} \times \mathbb{Z})/\approx$$

where

$$(t,n) \approx (t,m) \Leftrightarrow m = n \text{ or } t < 0$$

and hence G_N^N is the (path-connected) non-Hausdorff 1-manifold

(2.29)

Finally we move away from foliated bundles and consider the Reeb foliation of S^3 (cf. (2.5), (2.6)). Take a transversal $N = [-1,1]$ which starts near the closed leaf, tunnels through the solid snake in time $[-1,0)$, passes through the closed leaf at time 0, and tunnels through the other solid snake in $(0,1]$, stopping near (but not at) the closed leaf. Then N is a complete transversal. The corresponding equivalence relation for G_N^N is the following subset of the plane with the relative topology:

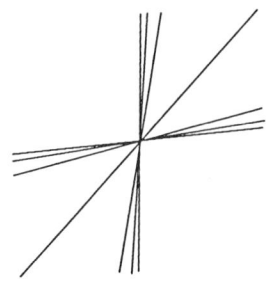

The point $(0,0)$ corresponds to the point $0 \in N$ which lies on the closed leaf ℓ_0. A sequence $\{(t,2^n t): t \in N-\{0\}\}$ corresponds to choosing a point $t \in N-\{0\}$, going around a corresponding holonomy path of degree n and returning to $\ell_t \cap [-1,1]$.

The map from G_N^N to the equivalence relation is a bijection except that $(0,0)$ has preimage $\mathbf{Z}^2$, corresponding to the fact that the closed leaf ℓ_0 has holonomy group $\mathbf{Z}^2$. So as a set fibred over the transversal N, G_N^N has structure

(2.31)

$$(o,s,[-1,0)) \cup \{(r,s)\} \cup (r,0,(0,1])$$

Write the lines for $t < 0$ as $(0,s,[-1,0))$ and the lines for $t > 0$ as $(r,0,(0,1])$. Then the set

$$(o,s,[-1,0)) \cup \{(r,s)\} \cup (r,0,(0,1])$$

is diffeomorphic to $[-1,1]$ in the topology of G_N^N. These sets serve as coordinate patches which exhibit G_N^N as a (non-Hausdorff) smooth topological groupoid. As an exercise, the reader is invited to show that the fundamental group $\pi_1(G_N^N)$ is the free group on two generators.

CHAPTER III: TANGENTIAL COHOMOLOGY

In this chapter we discuss certain cohomology groups associated with a foliated space, which we shall call tangential cohomology groups. It will be in these groups that invariants connected with the index theorem shall live. Similar groups have been considered before, for instance, by Kamber and Tondeur [KT2], Molino [Mol], Vaisman [V], Sarkaria [Sa1], Heitsch [He], El Kacimi-Alaoui [El], and Haefliger [Hae3] (whose work we discuss at the end of chapter IV). These cohomology groups are also related to those introduced by Zimmer in [Z2] for foliated measure spaces. The similarities and differences between the three situations are easy to describe; all involve differential forms which are smooth in the tangential direction of the foliation. The difference comes in the assumptions on the transverse behavior; for foliated manifolds, (Kamber-Tondeur, et al) forms are C^∞ in the transverse direction, while for foliated spaces (the present treatment), the forms are to be continuous in the transverse directions as that is all that makes sense, and finally for foliated measure spaces (Zimmer), the forms are to be measurable in the transverse direction, for again that is all that makes sense.

Thus let X be a metrizable foliated space with tangent bundle $FX \longrightarrow X$ as defined in Chapter II. The quickest and simplest way to introduce the tangential cohomology is via sheaf theory and sheaf cohomology, but for those readers who are not familiar with such notions we show how to define the groups via a de Rham complex and also show in an appendix how to give a completely algebraic definition. For details concerning sheaves and their cohomology, consult Godement [Gom], Wells [We].

We consider the sheaf on X, $\mathfrak{R}_\tau$ of germs of continuous real valued tangentially locally constant functions. Specifically, this sheaf assigns to each open set U of X the set of continuous real-valued functions on U that are locally constant in the tangential direction on the foliated space U (given the induced foliation from X). This is obviously a presheaf and it is immediate that the additional conditions defining a sheaf (Godement [Gom] p. 109) are satisfied.

<u>Definition</u> 3.1.   The *tangential cohomology* groups of the foliated space X are the cohomology groups of the sheaf $\mathcal{R}_\tau$, $H^*(X,\mathcal{R}_\tau)$, which we normally write as $H_\tau^*(X,\mathbb{R})$ or simply $H_\tau^*(X)$.

These cohomology groups can be defined by construction of resolutions ([Gom] p. 173) or perhaps more simply by the Čech method using cocycles defined on open covers of X ([Gom] p. 203) but for us the most useful and transparent way of dealing with them is via a de Rham complex.

<u>Definition</u> 3.2.   A *tangential differential k-form* at $x \in X$ is an alternating k-form on the tangent space FX_x at x i.e. an element of $\wedge^k(F^*X_x)$.   These fit together to yield a vector bundle denoted $\wedge^k F^*$ on X which is just the kth exterior power of the cotangent bundle, and it is quite evidently tangentially smooth in the sense of the previous chapter.

For each open set U, we assign to U the tangentially smooth sections $\Gamma_\tau(\underset{\sim}{\wedge}^k(F^*))$ (defined on U).   Just as before this gives a sheaf which we denote by $\underset{\sim}{\wedge}^k(F^*)$ - the *sheaf of germs of tangentially smooth k-forms*.

There is an obvious differential from $\Gamma_\tau(\underset{\sim}{\wedge}^k(F^*))$ to $\Gamma_\tau(\underset{\sim}{\wedge}^{k+1}(F^*))$ which can be defined in an elementary way in terms of local coordinates.   If $U \cong L^p \times N$ is a local coordinate patch with $x_1,...,x_p$ coordinates on the open ball L^p in $\mathbb{R}^p$, then a tangential differential k-form is an object that can be written locally as

$$(3.3) \qquad \omega = \sum_{(i_\ell)} a(x_1,...x_p,n)\, dx_{i_1} \wedge ... \wedge dx_{i_k}$$

with a and all of its derivatives with respect to the x_i continuous in all variables.   Then $d\omega$ is defined just as one does classically for a k-form with n playing the role of a parameter;

(3.4) $\quad d\omega = \sum_{(i_\ell), j} \dfrac{\partial a}{\partial x_j}(x_1,...,x_p,n)\, dx_j \wedge dx_{i_1} \wedge ... \wedge dx_{i_k}.$

We evidently have $d^2 = 0$ and so we have a sequence of sheaves with maps

(3.5) $\quad 0 \longrightarrow \mathcal{R}_{\mathcal{T}} \longrightarrow \underset{\sim}{\Lambda}^0(F^*) \overset{d}{\longrightarrow} \underset{\sim}{\Lambda}^1(F^*) \longrightarrow ... \longrightarrow \underset{\sim}{\Lambda}^p(F^*) \longrightarrow 0$

where the first map is the natural inclusion of tangentially locally constant functions into tangentially smooth functions. The Poincaré Lemma obviously holds in this context:

Proposition 3.6. The sequence (3.5) is an exact sequence of sheaves; that is, for sufficiently small open sets U, the kernel of each map on sections over U is the range of the map in one lower degree.

Proof. On an open set of the form $L \times N$, where L is a ball in $\mathbb{R}^p$, a k-form in the kernel of d is an expression $\omega(x,n)$ where for fixed n this is an ordinary closed k form with respect to the variable $x \in L$. (The variable n plays the role of a parameter with respect to which everything varies continuously.) By the usual Poincaré Lemma, one finds a k-1 form φ_n, one for each n, so that $d\varphi_n = \omega(\cdot,n)$ and what is at issue is that we can choose φ_n to be continuous as n varies over N so that φ_n defines a section of $\Lambda^{k-1}(F^*)$ over U. This is a routine exercise the details of which we omit. $\square$

Moreover, just as in the usual case, one sees that the sheaves $\underset{\sim}{\Lambda}^k(F^*)$ are fine [Gom, p. 157] and consequently by the general machinery of sheaf theory one can calculate the cohomology of the sheaf $\mathcal{R}_{\mathcal{T}}$ from this resolution - a de Rham type theorem. Let

$$\Omega_{\mathcal{T}}^k \equiv \Omega_{\mathcal{T}}^k(X) = \Gamma_{\mathcal{T}}(\underset{\sim}{\Lambda}^k(F^*))$$

denote the global tangentially smooth sections of the sheaf $\underset{\sim}{\Lambda}^k(F^*)$.

Proposition 3.7. There are isomorphisms

$$H_\mathcal{T}^k(X;\mathbb{R}) = H^k(X;\mathcal{R}_\mathcal{T}) \cong \frac{\ker\ (\Omega_\mathcal{T}^k \longrightarrow \Omega_\mathcal{T}^{k+1})}{\operatorname{Im}\ (\Omega_\mathcal{T}^{k-1} \longrightarrow \Omega_\mathcal{T}^k)} . \qquad \square$$

In particular, if $X = M$ foliated as one leaf then this is the usual identification of the de Rham cohomology groups $H^*(M;\mathbb{R})$. The analogous result (and indeed the entire chapter) holds with $\mathbb{R}$ replaced by $\mathbb{C}$ throughout.

From this proposition it is evident that $H_\mathcal{T}^k(X,\mathbb{R}) = 0$ for $k > p$ where p is the leaf dimension of the foliated space. However, these cohomology groups are in general going to be infinite dimensional, in contrast to the case of a compact manifold (with a foliation consisting of one leaf). Further, these groups, which are vector spaces, also inherit via the de Rham isomorphism the structure of (generally non Hausdorff) topological vector spaces. We topologize $\Omega_\mathcal{T}^k(X)$ by demanding that in all local coordinate patches we have uniform convergence of the functions $a(x_1,\dots,x_p)$ of (3.3) together with all their derivatives in the tangential direction on compact subsets of the coordinate patch. The differentials are clearly continuous with respect to these topologies and so $H_\mathcal{T}^n(X,\mathbb{R})$ is a topological vector space.

In general the image of d will fail to be closed and so $H_\mathcal{T}^n(X,\mathbb{R})$ will not be Hausdorff in these cases. It will be useful occasionally to replace the image of d by its closure, or equivalently replace $H_\mathcal{T}^n(X,\mathbb{R})$ by its quotient obtained from dividing by the closure of the identity; this is the largest Hausdorff quotient. We will denote this maximal Hausdorff quotient of $H_\mathcal{T}^k(X,\mathbb{R})$ by $\bar{H}_\mathcal{T}^k(X,\mathbb{R})$. The point of this is first that we shall usually only care about the image of a cohomology class $[\omega]$ of $H_\mathcal{T}^k(X,\mathbb{R})$ in this Hausdorff quotient rather than the class itself since $\int \omega d\nu$ depends only on the class of ω in $\bar{H}^p$, and second one at least has a chance of computing the groups $\bar{H}_\mathcal{T}^*$ in certain cases.

The tangential cohomology groups are related via natural maps to the usual cohomology groups of the compact metrizable space.

Since Čech cohomology and sheaf cohomology agree for such spaces ([Gom] p. 228) we shall simply write $H^*(X,A)$ for this cohomology for coefficients in an abelian group A. Specifically, the sheaf cohomology is defined to be the cohomology of the sheaf $\mathcal{R}$ of germs of locally constant real valued functions. As a sheaf $\mathcal{R}$ assigns to each open set U the locally constant real valued functions on U. The cohomology groups $H^*(X,\mathcal{R})$ are written as we indicated above as $H^*(X,\mathbb{R})$. But now there is a natural inclusion map of the sheaf $\mathcal{R}$ into the sheaf $\mathcal{R}_\tau$ of tangentially locally constant functions and we can complete this to a short exact sequence

$$(3.8) \qquad 0 \longrightarrow \mathcal{R} \xrightarrow{\ r\ } \mathcal{R}_\tau \longrightarrow \mathcal{R}_\nu \longrightarrow 0$$

where $\mathcal{R}_\nu$ is defined as the quotient sheaf of $\mathcal{R}_\tau$ by $\mathcal{R}$, ([Gom] p. 117). We obtain in particular an induced homorphism

$$(3.9) \qquad r_*\colon H^*(X,\mathbb{R}) = H^*(X,\mathcal{R}) \longrightarrow H^*(X,\mathcal{R}_\tau) = H^*_\tau(X,\mathbb{R})$$

which is a sort of "restriction" map from ordinary cohomology to tangential cohomology. Of course we also have a long exact sequence of cohomology corresponding to the short exact sequence of sheaves above, but we will not explore that further, and the sheaf $\mathcal{R}_\nu$ will not play any further role.

We should comment that the only reason for introducing sheaves was to obtain a natural definition of r_*. For we could have defined $H^*(X,\mathbb{R})$ as either Čech cohomology or equivalently Alexander-Spanier cohomology and we could have directly defined $H^*_\tau(X,\mathbb{R})$ as the cohomology of the tangential de Rham complex with no mention of sheaves. But then it is not at all apparent that there is a map r_* from the cohomology of X to the tangential cohomology of X. Actually as we have suggested before it is usually the composed map $\bar{r}_*$ from $H^*(X,\mathbb{R})$ to $\bar{H}^*_\tau(X,\mathbb{R})$ that is of more significance than r_*.

It might be helpful to look at an example and the simplest one is that of the Kronecker foliation of a two torus T^2 by parallel lines of a fixed irrational slope λ relative to given coordinate axes. The leaves are given in parametric form as $((e^{it}, e^{i\lambda t}c), t \in \mathbb{R})$ for some fixed $c \in T$. Tangential zero forms are simply real-valued functions which are tangentially smooth. Clearly the tangent bundle is a trivial bundle and in fact we can find an essentially unique tangential one form ω which is invariant under group translation on T^2. Then the most general tangential one form is easily seen to be of the type $f\omega$ where f is any tangentially smooth function. If θ is a group-invariant vector field on T^2 pointing in the tangential direction, then the differential d: $\Omega_{\tau}^0(F^*) \longrightarrow \Omega_{\tau}^1(F^*)$ is given by $d(g) = \theta(g)\omega$ for a suitable normalization of θ.

To investigate this more closely we expand g in double Fourier series

$$g(\xi_1, \xi_2) = \Sigma \, g_{n,m} \xi_1^n \xi_2^m$$

and for a second function f denote its Fourier coefficients by $f_{n,m}$. The condition for a function g to be tangentially smooth is easily seen to be that $g_{n,m}(n+\lambda m)^k$ should be the Fourier coefficients of a continuous function, for each $k = 0,1,\ldots$. The relation between f and g expressed by $d(g) = f\omega$ is simply that

(3.10) $\qquad\qquad f_{n,m} = (n + \lambda m)g_{n,m}.$

Quite clearly the kernel of d consists of constants so $H_{\tau}^0(X,\mathbb{R}) \simeq \mathbb{R}$. On the other hand if we are given $f\omega$ with f tangentially smooth, we have to find out when we can solve (3.10). An evident necessary condition is that $f_{0,0} = 0$ and indeed if $f_{0,0} = 0$ and if f is a trigonometric polynomial, we can find a trigonometric polynomial g solving the equation, as λ is irrational and $n + \lambda m$ is not zero unless $n = m = 0$. Since one easily sees that the set

$$\{f\omega \,|\, f_{0,0} = 0, \text{ f a trigonometric polynomial}\}$$

is dense in all $f\omega$ with $f_{0,0} = 0$ in the topology described above, one can conclude immediately that $\bar{H}_{\tau}^1(X,\mathbb{R}) \simeq \mathbb{R}$. It is interesting to note that the closed tangential cohomology $\bar{H}_\tau^*(X,\mathbb{R})$ in this case is the cohomology over $\mathbb{R}$ of a circle.

However $H_\tau^1(X,\mathbb{R}) \neq \bar{H}_\tau^1(X,\mathbb{R})$ and the former is in fact infinite dimensional, for (3.10) results in a classic "small denominator problem." Indeed, choose a sequence $(n(k), m(k))$ with $n(k) + \lambda m(k) = \xi(k)$ a summable sequence, and define $f_{n,m} = 0$ unless $(n,m) = (n(k), m(k))$ and $f_{n(k), m(k)}$ any sequence in k asymptotic to $\xi(k)^{-1}$. Then $f_{n,m}(n + \lambda m)^k$ is the Fourier series of a continuous function for each k as it holds for $k = 0$ and as $(n+\lambda m)$ is bounded (in fact tends to zero) where $f_{n,m} \neq 0$. But quite evidently $f_{n,m}(n+\lambda m)^{-1} = g_{n,m}$ is not the Fourier series of a continuous function as $g_{n,m}$ does not tend to zero. This shows that $H_\tau^1(X,\mathbb{R})$ is infinite dimensional and of course non-Hausdorff since $\bar{H}_\tau^1(X;\mathbb{R})$ is one dimensional.

We remark that if one uses differential forms which also are required to be C^∞ in the transverse direction then the result is quite different. Haefliger [Hae3] shows that for such forms the associated first cohomology groups of the Kronecker flow on the torus has either infinite dimension or dimension one, depending upon whether the irrational slope is Liouville or diophantine.

Let

$$L \longrightarrow X \longrightarrow B$$

be a fibration with X compact and with leaves L_b corresponding to preimages of points $b \in B$ assumed to be smooth. Then the tangential cohomology of X has a simple description. Form a vector bundle E over B with

$$E_b \cong H^*(L_b).$$

(It is locally trivial.) Then $H_\tau^*(X)$ is isomorphic to the continuous cross-sections of E. (This suggests that for more general fibrations $F \longrightarrow X \longrightarrow B$ of foliated spaces there should be a Serre spectral sequence of the type $H_\tau^*(B; H_\tau^*(F)) \Rightarrow H_\tau^*(X)$; we do not pursue this direction here.)

The next order of business is a more thorough study of $H_\tau^*(X)$. Let $\mathcal{F}$ denote the category of metrizable, (hence paracompact) foliated spaces and tangentially smooth (leaf-preserving) maps. Let $\mathbb{K} = \mathbb{R}$ or $\mathbb{C}$: we note once and for all that our results hold in both cases and that $\mathbb{R} \hookrightarrow \mathbb{C}$ induces an isomorphism $H_\tau^k(X;\mathbb{R}) \otimes_\mathbb{R} \mathbb{C} \cong H_\tau^k(X;\mathbb{C})$. Let $H_\tau^*(X) = H_\tau^*(X;\mathbb{K})$.

Proposition 3.11. H_τ^* is a contravariant functor from $\mathcal{F}$ to $\mathbb{Z}$-graded associative, graded-commutative topological $\mathbb{K}$-algebras and continuous homomorphisms, and $H_\tau^k(X) = 0$ for $k < 0$ or $k > p$ where p is the leaf dimension of X.

Proof: The wedge-product of forms yields a natural continuous map

$$\Omega_\tau^i(X) \otimes \Omega_\tau^j(X) \longrightarrow \Omega_\tau^{i+j}(X)$$

which supplies the product structure in the usual way. As $\Omega_\tau^k(X) = 0$ for $k < 0$ or $k > p = \dim X$, the cohomology groups also vanish. If $f: X \longrightarrow Y$ then by proposition 2.15 we may assume that f is tangentially smooth. Thus the induced map $\Omega_\tau^*(Y) \longrightarrow \Omega_\tau^*(X)$ is continuous and $f^*: H_\tau^*(Y) \longrightarrow H_\tau^*(X)$ is continuous. $\square$

Proposition 3.12.   If X is the topological sum of $\{X_j\}$ in $\mathcal{F}$, then

$$H_\tau^k(X) \cong \prod_j H_\tau^k(X_j) \qquad \square$$

Proposition 3.13. $H_\tau^0(X) = \{f \in C_\tau^\infty(X) \mid f\big|_\ell$ is a constant for any leaf $\ell\}$. In particular, if X has a dense leaf then $H_\tau^0(X) \cong \mathbb{K}$. $\square$

Definition 3.14.   Let $f,g: X \longrightarrow Y$ in $\mathcal{F}$. An $\mathcal{F}$-*homotopy* h between f and g is an $\mathcal{F}$-map $h: X \times \mathbb{R} \longrightarrow Y$ (where $X \times \mathbb{R}$ is foliated as (Leaf of X) $\times \mathbb{R}$) such that

$$h(x,t) = \begin{cases} f(x) & \text{for } t \leq 0 \\ g(x) & \text{for } t \geq 1 \end{cases}.$$

Homotopy in $\mathcal{F}$ is obviously an equivalence relation. The following technical proposition leads to homotopy-invariance of $H_\mathcal{T}^*$.

Proposition 3.15.   Let $J_t \colon X \longrightarrow X \times \mathbb{R}$ by $J_t(x) = (x,t)$. There is a $\mathbb{k}$-linear map

$$L \colon \Omega_\mathcal{T}^*(X \times \mathbb{R}) \longrightarrow \Omega_\mathcal{T}^*(X)$$

with the following properties:

   1) $L(\Omega_\mathcal{T}^k(X \times \mathbb{R})) \subset \Omega_\mathcal{T}^{k-1}(X)$
   2) $dL + Ld = J_1^* - J_0^*$

That is, L is a chain-homotopy from J_0 to J_1.

Proof: By a tangentially smooth partition of unity argument, we may assume that $X = \mathbb{R}^p \times N$. Then any k-form in $\Omega_\mathcal{T}^k(X \times \mathbb{R})$ with $k \geqslant 1$ may be written uniquely as a sum of monomials of the form

$$\alpha = a dx_I$$

or

$$\beta = b dx_I \wedge dt .$$

Define $L \big| \Omega_\mathcal{T}^0(X \times \mathbb{R}) \equiv 0$, $L(\alpha) = 0$, and

$$L\beta = \left[\int_0^1 b dt \right] dx_I$$

Then $L(\Omega_\mathcal{T}^k(X \times \mathbb{R})) \subset \Omega_\mathcal{T}^{k-1}(X)$. If $f \in \Omega_\mathcal{T}^0(X \times \mathbb{R})$ then

$$(dL + Ld)f = L\left[\frac{\partial f}{\partial t} dt\right] + L\left[\sum_i \frac{\partial f}{\partial x_i} dx_i\right]$$

$$= \int_0^1 \frac{\partial f}{\partial t} dt$$

$$= (J_1^* - J_0^*)f.$$

On forms of type α,

$$(dL + Ld)\alpha = Ld\alpha$$

$$= \left[\int_0^1 \frac{\partial a}{\partial t} \, dt\right] dx_I$$

$$= \left[J_1^* - J_0^*\right]\alpha.$$

On forms of type β, $J_1^* \beta = J_0^* \beta = 0$ and

$$dL\beta = -Ld\beta = \sum_{i=1}^p \left(\frac{\partial b}{\partial x_i} \, dt\right) dx_i \wedge dx_I \ .$$

This shows that $dL + Ld = J_1^* - J_0^*$ on monomials and hence in general. □

Theorem 3.16.   If $f,g: X \longrightarrow Y$ are $\mathcal{F}$-homotopic, then

$$f^* = g^*: H_{\mathcal{T}}^*(Y) \longrightarrow H_{\mathcal{T}}^*(X).$$

This is immediate from Proposition 3.15.   □

Corollary 3.17.   If $f: X \longrightarrow Y$ is a continuous leaf-preserving map then f induces a unique continuous map $f^*: H_{\mathcal{T}}^*(Y) \longrightarrow H_{\mathcal{T}}^*(X)$.

Proof: Say $f \simeq g$ and $f \simeq g'$ where $g,g' \in \mathcal{F}$.   Then $g^* = g'^*$ by Theorem 3.16, so declare $f^* = g^*$.   □

Corollary 3.18.   Say $X,Y \in \mathcal{F}$ and $f: X \longrightarrow Y$ is a leaf-preserving homeomorphism.   Then f^* is an isomorphism.   Thus $H_{\mathcal{T}}^*$ is a leaf-preserving topological invariant.

Proof: The map f^{-1} is also leaf-preserving.   By Corollary 3.17 the maps f,f^{-1} induce f^*, $(f^{-1})^*$, and clearly $(f^{-1})^* = (f^*)^{-1}$.   □

Corollary 3.19. (Homotopy-type invariance). Two foliated spaces having the same "leaf-preserving homotopy type" have isomorphic tangential cohomology.

Proof: Let $f: X \longrightarrow Y$ be a leaf-preserving homotopy equivalence with leaf-preserving homotopy inverse g, so that $fg \simeq 1_Y$, $gf \simeq 1_X$ via $\mathcal{F}$-homotopies. By Corollary 3.17, f^* and g^* exist, and then $f^* = (g^*)^{-1}$. $\square$

Next we introduce cohomology with compact supports. The *support* of a form $\omega \in \Omega^k_{\mathcal{T}}(X)$ is the closure of $\{x \in X \mid \omega(x) \neq 0\}$. Let $\Omega^k_{\mathcal{T}c}(X) \subset \Omega^k_{\mathcal{T}}(X)$ denote the forms of compact support. The groups $\Omega^k_{\mathcal{T}c}(X)$ form a complex, as $d: \Omega^k_{\mathcal{T}}(X) \longrightarrow \Omega^{k+1}_{\mathcal{T}}(X)$ decreases supports. Define *tangential cohomology with compact support* by

$$(3.20) \qquad H^k_{\mathcal{T}c}(X) = H^k(\Omega^*_{\mathcal{T}c}(X)).$$

The inclusion $\Omega^*_{\mathcal{T}c}(X) \hookrightarrow \Omega^*_{\mathcal{T}}(X)$ is the inclusion of a differential subalgebra, thus inducing a map of $\mathbb{Z}$-graded $\mathbb{K}$-algebras

$$(3.21) \qquad H^k_{\mathcal{T}c}(X) \longrightarrow H^k_{\mathcal{T}}(X)$$

which is an isomorphism if X is compact. The following proposition summarizes the elementary properties of $H^*_{\mathcal{T}c}$.

Proposition 3.22. 1) $H^*_{\mathcal{T}c}$ is a contravariant functor from $\mathcal{F}$ and *proper* $\mathcal{F}$-maps to $\mathbb{Z}$-graded associative, graded-commutative topological $\mathbb{K}$-algebras and continuous homomorphisms.

2) If X is the topological sum of $\{X_j\}$ in $\mathcal{F}$, then there is a natural isomorphism

$$H^k_{\mathcal{T}c}(X) \cong \oplus_j H^k_{\mathcal{T}c}(X_j).$$

3) $H_{\mathcal{T}c}^*(X)$ is a unital algebra if and only if X is compact, in which case $H_{\mathcal{T}c}^*(X) \cong H_{\mathcal{T}}^*(X)$ naturally.

4) $H_{\mathcal{T}c}^*(X)$ is a covariant functor with respect to inclusions of open sets $U \subset X$.

An $\mathcal{F}$-homotopy h: $X \times \mathbb{R} \longrightarrow Y$ is *proper* if $h \big| X \times I$ is proper.

Proposition 3.23.   Let $\Omega_x(X \times \mathbb{R}) = \{\omega \in \Omega_{\mathcal{T}}(X \times \mathbb{R}) \big| \ \text{supp}(\omega)$ is a compact projection}.   Then there is a linear map $L: \Omega_x^*(X \times \mathbb{R}) \longrightarrow \Omega_{\mathcal{T}c}^*(X)$ such that

$$L(\Omega_x^k(X \times \mathbb{R})) \subset \Omega_{\mathcal{T}c}^{k-1}(X)$$

and

$$dL + Ld = J_1^* - J_0^*. \qquad \square$$

Theorem 3.24.   If f,g are proper $\mathcal{F}$-homotopic continuous maps then $f^* = g^*$ on $H_{\mathcal{T}c}^*$. $\square$

Corollary 3.25.   Let h: $X \longrightarrow Y$ be a leaf-preserving homeomorphism. Then there is a map h^* and $h^*: H_{\mathcal{T}c}^*(Y) \xrightarrow{\cong} H_{\mathcal{T}c}^*(X)$. $\square$

Note that $H_{\mathcal{T}c}^*$ is **not** an invariant of homotopy type.   The space $\mathbb{R}^m$ has the homotopy type of the space $\mathbb{R}^p \times \mathbb{R}^m$ and $H_{\mathcal{T}c}^0(\mathbb{R}^p \times \mathbb{R}^m) = 0$, since no constant function on a leaf has compact support, but $H_{\mathcal{T}c}^0(\mathbb{R}^m) = C_c^0(\mathbb{R}^m)$.

In preparation for the Thom isomorphism theorem, we introduce a third sort of cohomology which best suits the total space of vector bundles.   (Bott-Tu [BT] is an excellent general reference.)   Suppose that $\pi: E \longrightarrow X$ is a tangentially smooth real vector bundle over X with E foliated by leaves which locally are of the form

$$(\text{leaf of } X) \times E_x$$

where $E_x = \pi^{-1}(x) \cong \mathbb{R}^n$. Let $\Omega^k_{\mathcal{T}v}(E)$ be those forms $\omega \in \Omega^k_{\mathcal{T}}(E)$ which are compactly supported on each fibre E_x. Then $\Omega^*_{\mathcal{T}v}(E)$ is a subcomplex; let $H^*_{\mathcal{T}v}(E)$ be the associated cohomology groups. We shall refer to them as *tangential compact vertical* or more simply as *tangential vertical* cohomology groups. If X is compact then $\Omega^*_{\mathcal{T}v}(E) = \Omega^*_{\mathcal{T}c}(E)$ and so $H^*_{\mathcal{T}v}(E) = H^*_{\mathcal{T}c}(E)$; in general these groups differ.

Theorem 3.26. (Mayer-Vietoris) Let U, V be open subspaces of the foliated space X. Then the Mayer-Vietoris sequence (with usual maps)

$$0 \longrightarrow \Omega^*_{\mathcal{T}}(U \cup V) \longrightarrow \Omega^*_{\mathcal{T}}(U) \oplus \Omega^*_{\mathcal{T}}(V) \longrightarrow \Omega^*_{\mathcal{T}}(U \cap V) \longrightarrow 0$$

is exact. Hence there is a long exact sequence

$$\dots \longrightarrow H^k_{\mathcal{T}}(U \cup V) \longrightarrow H^k_{\mathcal{T}}(U) \oplus H^k_{\mathcal{T}}(V) \longrightarrow H^k_{\mathcal{T}}(U \cap V) \longrightarrow H^{k+1}_{\mathcal{T}}(U \cup V) \longrightarrow \dots$$

Similarly, if $\pi: E \longrightarrow X$ is a tangentially smooth bundle over X, X_1 and X_2 are open sets in X with $X_1 \vee X_2 = X$, $U = \pi^{-1}(X_1)$, $V = \pi^{-1}(X_2)$, then the sequence

$$0 \longrightarrow \Omega^k_{\mathcal{T}v}(E) \longrightarrow \Omega^k_{\mathcal{T}v}(U) \oplus \Omega^k_{\mathcal{T}v}(V) \longrightarrow \Omega^k_{\mathcal{T}v}(U \cap V) \longrightarrow 0$$

is exact and so there is a long exact sequence

$$\longrightarrow H^k_{\mathcal{T}v}(E) \longrightarrow H^k_{\mathcal{T}v}(U) \oplus H^k_{\mathcal{T}v}(V) \longrightarrow H^k_{\mathcal{T}v}(U \cap V) \longrightarrow H^{k+1}_{\mathcal{T}v}(E) \rightarrow \dots$$

Proof: Let $\omega_U \in \Omega^k_{\mathcal{T}}(U)$, $\omega_V \in \Omega^k_{\mathcal{T}}(U)$, and suppose that $\omega_U|_{U \cap V} = \omega_V|_{U \cap V}$. Define $\omega \in \Omega^k_{\mathcal{T}}(U \cup V)$ by

$$\omega(x) = \begin{cases} \omega_U(x) & \text{if } x \in U \\ \omega_V(x) & \text{if } x \in V \end{cases}$$

Then ω maps to (ω_U, ω_V). This shows exactness of the first sequence. The other verifications are as trivial. $\square$

Proposition 3.27. There is a natural associative continuous external product pairing

$$\alpha: H^i_{\mathcal{T}}(X) \otimes H^j_{\mathcal{T}}(Y) \longrightarrow H^{i+j}_{\mathcal{T}}(X \times Y)$$

and similarly

$$\alpha_c: H^i_{\mathcal{T}c}(X) \otimes H^j_{\mathcal{T}c}(Y) \longrightarrow H^{i+j}_{\mathcal{T}c}(X \times Y)$$

respecting $H^*_{\mathcal{T}c} \longrightarrow H^*_{\mathcal{T}}$.

Proof: Let $X \xleftarrow{\pi_x} X \times Y \xrightarrow{\pi_y} Y$ be the projections. Define the first pairing by

$$H^i_{\mathcal{T}}(X) \otimes H^j_{\mathcal{T}}(Y) \xrightarrow{\pi_x^* \otimes \pi_y^*} H^i_{\mathcal{T}}(X \times Y) \otimes H^j_{\mathcal{T}}(X \times Y) \xrightarrow{\text{multiply}} H^{i+j}_{\mathcal{T}}(X \times Y)$$

and similarly for $H^*_{\mathcal{T}c}$. $\square$

Proposition 3.28. Let X be a foliated space and let M be a smooth manifold. Foliate $X \times M$ as (leaf of X) $\times M$. Then α induces natural continuous isomorphisms

$$H^*_{\mathcal{T}}(X) \otimes H^*(M) \longrightarrow H^*_{\mathcal{T}}(X \times M)$$

and

$$H^*_{\mathcal{T}v}(X) \otimes H^*_c(\mathbb{R}^n) \longrightarrow H^*_{\mathcal{T}v}(X \times \mathbb{R}^n).$$

Proof: There is a natural isomorphism of sheaves

$$\mathcal{R}_{\mathcal{T}}(X) \otimes \mathcal{R}(M) \cong \mathcal{R}_{\mathcal{T}}(X \times M)$$

corresponding to the pairing α. This is clear since (on a local patch and hence globally) a function on $X \times M$ which is constant on leaves corresponds uniquely to a function on X which is constant on leaves and a constant function on M. This proves the first isomorphism.

For the second we regard $X \times \mathbb{R}^n$ as a trivial bundle over X. The homotopy inverse to the Künneth pairing above is given by integration along the fibre

$$H^k_{\mathcal{T}v}(X \otimes \mathbb{R}^n) \longrightarrow H^{k-n}_{\mathcal{T}}(X) \cong H^{k-n}_{\mathcal{T}}(X) \otimes H^n_c(\mathbb{R}^n)$$

as in Bott-Tu [BT], page 61. □

Corollary 3.29.   Let $u_n \in H^n_c(\mathbb{R}^n) \cong \mathbb{R}$ be the canonical generator and for a foliated space X, let $u^X_n = \alpha(1^X \otimes u_n) \in H^n_{\mathcal{T}v}(X \times \mathbb{R}^n)$.   Then

1) $H^*_{\mathcal{T}v}(X \times \mathbb{R}^n)$ is a free continuous $H^*_{\mathcal{T}}(X)$-module on u^X_n.

2)   If f: $X \longrightarrow Y$ in $\mathcal{F}$ then

$$(f \times 1)^* u^Y_n = u^X_n$$

3) $\bar{\alpha}(u^X_n \otimes u^Y_m) = u^{X \times Y}_{n+m}$ (explained below)

The class u^X_n is the Thom class of the trivial bundle $X \times \mathbb{R}^n \longrightarrow X$. The map $\bar{\alpha}$ is the composition

$$H^n_{\mathcal{T}v}(X \times \mathbb{R}^n) \otimes H^m_{\mathcal{T}v}(X \times \mathbb{R}^m) \xrightarrow{\ \alpha\ } H^{n+m}_{\mathcal{T}v}(X \times \mathbb{R}^n \times Y \times \mathbb{R}^m)$$

with a diagonal map labeled $\bar{\alpha}$ going down to

$$\Big\downarrow \cong (\tilde{t} \text{ defined below})$$

$$H^{n+m}_{\mathcal{T}v}(X \times Y \times \mathbb{R}^{n+m})$$

Proof: Apply Proposition 3.28 to $X \times \mathbb{R}^n$. Then

$$H^{k+n}_{\mathcal{T}v}(X \times \mathbb{R}^n) \cong \oplus_j H^j_{\mathcal{T}}(X) \otimes H^{k+n-j}_c(\mathbb{R}^n)$$

$$\cong H^k_{\mathcal{T}}(X) \otimes H^n_c(\mathbb{R}^n)$$

As $H^n_c(\mathbb{R}^n) \cong \mathbb{R}$ on u_n, the class $u^X_n = 1^X \otimes u_n$ generates

$H_{\tau v}^{k+n}(X \times \mathbb{R}^n)$ as an $H_{\tau}^k(X)$–module.   This proves 1).

For 2), we compute:

$$(f \times 1)^* u_n^Y = (f \times 1)^* (1^Y \otimes u_n)$$

$$= f^*(1^Y) \otimes u_n$$

$$= 1^X \otimes u_n$$

$$= u_n^X \quad \text{as required.}$$

For 3), consider the graded commutative diagram below, where t generically denotes twist maps:

Then

$$\bar{a}(u_n^X \otimes u_m^Y) = \tilde{t}\alpha(u_n^X \otimes u_m^Y)$$

$$= \tilde{t}\alpha(\alpha \otimes \alpha)(1^X \otimes u_n \otimes 1^Y \otimes u_m)$$

$$= \alpha(\alpha \otimes \alpha)(1 \otimes t \otimes 1)1^X \otimes u_n \otimes 1^Y \otimes u_m)$$

$$= \alpha(\alpha \otimes \alpha)(1^X \otimes 1^Y \otimes u_n \otimes u_m)$$

$$= \alpha(1^{X \times Y} \otimes u_{n+m}) \quad \text{since } \alpha(u_n \otimes u_m) = u_{n+m}$$

$$= u_{n+m}^{X \times Y} \quad \square$$

Thom Isomorphism Theorem (3.30). For each compact foliated space X and for each tangentially smooth real oriented n-plane bundle p: E ⟶ X there is a unique Thom class $u_E \in H^n_{\mathcal{T}c}(E)$ with the following properties:

1) If f: X ⟶ X' in $\mathcal{F}$ and E' is a real oriented bundle over X', then

$$f^* u_{E'} = u_E.$$

2) Let $x \in X$ and let E_x denote the fibre over X in E. Then the inclusion $E_x \subset E$ induces a map $H^n_{\mathcal{T}c}(E) \longrightarrow H^n_c(E_x) \cong H^n_c(\mathbb{R}^n)$ under which u_E is sent to u_n.

The $\mathbb{K}$-algebra $H^*_{\mathcal{T}c}(E)$ is a free $H^*_{\mathcal{T}}(X)$-module on the Thom class u_E; precisely, there is a continuous Thom isomorphism

(3.31) $$\Phi_{\mathcal{T}}: H^k_{\mathcal{T}}(X) \xrightarrow{\cong} H^{n+k}_{\mathcal{T}c}(E)$$

given by

(3.32) $$\Phi_{\mathcal{T}}(\omega) = u_E \omega.$$

Further, if E and E' are bundles over X as above then

$$u_{E \oplus E'} = u_E u_{E'}.$$

Proof: We shall establish the Thom theorem in somewhat greater generality than stated above. Let us say that a bundle E over X is r-trivial if there exist a finite open cover $X_1, \ldots, X_r$ of X such that the bundle E is trivial when restricted to any X_i. We shall prove that the Thom theorem in the form

$$\Phi_{\mathcal{T}}: H^k_{\mathcal{T}}(X) \xrightarrow{\cong} H^{n+k}_{\mathcal{T}v}(E)$$

holds for X a foliated space (compact or not) and $E \longrightarrow X$ any oriented r-trivial bundle. This implies Theorem (3.30) since if X is compact then every vector bundle over X is r-trivial for some r and also $H^*_{\tau v}(E) \cong H^*_{\tau c}(E)$.

We proceed by induction on r. If $r = 1$ then E is a trivial bundle. Corollary 3.29 establishes the existence and uniqueness of the classes $u^X_n \equiv u_E$ with the properties 1) and 2) compatible with orientation. Suppose inductively that for all bundles E' which are k-trivial for some $k < r$ we have shown uniqueness of the Thom class $u_{E'}$ compatible with orientations. Let $\pi: E \longrightarrow Y$ be an oriented bundle which is oriented r-trivial, via open sets $X_1, ..., X_r$. Let $U = \pi^{-1}(X_1, \cup...\cup X_{r-1})$ and $V = \pi^{-1}(X_r)$. Then U and V are open sets in E with $E = U \cup V$. The Mayer-Vietoris sequence (3.26) yields the long exact sequence

$$\longrightarrow H^{n-1}_{\tau v}(U \cap V) \longrightarrow H^n_{\tau v}(E) \longrightarrow H^n_{\tau v}(U) \oplus H^n_{\tau v}(V) \longrightarrow H^n_{\tau v}(U \cap V).$$

Since $U \cap V \subset \pi^{-1}(X_r)$, we have

$$U \cap V \cong \left[(X_1 \cup...\cup X_{r-1}) \cap X_r \right] \times \mathbb{R}^n.$$

Thus $H^{n-1}_{\tau v}(U \cap V) = 0$ by Corollary 3.25. By induction we have existence and uniqueness of the classes

$$u^U_n \in H^n_{\tau v}(U)$$

$$u^V_n \in H^n_{\tau v}(V)$$

and each of these restrict to the class

$$u^{U \cap V}_n \in H^n_{\tau v}(U \cap V)$$

using uniqueness and orientability. Exactness of the Mayer-Vietoris sequence implies that there is a unique class $u_E \in H^n_{\tau v}(E)$ which maps to (u_n^U, u_n^V). This establishes the existence and uniqueness of the Thom classes u_E for all r-trivial bundles E. Define

$$\Phi_\mathcal{T}: H_\mathcal{T}^k(X) \longrightarrow H_{\mathcal{T}\,v}^{n+k}(E)$$

by $\Phi_\mathcal{T}(\omega) = u_E\omega$. This is a continuous isomorphism on trivial bundles, by Corollary 3.29. A Mayer-Vietoris argument which we omit (cf. Bott-Tu [BT], p. 64) implies that $\Phi_\mathcal{T}$ is an isomorphism for all r-trivial bundles. □

Remark: Let $z: X \longrightarrow E$ be the zero-section. Define $\tilde{e}_\mathcal{T}(E) \in H_\mathcal{T}^n(X)$ by $\tilde{e}_\mathcal{T}(E) = z^*u_E$. The class $\tilde{e}_\mathcal{T}$ is the tangential Euler class. Similarly we may construct Chern classes and the tangential Chern character in this manner. Since u_E lies in cohomology with real/complex (but not integer) coefficients, characteristic classes constructed in this manner are not visibly integral classes.

Our final topic in this chapter is the introduction of tangential homology. Recall that the tangential forms $\Omega_\mathcal{T}^k(X)$ and $\Omega_{\mathcal{T}\,c}^k(X)$ have been topologized by demanding that in all coordinate patches we have uniform convergence of the functions $a(x_1,..,x_p,n)$ together with all their derivatives in the tangential direction on compact subsets of the coordinate patch. The differential d is continuous. Define

(3.33) $$\Omega_k^\mathcal{T}(X) = \mathrm{hom}(\Omega_{\mathcal{T}\,c}^k(X), \mathbb{R})$$

where hom denotes *continuous* homomorphisms. Elements $c \in \Omega_k^\mathcal{T}$ are called *currents*. The natural differential

$$d_*: \Omega_k^\mathcal{T} \longrightarrow \Omega_{k-1}^\mathcal{T}$$

is given by

$$\langle\omega, d_*c\rangle = (-1)^{k-1}\langle d\omega, c\rangle$$

where $c \in \Omega_k^\mathcal{T}$, $\omega \in \Omega_\mathcal{T}^{k-1}$, and $\langle\ ,\ \rangle$ denotes the evaluation of a current $c \in \Omega_*^{\mathcal{T}}$ on a form ω. (Our sign convention dictates that forms are placed on the left, currents on the right, in $\langle\omega,c\rangle$.) It is immediate that $d_*^2 = 0$. Let $\Omega_k^{\mathcal{T}c} \subset \Omega_k^{\mathcal{T}}$ be those currents which have compact support in the obvious sense. This is a differential submodule. Define *tangential homology* by

$$(3.34) \qquad H_k^{\mathcal{T}}(X;\mathbb{R}) = \frac{\ker\, d_*:\ \Omega_k^{\mathcal{T}} \longrightarrow \Omega_{k-1}^{\mathcal{T}}}{\mathrm{Im}\, d_*:\ \Omega_{k+1}^{\mathcal{T}} \longrightarrow \Omega_k^{\mathcal{T}}}$$

and similarly for $H_*^{\mathcal{T}c}(X;\mathbb{R})$.

Proposition 3.35.

1) Each $H_k^{\mathcal{T}}$ is a covariant functor from foliated spaces and tangentially smooth maps to $\mathbb{R}$ (resp. $\mathbb{C}$)-vector spaces and continuous homomorphisms.

2) If f and g are tangentially homotopic maps $X \longrightarrow Y$ then

$$f_* = g_*: H_*^{\mathcal{T}}(X;\mathbb{R}) \longrightarrow H_*^{\mathcal{T}}(Y;\mathbb{R}).$$

3) The pairings

$$\langle\ ,\ \rangle:\ \Omega_{\mathcal{T}}^* \otimes \Omega_*^{\mathcal{T}c} \longrightarrow \mathbb{R}$$

and

$$\langle\ ,\ \rangle:\ \Omega_{\mathcal{T}c}^* \otimes \Omega_*^{\mathcal{T}} \longrightarrow \mathbb{R}$$

induce continuous pairings

$$\langle\ ,\rangle:\ H_{\mathcal{T}}^*(X;\mathbb{R}) \otimes H_*^{\mathcal{T}c}(X;\mathbb{R}) \longrightarrow \mathbb{R}$$

$$\langle\ ,\rangle:\ H_{\mathcal{T}c}^*(X;\mathbb{R}) \otimes H_*^{\mathcal{T}}(X,\mathbb{R}) \longrightarrow \mathbb{R}.$$

and an isomorphism

$$H_k^{\mathcal{T}}(X) \cong \mathrm{Hom}_{\mathrm{cont}}(H_{\mathcal{T}c}^k(X),\mathbb{R}).$$

4)   If $X = M$ foliated as one leaf then

$$H_*^{\mathcal{T}}(X;\mathbb{R}) \cong H_*(X;\mathbb{R}).$$

Proof: Only 3) requires comment.   Compactness (on one side or the other) guarantees that the pairings exist at the chain level.   There is a natural continuous pairing

$$(\mathrm{cocycles}) \otimes (\mathrm{cycles}) \longrightarrow \mathbb{R}$$

given by evaluation.   If $\omega = d\omega'$ then

$$\langle\omega,c\rangle = \langle d\omega',c\rangle$$

$$= \pm \langle\omega',d_*c\rangle \qquad d_*c = 0 \text{ since c is a cycle}$$

$$= 0$$

and similarly, if c is a boundary then

$$\langle\omega,c\rangle = \langle\omega,d_*c'\rangle$$

$$= \pm \langle d\omega,c'\rangle \qquad (d\omega = 0)$$

$$= 0.$$

Thus there are pairings as indicated.   The isomorphism comes on purely algebraic grounds from the fact that $\mathrm{Hom}_{\mathrm{cont}}(-,\mathbb{R})$ is an exact functor. $\square$

APPENDIX

In this addendum we want to rephrase the construction of the tangential de Rham complex so as to make the algebraic essentials of the construction clear and to show the essential unity of this construction with ordinary Lie algebra cohomology. In this we will be following what is folklore. We start with a pair consisting of a commutative associative algebra A over a field k and a Lie algebra L also over k. We assume that we have a representation of L as a Lie algebra of derivations of A, and we write the action of $\Theta \in L$ on an element $a \in A$ as just $\Theta(a)$. We further assume that L as a linear space is a module over A (but not that L is a Lie algebra over A). Rather one assumes

(3A.1) $\quad [\Theta, b\psi] = b[\Theta, \psi] + \Theta(b)\psi$

where $a\psi$ is left multiplication of $a \in A$ on $\psi \in L$. We call (A,L) a *Lie-associative pair*. Note that if $\Theta(b) = 0$ for all Θ and b then L is a Lie algebra over A. A module M for the pair (A,L) is simply a vector space over k with a module structure for A and with a representation of L on M (as vector space) satisfying

(3A.2) $\quad \Theta(am) = a \cdot \Theta(m) + \Theta(a) \cdot m$

(3A.3) $\quad (a\Theta)(m) = a(\Theta(m)) \qquad$ for $a \in A, \ \Theta \in L, \ m \in M,$

where $\Theta(m)$ is the Lie algebra action and $b \cdot m$ is the left module action. We observe the following fact.

Proposition 3A.4. The algebra A itself, given the structure of A module by left multiplication, and the defining representation of L as derivations of A, is an (A,L) module.

Proof: That the key identities (3A.2) and (3A.3) are satisfied in the first case is just the fact that L is given to act as derivations of A. □

As noted previously this structure already contains ordinary Lie algebras by taking for instance $A = k$ in which case modules as defined above are ordinary L modules. However, the motivating example for this is given by a C^∞ manifold X, with $A = C^\infty(X)$ and L the C^∞ vector fields. More generally X could be a foliated space with A the tangentially smooth functions $C^\infty_\tau(X)$ and L the tangentially smooth sections of the tangent bundle $\Gamma_\tau(FX)$. That L is a Lie algebra under commutator brackets is immediate.

The immediate point here is to define cohomology groups $H^*(M)$ for any (A,L) module M so that in the first example above (A = k) one obtains usual Lie algebra cohomology while in the second example, one obtains the usual de Rham cohomology of X and in the third example one obtains the tangential cohomology. The construction is patterned exactly on the classical Koszul complex (cf. MacLane [Mac]) and the construction of differential forms. If M is an (A,L) module we let $C^p_A(L,M)$ or for short $C^p(M)$ denote the space of all alternating p linear, <u>A-linear</u> maps from L to M. A differential $C^p(M) \longrightarrow C^{p+1}(M)$ is defined as usual by

(3A.5)
$$d\phi(\Theta_0,...,\Theta_p) = \Sigma^p_{i=0}(-1)^i\Theta_i(\phi(\Theta_1,...,\hat{\Theta}_i,...,\Theta_p))$$
$$+ \sum_{i<j} (-1)^{i+j}\phi([\Theta_i,\Theta_j],\Theta_1,...,\hat{\Theta}_i,...,\hat{\Theta}_j,...,\Theta_p).$$

What requires checking is that $d\phi$ is actually A linear since none of the individual terms on the right are A linear. Use of the basic identities (3A.1) and (3A.2) for A, L, and M produces the necessary cancellations. We omit the details. It is also evident that $d^2 = 0$ and so one as usual defines cohomology groups $H^p_A(L,M)$ or for short $H^p(M) = Z^p(M)/B^p(M)$ where $Z^p(M)$ is the kernel of d in $C^p(M)$ and $B^p(M)$ is the image of d from $C^{p-1}(M)$.

It is an easy matter to check that in the case A = k, this yields the usual Lie algebra cohomology of the module M as the formulas (3A.5) are the standard ones, cf. MacLane [Mac]. It is equally easy to see that if $A = C^\infty(X)$, and L the C^∞ vector fields on X, then the cohomology $H^*_A(L,A) = H^*(A)$ is the usual de Rham cohomology of X because $C^*(A)$ is visibly the de Rham complex of differential forms with its usual differential. In the same way, when $A = C_\tau^\infty(X)$ is the tangentially smooth functions on a foliated space and L is the C^∞ tangentially smooth tangential vector fields, then $H^*(A)$ is also visibly the tangential cohomology as defined in Chapter III.

CHAPTER IV: TRANSVERSE MEASURES

In this chapter we concentrate upon the measure theoretic aspects of foliated spaces, including especially the notion of transverse measures.

We begin with a general study of groupoids, first in the measurable and later in the topological context. Our examples come from the holonomy groupoid of a foliated space (2.20) and a discrete version corresponding to a complete transversal. We introduce transverse measures ν with a given modulus and discuss when these are invariant.

Next we look in the tangential direction, defining a tangential measure λ to be a collection of measures λ^X (one for each leaf in the case of a foliated space) which satisfies certain invariance and smoothness conditions. For instance, a tangential, tangentially elliptic operator D yields a tangential measure ι_D as follows. Restrict D to a leaf ℓ. Then from Chapter I Ker D_ℓ and Ker D_ℓ^* are locally finite dimensional and hence the local index ι_{D_ℓ} is defined as a signed Radon measure on ℓ. [A priori it would seem that ι_D depends on the domains $\mathfrak{Dom}(D_\ell)$ but in Chapter VII we shall demonstrate that ι_D is well-defined.] Then $\iota_D = (\iota_{D_\ell})$ is a tangential measure. Tangential measures, suitably bounded, correspond to integrands: if λ is a tangential measure and ν is an invariant transverse measure then $\lambda d\nu$ is a measure on X and $\int \lambda d\nu \in \mathbb{R}$ is defined.

Next we specialize to topological groupoids and continuous Radon tangential measures. In the case of a foliated space we recount the Ruelle–Sullivan construction of a current associated to a transverse measure and we show that the current is a cycle if and only if the transverse measure is invariant.

Finally we prove a Riesz representation theorem for (signed) invariant transverse measures on a foliated space X; they correspond precisely to the topological vector space $(\bar{H}_\tau^p(X))^*$.

A groupoid, whose main feature is a partially defined associative multiplication, is best understood by two extreme special cases – a group on the one hand, and an equivalence relation on the other.   We need say no more about groups, but if $\mathcal{R}$ is an equivalence relation on a set X so that $\mathcal{R}$ is a subset of X×X, one can construct a partially defined associative multiplication on $\mathcal{R}$ so that it becomes a groupoid.   Specifically if u = (x,y) and v = (w,z) are two elements of $\mathcal{R}$, the product uv is defined exactly when y = w, and then uv = (x,z).   It is suggestive to define the range of an element u = (x,y), denoted r(u), to be the first coordinate x and the source of u, denoted s(u) to be the second coordinate y.   Then uv is defined precisely when r(v) = s(u).   Intuitively one might think of the pair (x,y) as something starting at y and going to x so that multiplication is in some way a kind of composition.

If X is a foliated space, there is an obvious equivalence relation on X defined by the leaves, but as we saw in Chapter II a foliated space has associated to it something more, namely its graph G(X) or as it is also called, the holonomy groupoid of X.   This is but one example where a Borel or topological groupoid presents itself naturally – another is when one has a topological group acting on a space where the action is not necessarily free.   Thus we are led to the notion of a groupoid:

Definition 4.1.   A *groupoid* G with unit space X consists of the sets G and X together with maps

(1) Δ: X $\longrightarrow$ G (the diagonal or identity map)

(2)   An involution i: G $\longrightarrow$ G, called inversion and written i(u) = u^{-1}

(3)   range and source maps

$$r: G \longrightarrow X \text{ and } s: G \longrightarrow X$$

(4)   an associative multiplication m defined on the set G' of pairs (u,v), u,v $\in$ G, with r(v) = s(u); one writes m(u,v) = u·v or just uv.

93

In addition to the properties already listed one needs the obvious extra conditions

(5) $r(\Delta(x)) = x$, $s(\Delta(x)) = x$ and $u \cdot (\Delta(s(u))) = u$,
       $\Delta(r(u)) \cdot u = u$

(6) $r(u^{-1}) = s(u)$ and $m(u, u^{-1}) = \Delta(r(u))$

Alternatively, one could define a groupoid as a small category where every map has an inverse. At all events, if X is reduced to a single point, G is simply a group with identity element $\Delta(x)$, X = $\{x\}$. In general the maps r and s together yield a map Φ: u $\longrightarrow$ $(r(u), s(u))$ of G into X×X. The image of this map is an equivalence relation on X in view of the axioms above. If this map is injective, then G as a groupoid is (isomorphic to) this equivalence relation; G is called *principal* if this happens. In any case this shows that associated to any groupoid there is always a principal groupoid (i.e. an equivalence relation). A general groupoid can be viewed as a mixture or combination of this equivalence relation ℜ and the other extreme case of a groupoid, namely a group.

Specifically, we let

$$G_x = \{u: s(u) = x\},$$
$$G^y = \{u: r(u) = y\},$$
$$G^y_x = G_x \cap G^y$$

$$G^Z_Y = \{u \mid r(u) \in Z, s(u) \in Y\}$$

more generally. Then G^x_x is immediately seen to be a group with identity element $\Delta(x)$. The sets G^x_y for $(x,y) \in$ ℜ are principal homogeneous spaces for G^x_x and G^y_y with G^x_x acting on the left and G^y_y acting on the right. In particular for $(x,y) \in$ ℜ, G^x_x and G^y_y are isomorphic. Thus the groupoid G appears as a kind of fibre space over the equivalence relation ℜ as base and with the group-like objects G^y_x as fibres. This is exactly the geometric structure that the holonomy groupoid of Chapter II displayed. Indeed we will often refer to the groups G_x^x as holonomy groups. They can also be thought of as "isotropy groups" because of another important example of groupoids coming from group actions. If a group H acts as a group of transformations on a space X, the set $G = H \times X$ becomes, as George Mackey has emphasized in his seminal papers [Ma3], [Ma5], a groupoid. The unit space is X; $\Delta(x) = x$; the range and source maps are $s(h,x) = x$, $r(h,x) = h \cdot x$ where $h \cdot x$ is the result of the group element h acting on the point x. The inverse of (h,x) is $(h^{-1}, h \cdot x)$. Two points (g,y) and (h,x) are multipliable when $y = h \cdot x$ and then $(g,y) \cdot (h,x) = (gh,x)$. Finally the holonomy group G_x^x is visibly just the isotropy group $(h: h \cdot x = x)$ of the action at x.

Now that the purely algebraic structure of groupoids has been described, we impose the extra conditions appropriate for the analytic and geometrical applications.

Definition 4.2. A (standard) *Borel groupoid* (cf. Mackey [Ma5] or Ramsay [Ra]) is a groupoid G so that G and its unit space are Borel spaces - that is, come equipped with a σ-field of sets - so that the defining maps Δ, r, s, i and m are Borel.

The set X becomes, via the diagonal map Δ, a subset of G, and it will have the relative Borel structure because Δ and r are Borel maps. (In principle it would not be necessary to separately assume that X was a Borel space.) The subset G' of $G \times G$, where m is defined, is given the product Borel structure. We will be assuming throughout that the Borel space G is a standard Borel space. This means that G with its Borel σ-field is isomorphic to a Borel subset of a complete and separable metric space given its Borel σ-field. The reader is referred to Mackey [Ma5], Arveson [Ar], Bourbaki [Bou] for further discussion of this important and pervasive regularity condition for Borel spaces. It is a condition that can be easily checked in the examples to be treated.

Proposition 4.3. The graph G(X) of a foliated space (cf. Definition 2.20 in Chapter II) with the σ-field generated by the open sets is a standard Borel groupoid.

Proof: In case the graph G(X) is Hausdorff, this is obvious for it is a locally compact Hausdorff second countable space and can be given a separable complete metric. In general G(X) may be covered by a countable number of open sets U_i each of which is locally compact, Hausdorff and second countable. It is easy to see that a subset E of G(X) is Borel if and only if $E \cap U_i$ is Borel for all i. Since each U_i is standard as a Borel space, it follows easily that G(X) is standard. □

We will impose two further conditions on our standard Borel groupoid G, both of which are very natural and immediate in the context of foliated spaces. First we shall assume that *each holonomy group* G_x^x *is countable*. The second condition revolves around the notion of a transversal.

Definition 4.4. If G is a standard Borel groupoid with unit space X and associated equivalence relation $\mathcal{R}$, a Borel subset S of X is called a *transversal* if S intersects each equivalence class of $\mathcal{R}$ in a countable set. (For us countable shall mean finite or countably infinite.) A transversal is *complete* if it meets every equivalence class.

If $\mathcal{R}$ is a countable standard Borel equivalence relation in the sense of Feldman-Moore [FMI] (that is, the equivalence classes are countable), then of course any Borel subset is a transversal. As we shall see, the existence of a complete Borel transversal for a general G ensures that it can be built up in a simple way from a countable standard equivalence relation.

We shall now forthwith assume that $\mathcal{R}$ *always has a complete (Borel) transversal*. Note that for foliated spaces the existence of such sets is an immediate consequence of the definitions. In dealing analytically with transversals, we will have need of a very helpful result about Borel spaces that is not too well known.

Theorem 4.5. Let X and Y be standard Borel spaces and let f be a Borel map from X into Y with property that $f^{-1}(y)$ is countable for each y. Then X can be written as the disjoint union of Borel subsets U_i so that f is injective on each U_i. Moreover f(X) is a Borel subset of Y.

We shall not include a proof. The reader is referred to the discussion in Kuratowski [Kur] in section 35,VII. A proof may be found in Hahn [Hah] p. 381. See also Purves [Pu].

We list some consequences of this result that will be relevant for us.

Proposition 4.6. Let G be a standard Borel groupoid with countable holonomy groups.

(1)   The equivalence relation $\mathfrak{R}$ is a Borel subset of X×X, hence a standard space.

(2)   If S is a Borel transversal, the saturation $\mathfrak{R}(S)$ of S with respect to the equivalence relation $\mathfrak{R}$ is a Borel subset of X.

(3)   If S is as in (2), there is a Borel map f from $\mathfrak{R}(S)$ to S with $f(x) \sim x$.

Proof: (1) The map G $\longrightarrow$ X×X given by u $\longrightarrow$ (r(u),s(u)) is Borel and countable to one. Hence its image $\mathfrak{R}$ is Borel.

(2)   Recall that the saturation of a set S is the set of all points equivalent to a member of S. Let W be the subset of X×X given by (S×X) $\cap$ $\mathfrak{R}$. By (1) W is a Borel set. Now let p be the projection map to the second factor. The image of W under p is nothing else but $\mathfrak{R}(S)$, and since S is a transversal p on W is countable to one. Hence p(W) is Borel by Theorem 4.5.

(3)   By the first part of the theorem, we may with a little cutting and pasting construct a subset U of W above so that p is injective on U and p(U) = p(W) = $\mathfrak{R}(S)$. Then define a map f of $\mathfrak{R}(S)$ into S by the condition that f(x) is the unique point so that (f(x),x) $\in$ U. The graph of f is a Borel function, and it follows (cf. Auslander-Moore Ch. I [AM]) that f itself is Borel. This is the desired function. $\square$

The existence of a complete Borel transversal in a standard Borel groupoid G guarantees by part (3) of the proposition above that the equivalence relation ℛ of G is built up in a very simple way from a countable standard equivalence relation. To see this note that a complete Borel transversal S in X defines a countable standard equivalence $ℛ_S$ on S itself; $ℛ_S$ = (S×S) ∩ ℛ. Then the map f guaranteed by (3) of Proposition 4.6 from X to S with f(x)~x displays X as a fibre space over S so that ℛ is also fibred over $ℛ_S$ in the sense that two points of X are ℛ-equivalent if and only if their images under f are $ℛ_S$-equivalent. We shall exploit this structural representation heavily in our discussion of transverse measures.

We observe that not every standard Borel groupoid ℛ satisfies our condition on the existence of a complete Borel transversal. Indeed let X be a Borel subset of the plane $ℝ^2$ whose projection to the first axis is not a Borel set (Kuratowski [Kur]), and define an equivalence relation on X by declaring that two points are equivalent if their first coordinates are the same. This is clearly a standard Borel groupoid (with no holonomy), but there is no complete Borel transversal. For if S is such a transversal, it would follow by Theorem 4.5 that the projection of S to the first axis, which is the same as the projection of X to that axis, would be a Borel set.

Mackey [Ma5], followed by many others (cf. [Ra], [Ra2]), has introduced and studied the notion of a measured groupoid; these are by definition standard Borel groupoids with one more additional datum, namely a Borel measure or better an equivalence class of Borel measures on the groupoid. This class of measures has to satisfy an invariance property that reduces in the case of a principal groupoid (an equivalence relation) to the condition that $θ_*μ$ be equivalent to $μ$ where $μ$ is any measure in the class, and $θ$ is the flip $θ(x,y) = (y,x)$ on the equivalence relation and $θ_*μ$ is the image of the measure $μ$ under the map $θ$. The condition in general is somewhat more complicated but basically the same.

Definition 4.7. A measure $μ$ on a standard Borel groupoid G is *quasi invariant* if $ϕ_*μ$ is quasi invariant on ℛ where ℛ is the principal groupoid associated to G and ϕ is the projection map $G \longrightarrow \mathcal{R}$ and if when μ is disintegrated over $\phi_*(\mu)$ into measures μ_x^y on the fibres G_x^y of the maps ϕ, then for almost all pairs (x,y) in $\mathcal{R}$, μ_x^y should be quasi invariant under the action of the groups G_x^x and G_y^y.

In the present case when the holonomy groups G_x^x are countable this last condition can be rephrased more simply as the condition that for almost all (x,y) μ_x^y gives positive mass to each point in the countable set G_x^y. Note that $r_*(\mu)$ is equivalent to $s_*(\mu)$ and defines an equivalence class of measures on the unit space of X.

Although we have seen that a standard Borel groupoid may fail to have a complete transversal, an important result of Ramsay [Ra] shows that a standard measure groupoid does have such a transversal up to null sets.

Theorem 4.8.   Let G be a standard measured Borel groupoid with unit space X.   Then there is a Borel subset Y of X conull for the natural measure (class) on X defined by the measure μ on G and a subset T of Y which is a complete Borel transversal for the groupoid $G_Y = r^{-1}(Y) \cap s^{-1}(Y)$.   That is, T is a transversal for the original equivalence relation on X and meets every equivalence class of that relation which has a non empty intersection with the conull subset Y.

Thus while the results to follow concerning transverse measures which all assume the existence of a complete transversal on the groupoid do not strictly apply to a measured groupoid, they will apply after one deletes an inessential null set from the unit space.   Our point of view in that discussion is that *all* the points count and that one cannot delete or ignore null sets, especially when one is dealing, as we shall later, with locally compact groupoids.

The discussion to follow concerning transverse and tangential measures can be interpreted as an analysis of a measure on a groupoid into a product (in a Fubini type sense) of a part tangential to the orbits of the groupoid times a part transverse to the orbits.   The transverse part is thus in some vague sense a measure on the space

99

of orbits of the groupoid.

Let us now turn to the crucial topic of transverse measures. If G is a standard Borel groupoid (or more particularly an equivalence relation $\mathcal{R}$ on X) a transverse measure provides, at least intuitively, a method of integrating some kind of object over the set of equivalence classes of the principal groupoid $\mathcal{R}$ associated to G – that is over the quotient space $X/\mathcal{R}$. This quotient space is in general a very pathological space from the point of view of measure theory, containing subsets like $\mathbb{R}/\mathbb{Q}$, the real numbers mod the rational numbers. If the quotient space $X/\mathcal{R}$ with its quotient Borel structure were a standard or even analytic Borel space (e.g. if G were to come from a foliation given by the fibres of a fibre bundle) then transverse measures would be really just ordinary measures on $X/\mathcal{R}$. For general foliations transverse measures suitably defined have played an important role for years. In addition, as Connes points out [Co3], one has to rethink one's concept of what sort of functions are suitable integrands for integrating against a transverse measure.

We will treat something a bit more general than what traditionally in the theory of foliations is called a transverse measure; transverse measures here will involve a modular function analogous to the modular function on a locally compact group. When this modular function is identically one, as is traditional in foliation theory, the transverse measure will be called invariant. Hence what in foliations is called a transverse measure, we shall call an invariant transverse measure.

The modular function in question above is simply a homomorphism from the groupoid to the group of positive real numbers $\mathbb{R}^+$ under multiplication. A homomorphism of a groupoid G to a group H (or indeed to another groupoid) is a map ϕ from G to H so that when uv is defined $\phi(u)\phi(v)$ is defined and is equal to $\phi(uv)$. When G and H are standard Borel groupoids, one insists naturally that ϕ be a Borel map. For the purposes at hand we fix a Borel homomorphism, denoted by Δ, of G into the group $\mathbb{R}^+$. We further assume that Δ is holonomy invariant in that $\Delta(u)$ depends only on r(u) and s(u). Put another way, there is a homomorphism Δ' of the principal groupoid (equivalence relation) $\mathcal{R}$ associated to G so that

$\Delta(u) = \Delta'(p(u))$ where $p(u) = (r(u), s(u))$ is the projection of G onto the equivalence relation $\mathcal{R}$.

Let us now consider the case when G (or $\mathcal{R}$) has countable equivalence classes. Then in view of our standing hypothesis that all holonomy groups are countable, the range and source maps r and s are countable to one. In [Ma5], the notion of a quasi invariant measure on X with given Radon-Nikodym derivative (or modulus) Δ is discussed, at least in the case of trivial holonomy groups, see also [FM1]. The discussion extends without change; namely we start with a measure ν on X quasi invariant under $\mathcal{R}$ in the sense that a subset E of x is ν-null if and only if its $\mathcal{R}$ saturation – again a Borel set by Proposition 4.6 – is also ν-null. As r is countable to one, there is a unique measure ν_r on $\mathcal{R}$ which is the integral of the counting measures on the fibres of the map r over the base X. Specifically if $|C|$ is the cardinality of C, then

$$\nu_r(S) = \int_X |S \cap r^{-1}(x)| \ d\nu(x).$$

There is a similar measure ν_s defined using the source map s instead of r. As in Feldman-Moore [FMI], these measures are mutually absolutely continuous and the Radon-Nikodym derivative $d\nu_r/d\nu_s = \Delta$ is called the *modulus* of ν. This function on G is readily seen to depend only on the projection of G onto $\mathcal{R}$. As a function on G or $\mathcal{R}$ the modulus Δ is a homomorphism up to null sets in that $\Delta(uv) = \Delta(u)\Delta(v)$ for almost all u and v in the obvious sense.

We return to the case of a general standard Borel groupoid with countable holonomy groups and a complete Borel transversal. We observe that the set of all Borel transversals $\mathcal{S}$ is indeed a σ-ring, but not in general a σ-field, of subsets of X. A transverse measure will be simply a measure on this σ-ring. For each $S \in \mathcal{S}$ we can form the restriction of G to S, $G_S^S = \{u \in G, r(u), s(u) \in S\}$. This is a groupoid with countable orbits and countable holonomy groups of the kind discussed a moment ago.

There is a subtle point here about whether Δ is or is not constant on the holonomy groups G_x^x. There will be instances later on when we specifically will want to allow Δ to be non constant on some holonomy groups G_x^x; the point is that this cannot happen for too many x's, for it follows from the fact that Δ is equal almost everywhere to the Radon-Nikodym derivative of $\nu\big|_S$ on G_S^S and some null set manipulations, that for $\nu\big|_S$ almost all $x \in S$, Δ is constant on G_x^x.

Definition 4.9.   A *transverse measure* with modulus Δ on a standard Borel groupoid is a measure ν on the σ-ring of Borel transversals $\mathcal{J}$ so that $\nu\big|_S$ is σ-finite for each $S \in \mathcal{J}$ and so that $\nu\big|_S$ is quasi invariant on G_S^S with modulus equal to Δ almost everywhere on G_S^S. If $\Delta = 1$, one says that ν is an *invariant transverse measure*.

A transverse measure allows one to talk consistently about what it means for a set L of equivalence classes of the equivalence relation $\mathcal{R}$ to have measure zero.   The condition is that the intersection of the leaves in L with each Borel transversal S should be contained in a $\nu\big|_S$ Borel null set, or equivalently that this should happen for a single complete Borel transversal. Since the modulus of a quasi invariant measure is constant on holonomy groups, we conclude from this discussion that Δ is constant on the holonomy groups of almost all leaves.

As an example of a transverse measure we consider the Kronecker foliation on the two-torus T^2 where the leaves are of the form

$$\{(\exp(2\pi i(x + x_0)), \exp(2\pi i\lambda x), x \in \mathbb{R}\}$$

λ irrational. Regard the two-torus as the square

$$\{(x,y) \mid 0 \leqslant x < 1, 0 \leqslant y < 1\},$$

and for each ρ, $-\lambda < \rho \leqslant 1$, let ℓ_ρ be the part of the line $y = \lambda x + \rho$ inside the square described above.

If S is any Borel transversal, then S must meet each ℓ_ρ in at most a countable set. If $n(\rho)$ is the cardinality of $S \cap \ell_\rho$, then n is obviously a Borel function. We define ν by the formula

$$\nu(S) = \int_{-\lambda}^{1} n(\rho) d\rho.$$

It is not difficult to verify that this produces an invariant transverse measure for the graph of this foliation. If instead one defined

$$\nu(S) = \int_{-\lambda}^{1} n(\rho) f(\rho) d\rho$$

for some positive Borel function f, the result would be a transverse measure with modular function

$$\Delta((x_1, y_1), (x_2, y_2)) = f(y_1 - \lambda x_1)/f(y_2 - \lambda x_2).$$

Recall that given a diffeomorphism ϕ of F then one may form its suspension $M = \mathbb{R} \times_{\mathbb{Z}} F$ (cf. (2.3)) which is foliated with leaves of dimension 1. An invariant transverse measure for M corresponds to a ϕ-invariant measure on F. More generally, in the situation (2.2) of a manifold with discrete structural group

$$M = \tilde{B} \times_{\pi_1(B)} F$$

an invariant transverse measure on M corresponds to a measure on F which is invariant under the action $\pi_1(B) \longrightarrow \text{Homeo}(F)$.

Also let us consider the very special case when $G = \mathcal{R}$ is an equivalence relation coming from a Borel map p of X onto a standard Borel space B - in other words, a fibration. Here $x \sim y$ if and only if $p(x) = p(y)$. We let $\Delta = 1$ so we are looking for invariant transverse measures. If $\tilde{\nu}$ is a measure on the base B then if N is transversal we define

$$\nu(N) = \int |N \cap p^{-1}(b)| \, d\tilde{\nu}(b).$$

103

It is clear that ν so defined on the σ-ring of transversals is an invariant transverse measure. Conversely we claim every such ν is of this form. To see this, observe that the assumed existence of a complete transversal yields by Proposition 4.6 the existence of a Borel cross section S for the map p. It follows that p maps S bijectively onto B; by Kuratowski [Kur] it is therefore a Borel isomorphism. If ν is an invariant transverse measure on X, ν gives in particular an ordinary measure on S. This may be transported to B via $p \mid S$ to give a measure $\tilde{\nu}$ on B. Then it is an easy exercise to see that ν on any transversal N is given by the formula above in terms of $\tilde{\nu}$ and $\mid N \cap p^{-1}(b) \mid$.

It is well to extend the remarks above a bit to observe that ν is determined by what happens on any complete transversal.

Proposition 4.10. For any standard Borel groupoid with countable holonomy and unit space X, a transverse measure ν of modulus Δ is completely determined by $\nu \mid_N$ where N is any complete transversal. Conversely if ν_N is a transverse measure on N with modulus $\Delta \mid_N$, then there exists a (unique) transverse measure on X with modulus Δ.

Proof. By Proposition 4.6 we construct a Borel map t from X to N with $f(x) \sim x$. If S is any transversal, f restricted to S, $f \mid_S$, is a countable to one map of S to N; then assuming we know $\nu \mid_N = \nu_N$ for some transverse measure, we can immediately calculate ν on S given the invariance properties in terms of $\Delta(f(s),s)$ as follows:

$$\nu(S) = \int_N (\Sigma_s \Delta(t,s)) d\nu_N(t)$$

where for each t the sum is taken over all s with $f(s) = t$. This shows that ν on N determines ν altogether.

Conversely if we are given a transverse measure ν_N on N, we use the same formula to extend ν_N to all transversals. It is a simple calculation to show that the result is a transverse measure on X. □

If X is a foliated manifold with oriented transverse bundle, we remark that there is a canonical transverse measure class given by the volume element on q-dimensional transverse submanifolds. This may or may not be an invariant transverse measure (class).

A transverse measure on a general groupoid in this formulation is really an ordinary measure but is defined on a σ-ring $\mathcal{S}$ instead of a σ-field. The measure could of course be extended to the σ-field generated by $\mathcal{S}$ but this extension would in general be impossibly non σ-finite as a measure on the entire space. (If the entire groupoid has countable orbits then a transverse measure is just an ordinary (σ-finite) measure on the unit space.) These facts make a huge difference in the type of object that can be integrated in general against a transverse measure.

We insert here several diverse examples of foliated spaces which yield interesting classes of (primarily Type III) von Neumann algebras. In Chapter VI we shall consider the question of exactly which von Neumann algebras may be realized as the von Neumann algebras of foliated spaces.

Let $G = SL_2(\mathbb{R})$, let Γ be a discrete cocompact subgroup, and let $M = G/\Gamma$ be the resulting compact 3-dimensional manifold. Foliate M by the left action of the triangular subgroup

$$B = \begin{bmatrix} a & 0 \\ b & a^{-1} \end{bmatrix} \qquad a > 0.$$

The orbits are two dimensional, hence this is a codimension 1 foliation of M. Each leaf is dense. In fact, if one lets

$$N = \begin{bmatrix} 1 & 0 \\ b & 1 \end{bmatrix}$$

act instead [this is called the horocycle flow] then each leaf is still dense. The foliation arising from N has an invariant transverse measure. However, there is no invariant transverse measure at all for the foliation which arises from the action of B. The associated von Neumann algebra is a III_1 factor; cf. Bowen [Bow].

Here is an example of a 1-dimensional foliation of a

3-dimensional manifold which is due to Furstenberg [Fu]. It is built by first defining a $\mathbb{Z}$ action on a 2-manifold and then suspending it to make an $\mathbb{R}$ action on a 3-manifold.

Let $M = \mathbb{T} \times \mathbb{T}$ be the 2-torus and let $\mathfrak{s}$ be an irrational number. Let

$$\phi(x,y) = (e^{2\pi i \mathfrak{s}} x, g(x)y)$$

where x,y are complex numbers of absolute value 1 and where $g: \mathbb{T} \longrightarrow \mathbb{T}$ is a function at our disposal. We construct g as follows by first defining

$$h(x) = \sum_{k \neq 0} \frac{1}{k} e^{2\pi i n_k \mathfrak{s}}$$

where n_k is a sequence of integers tending to ∞ at our disposal. Observe that

$$k(x) = h(e^{2\pi i \mathfrak{s}} x) - h(x) =$$

$$= \sum_{k \neq 0} \frac{1}{k} (e^{2\pi i n_k \mathfrak{s}} - 1) e^{2\pi i n_k \mathfrak{s}}.$$

Now pick $\mathfrak{s}$ and n_k such that, say,

(*) $$|e^{2\pi i n_k \mathfrak{s}} - 1| < r^{n_k} \quad \text{some } r < 1.$$

This is possible for suitable $\mathfrak{s}$, but such $\mathfrak{s}$'s are not very common — they are highly Liouville; alternatively one could make

(**) $$|e^{2\pi i n_k \mathfrak{s}} - 1| = 0(n_k^{-r}) \quad \text{for all } r > 0.$$

Then consider

$$g(x) = e^{it(h(e^{2\pi i \mathfrak{s}} x) - h(x))}$$

for suitable t as our g. First of all, if $k(x) = h(\mathfrak{s}x) - h(x)$, then k is real analytic under (*) and C^∞ under (**). Hence $\phi(x,y)$ is real analytic, respectively C^∞. It is a theorem of Furstenberg [Fu] that for any ϕ of the form above,

(1) ϕ is minimal iff one cannot factor any power of g as

$$g^m(x) = u(e^{2\pi i \zeta}x)/u(x)$$

for a continuous function u: $\mathbb{T} \longrightarrow \mathbb{T}$, and

(2) ϕ is ergodic with respect to Lebesgue measure if one cannot factor any power of g as

$$g^m(x) = u(e^{2\pi i \zeta}x)/u(x)$$

for a measurable function u: $\mathbb{T} \longrightarrow \mathbb{T}$.

The proofs are not hard.

Now the g is cooked up so that

$$g(x) = e^{ith(e^{2\pi i \zeta}x)}/e^{ith(x)}$$

so that for all t the transformation is not ergodic, hence not unqiuely ergodic. However, if one could factor as above, then the factorization would be unique up to a constant, as ζ is irrational and rotation by ζ is ergodic on $\mathbb{T}$. So if one could factor g (or any power of g) then the factorization would have the same form as above. Hence ϕ will be minimal for given t provided that we can be assured that

$$e^{ith(x)}$$

is *not* continuous. If this is continuous for all t, it is easy enough to see that h(x) is continuous (and conversely, of course). But h is *not* continuous because the Fourier series of h(x) would then be Cesaro summable to h for every x, by advanced calculus. Then we would have

\

$$h(1) = \text{C.S.} \sum_{k \neq 0} \frac{1}{k}$$

which is nonsense, since this is a sum of positive terms.

So ϕ is minimal – each orbit is dense, but (for suitable choice of g and $\mathfrak{s}$) ϕ is not ergodic. Now form the suspension of ϕ to obtain a one dimensional foliation of the 3-torus which has corresponding properties. A transversal is of course M with the equivalence relation induced by powers of ϕ. This is a real analytic foliation. There are a continuum of ergodic invariant transverse measures of this foliation – in fact they are indexed by the circle. Each is singular with respect to Lebesgue measure and in the foliation case live on a measurable *but* *not* topological 2-torus inside the manifold. Measure theoretically this foliation looks like a Kronecker foliation on the 2-torus with angle $\mathfrak{s}$ crossed with a circle – nothing happening in the transverse direction here. The invariant ergodic transverse measures are just the measures on the copies of the Kronecker torus in this product structure.

Here is an example given by Connes [Co2, p. 150] of a foliation whose von Neumann algebra is of type III_λ for some fixed λ with $0 < \lambda < 1$. Let S be a circle of length s, let $X = SL(2,\mathbb{R})/\Gamma$ for a discrete cocompact subgroup Γ, and let $Y = S \times X$. Act on Y by the group of matrices of the form

$$\begin{bmatrix} e^t & 0 \\ b & e^{-t} \end{bmatrix}$$

for $t, b \in \mathbb{R}$, where

$$\begin{bmatrix} 1 & 0 \\ b & 1 \end{bmatrix}$$

acts trivially on S and by the horocycle action on X, and

$$\begin{bmatrix} e^t & 0 \\ 0 & e^{-t} \end{bmatrix}$$

acts by a rotation of speed 1 on S by the geodesic flow on X. The resulting foliation has a von Neumann algebra of type III_λ where $\lambda = e^{-s}$. If S is replaced by a space K of dimension at least 2 with an ergodic action then a III_0 factor results.

With these examples in hand, we return to the general development. The next order of business is the introduction of appropriate integrands to pair with transverse measures.

By way of motivation to see what the appropriate integrands are, we consider the case when G is an equivalence relation on X with each equivalence class consisting of one point. As noted, a transverse measure ν is an ordinary measure; one uses a (non negative) Borel function f on X as integrand. Instead of looking at f as a function on X, we view f as an assignment to each equivalence class of G (i.e. each point of X), of a measure living on that equivalence class. The measure attached to $\{x\}$ is of course $f(x)\delta_x$ where δ_x is the Dirac measure at x. Moreover, we regard the process of integration as first passing from the integrand f to the measure $f \cdot \nu$ on X, where

$$f \cdot \nu(E) = \int_E f d\nu,$$

and then passing to the total mass of $f \cdot \nu$ to obtain a real number – the integral of f.

This point of view guides us in the general case: the proper integrand for a transverse measure ν on a standard Borel groupoid G will be a family of measures $\{\lambda^\ell\}$ one on each "leaf" of the groupoid. If G is an equivalence relation this is simply an assignment $\ell \longrightarrow \lambda^\ell$ of a (non negative, σ-finite) measure λ^ℓ on each equivalence class ℓ of the equivalence relation. This map $\ell \longrightarrow \lambda^\ell$ should be Borel in an obvious but tedious sense that we shall not write down.

If for example G has countable orbits, there is a very natural such family of measures; namely, λ^ℓ is counting measure on the (countable) set ℓ. For the Kronecker foliations discussed above, each orbit is an affine real line; that is, the real line without an origin specified. On such affine lines we can simultaneously normalize Haar measure to obtain a family $\{\lambda^\ell\}$ of the type described. Finally if $\mathcal{R}$

is any standard equivalence relation and if S is a Borel transversal, we can define a family $\{\lambda^\ell\}$ by letting λ^ℓ be the counting measure on the countable set $T \cap \ell$, viewed as a measure on ℓ.

Definition 4.11. A *tangential measure* $\lambda = \{\lambda^\ell\}$ for an equivalence relation $\mathcal{R}$ is an assignment of (non-negative) measures $\ell \to \lambda^\ell$ as above.

This was all for a principal groupoid (i.e. equivalence relation). The presence of holonomy complicates matters a bit, but the complication is largely notational. Recall that the inverse images of the range map r are denoted $G^x = r^{-1}(x)$. A tangential measure on G is first an assignment of a (σ-finite) measure λ^x on G^x for each x in a Borel fashion subject to an invariance condition. For equivalence relations, when $x \sim y$, $r^{-1}(x)$ is actually the same as $r^{-1}(y)$ and we demanded $\lambda^x = \lambda^y$ (= λ^ℓ) where ℓ is the common equivalence class of x and y. In general the requirement is that

$$\int f(uu')d\lambda^x(u') = \int f(u')d\lambda^y(u')$$

for every $u \in G^y_x$ (= $r^{-1}(y) \cap s^{-1}(x)$) and every non negative Borel function f on G. Since the meaning of this formula is not immediately transparent, we rephrase it more geometrically. Each set G^x is acted upon by the group G^x_x which acts freely from the left by groupoid multiplication. The quotient space $G^x_x \backslash G^x$ is canonically identified to the equivalence class $\ell(x)$ of the corresponding principal groupoid (equivalence relation) associated to G. If $y \sim x$ with respect to $\mathcal{R}$, then each element of G^y_x defines a bijection of G^x onto G^y; moreover G^y_x is acted upon freely by G^x_x on the right and G^y_y on the left again using groupoid muliplication. By associativity, the transformations of G^y_x intertwine the actions of G^x_x on G^x and G^y_y on G^y so that the quotient spaces $G^x_x \backslash G^x$ and $G^y_y \backslash G^y$ can be identified. The identification is independent of the element in G^y_x and when these two sets are further identified with $\ell(x)$ and $\ell(y)$ respectively, the mapping becomes the identity map between $\ell = \ell(x)$ and $\ell = \ell(y)$, the common equivalence classes of x and y under $\mathcal{R}$.

The invariance condition expressed by the integral formula

above says first of all that λ^x on G^x is invariant under the left action of G_x^x for all x, and furthermore that elements of G_x^y, viewed as mappings from G^x to G^y, carry λ^x onto λ^y. We see now that the standing hypothesis of countability of G_x^x allows us to simplify the situation. By a choice of cross section, G^x can be viewed as the product $G_x^x \times \ell(x)$, and using counting measure on G_x^x, there is a bijection between measures on $\ell(x)$ and G_x^x-invariant measures on G^x which is independent of the cross section. Thus if λ^x is a choice of G_x^x invariant measures on G^x, $x \in X$, with corresponding measures $\tilde{\lambda}^x$ on $\ell(x)$, the further invariance under G_x^y for a tangential measure means that $\tilde{\lambda}^x = \tilde{\lambda}^y$ if $x \sim y$. Hence $\tilde{\lambda}^x = \tilde{\lambda}^\ell$ if $x \in \ell$ defines a tangential measure on $\mathcal{R}$, the associated principal groupoid. Summarizing, we obtain the following observation which allows us better to understand tangential measures in general.

Proposition 4.12. If G is a standard Borel groupoid with countable isotropy groups and $\mathcal{R}$ is the corresponding equivalence relation, then the map $\lambda \longrightarrow \tilde{\lambda}$ defined above is a bijection from tangential measures on G to tangential measures on $\mathcal{R}$. □

To illustrate further the notion of a tangential measure when there is holonomy, consider the example of a groupoid G coming from the action of a locally compact group H on a Borel space X. Recall that elements of G are pairs (h,x) and that the range map is r(h,x) = h·x. If we fix a point $x_0 \in X$, then $r^{-1}(x_0)$ can be represented as the set $\{(h, h^{-1}x_0), h \in H\}$, and we use the first coordinate to parametrize this set. If y_0 is equivalent to x_0, so that $y_0 = h_0 \cdot x_0$, then $r^{-1}(y_0)$ can be represented as the set $\{(k, k^{-1}y_0), k \in H\}$, and elements of $G_{x_0}^{y_0}$ are of the form $(h_0 h_1, x_0)$ where h_1 is in the isotropy group of x_0. Groupoid multiplication shows that the map from G^{x_0} to G^{y_0} is $(h, h^{-1}x_0) \longrightarrow (h_0 h_1 h, h^{-1}h_1^{-1}h_0^{-1}y_0)$. Hence in terms of the parameters on these spaces the map is left translation. Therefore a suitable choice of tangential measure would be λ^x equal to left Haar measure on H transported over to $r^{-1}(x)$ as indicated above.

There are evidently many other choices also.

Ultimately we will want to consider tangential measures of mixed sign. In outline the notion is clear, but there are technical difficulties because generically the measures λ^ℓ (or λ^x) will be infinite measures. Indeed one sees easily that for an equivalence relation $\mathcal{R}$, the existence of a tangential measure with λ^ℓ finite for each ℓ implies that the equivalence relation $\mathcal{R}$ is smooth; that is, the quotient space $X/\mathcal{R}$ is an analytic Borel space, [Ar, p. 71]. Since infinite signed measures cause problems in this general context, one would only want to discuss tangential measures of mixed sign in the presence of some topological assumptions.

Let us now turn to the integration process, which is related to the Ruelle-Sullivan pairing [RuS]. Begin with a standard Borel groupoid G together with a transverse measure ν with modulus Δ and a tangential measure λ. The integration process is going to produce first a measure μ, written $d\mu = \lambda d\nu$, on the unit space X whose total volume $\mu(X) = \int \lambda d\nu$ will be the integral of the tangential measure with respect to the transverse measure. To define these objects we first fix a complete Borel transversal S, which exists by our standing hypothesis. By Proposition 4.6 we find a Borel function f from X to S with $f(x) \sim x$. Next we observe by Proposition 4.12 that we may as well assume that $G = \mathcal{R}$ is principal. Then for each point $s \in S$ we define a measure ρ_s on $f^{-1}(s)$ as the restriction of $\lambda^{\ell(s)}$ to $f^{-1}(s) \subset \ell(s)$, the equivalence class of s. The modular function Δ of ν comes to us as a function on G_S^S, but we have observed that if we stay away from a ν-null set of equivalence classes of the relation $\mathcal{R}$, then Δ is constant on holonomy groups, and is almost everywhere really a function on $\mathcal{R}$. In the present context, this means that there is a saturated null set N of S so that for $s \notin N$, $\Delta(s,x)$ is well defined. That is the meaning of the function which appears in the integral below which defines the measure $\mu = \int \lambda d\nu$ on X, the result of integrating λ against the transverse measure ν:

$$\int_E \lambda d\nu = \mu(E) = \int \left[\int_{f^{-1}(s)} \Delta(s,x) \chi_E(x) d\rho_s(x) \right] d\nu(s)$$

for any Borel set E in X. The first remark is that this is independent of the choice of the complete transversal S and of the function f from X to S. The presence of the modular function Δ in the above formula is exactly what is needed to achieve this, and we omit the simple calculation.

The resulting measure μ on X is thus well defined and depends only on the data given, the transverse measure ν of modulus Δ and the tangential measure λ. Its total mass is written as

$$\mu(X) = \int_X \lambda d\nu.$$

In the most primitive special case of an equivalence relation on X given by a fibration p of a space X over a base space B, we have seen already that a transverse measure ν with modulus $\Delta = 1$ is exactly a measure on the base B, and that a tangential measure is a family of measures $\{\lambda^b\}$, one on each fibre $p^{-1}(b)$. The integral $\lambda d\nu$ is the usual construction of a measure on the total space X from a measure on the base and measures on the fibres. The formula given above in the general case makes the general situation very similar intuitively to the fibration case. Indeed the total space X is fibred measure theoretically over the transversal S, instead of a base space B; the picture is quite similar:

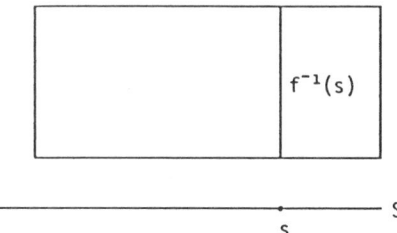

In accord with the notion that a transverse measure ν on G is in some sense a measure on the orbit space X/G, we have already remarked that it is possible to say what it means for a Borel set of orbits to be a null set of orbits. This is clear for a Borel set of orbits corresponds to a Borel set E in the unit space which is saturated or invariant with respect to the equivalence relation R of G.

Definition 4.13. An invariant Borel set E in X is a ν-null set if every transversal in E has ν measure zero.

Using this definition it is easy to define an ergodic transverse measure; namely if $X = E_1 \cup E_2$ where E_i are invariant Borel sets, one of them is a ν-null set. In addition one has as usual a type classification of ergodic transverse measures into types I, II, and III. Indeed if N is a complete transversal then $(\mathcal{R}_N, N, \nu \mid_N)$ is an ergodic countable standard measured equivalence relation which has a type classification (Feldman-Moore [FM]). In the type II case, one may have different transversals where one is type II_1 while another is type II_∞. Hence there is no meaningful distinction between these types and one has one class of type II transverse measures. As usual one may further divide the type III case into the III_λ $0 \leqslant \lambda \leqslant 1$ subtypes by the type classification of the discrete versions $(\mathcal{R}_N, N, \nu \mid N)$. For some examples of type III_λ factors, cf. Connes [Co2], pp. 149-150.

Further, a general transverse measure ν can be displayed as a continuous sum of ergodic components. To see this, one makes an ergodic decomposition of $(N, \nu \mid N)$ and then uses the projection map p of Proposition 4.6 of all of X on N to decompose ν itself. By construction, all of the groupoids appearing as disintegration products will have complete transversals.

Throughout this entire discussion the modular function Δ has remained fixed. If we change the modular function to a new one Δ' which is however in the same cohomology class, that is

$$\Delta'(u) = \Delta(u) \ b(r(u))b(s(u))^{-1}$$

where b is some Borel function on X into the strictly positive real numbers, then there is no essential difference between transverse measures of modulus Δ and transverse measures of modulus Δ'. Indeed if ν is a transverse measure of modulus Δ, then $b \cdot \nu$, where multiplication of a (transverse) measure by a positive Borel function has the usual meaning, is by a simple computation (cf. Feldman-Moore I [FM1], p. 291) a transverse measure of modulus Δ'

where Δ' is as above, and conversely.

Since most of the groupoids we shall meet carry not just a Borel structure, but also a topology, we shall now discuss briefly the notion of a topological groupoid. Following Renault [Ren1], we impose the following conditions.

Definition 4.14. A groupoid G with unit space X is a *topological groupoid* if G and X are topological spaces and

(1)   The set where the partially defined multiplication is defined is closed in G×G and multiplication is continuous.

(2)   The range and source maps are open and continuous.

(3)   The inversion map is a homeomorphism.

For our discussion G and X will be assumed to be locally compact in which case we will say that G is a locally compact (topological) groupoid. Ordinarily one would automatically assume that G and X are Hausdorff and most of the time in the sequel we will have this as a standing assumption. However the reader should be aware that there are a number of interesting, natural, and significant examples where a non-Hausdorff structure is forced upon one. The graph of the Reeb foliation discussed in Chapter II is one such example. All interesting examples known to us satisfy the following condition that could be used in place of the Hausdorff condition:

(4)   X is Hausdorff and G has a cover consisting of open sets each of which is Hausdorff.

If (4) is satisfied we say that G is *locally Hausdorff*.

We remark that if G = $\mathcal{R}$ is an equivalence relation, then $\mathcal{R}$ is a subset of X×X; yet the topology of $\mathcal{R}$ will not be the relative topology from X×X. For instance if we consider the Kronecker equivalence relation $\mathcal{R}$ on the circle $\mathbb{T}^1$ given by ξ ~

ξ exp(2πinλ), n $\in$ η, λ irrational, then $\mathcal{R}$ as a subset of $\mathbb{T}^1 \times \mathbb{T}^1$ is a line of irrational slope in the two torus. To make it a locally compact groupoid one has to give $\mathcal{R}$ the usual topology of the real line.

The prime example we have in mind is the graph of a foliation, at least when it is Hausdorff, as described in Chapter II. If H is a locally compact group acting as a topological transformation group on a locally compact Hausdorff space X, then the groupoid H$\times$X described earlier in this chapter becomes a locally compact topological groupoid.

Finally, the following simple example displays for us in a discrete context the need for introducing the graph of a foliation. On the real line $\mathbb{R}$ consider the equivalence relation $\mathcal{R}$ where $x \sim 2^{-n}x$ for all n $\in$ $\mathbf{Z}$. In spite of the simplicity of this, the equivalence relation $\mathcal{R}$ does not admit any reasonable locally compact topology. The trouble comes near (0,0) = ρ_0 where $\mathcal{R}$ appears to have an infinite number of line segments all passing through this point.

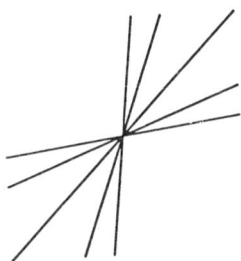

If however we introduce points p_n which are formally the limits of $(x, 2^{-n}x)$ as $x \longrightarrow 0$ with n fixed, then we can visualize this new object G as an infinite set of (parallel) real lines

.
.
p_1
—————————— = $\{(x, s/2)\}$

—————————— = $\{(x, x)\}$
p_0

—————————— = $\{(x, 2x)\}$
p_{-1}
.

It is easy to see that G may be turned into a locally compact topological groupoid. Indeed it is a discrete version of the graph construction for a foliation. We remark that if we modify the equivalence relation $\mathcal{R}$ by saying that $x \sim x$ for $x < 0$ and $x \sim 2^{-n}x$ all n, for $x \geq 0$, then this construction leads to a non-Hausdorff graph-like object

$$
\underset{\displaystyle \underline{}_{\;p_{-1}}}{\underset{\displaystyle \underline{}_{\;p_0}}{\overline{}^{\;p_1}}} = \{(x,x)\}
$$

where a neighborhood of p_1 is a small interval containing p_1 and extending to the right of p_1 plus a small interval to the left of p_0 (but not including p_0) which has already arisen in Chapter II (2.29).

In both examples it is clear that Lebesgue measure is a quasi invariant measure. It would be natural to hope that the modular function Δ could be fixed up to be continuous. A simple calculation shows that on the nth horizontal line in these examples Δ is almost everywhere equal to 2^n. Hence in the first example we can make Δ continuous, but then over the point 0 it is non-constant on the holonomy group G_0^0. This happens only on a null set-one point, in accord with our earlier discussion. In the second example we see that Δ cannot be constructed so as to be continuous.

Another class of examples of interest of topological groupoids are ones that arise from the holonomy of a single leaf of a foliation. (Compare with the bundle construction in Chapter II; cf. 2.25.) Let M be a manifold and let Γ be a quotient group of $\pi_1(M)$. Then there is a covering $\tilde{M}$ of M with deck group Γ, and we identify M as the orbit space $\tilde{M}/\Gamma$. We form $G = (\tilde{M} \times \tilde{M})/\Gamma$ where Γ is acting diagonally. Two Γ-orbits $\Gamma \cdot (x,y)$ and $\Gamma \cdot (z,w)$ are multipliable if $\Gamma \cdot y = \Gamma \cdot z$; we define their product to be $\Gamma \cdot (\gamma_1 x, \gamma_2 \cdot w)$ where γ_1 and γ_2 are elements of Γ so that $\gamma_1 \cdot y = \gamma_2 \cdot z$. The unit space is the original manifold M, and the range and source maps are $r(\Gamma \cdot (x,y)) = \Gamma \cdot x \in M$, and $s(\Gamma \cdot (x,y)) = \Gamma \cdot y \in M$. It is not difficult to see that this produces a topological groupoid with Γ

as constant holonomy group. One easily sees that as a Borel groupoid this groupoid is simply the product of the group Γ and the equivalence relation on M where all points are equivalent, but it is not the product as a topological groupoid.

In the context of topological groupoids, homomorphisms of a groupoid to a group or another groupoid should be assumed to be continuous. Transverse measures considered in this context will be assumed to have continuous modular functions.

In the context of topological groupoids there is a special kind of tangential measure of interest. If we recall that tangential measures are objects to be integrated against transverse measures and hence are analogues of functions, it makes sense to try to define, in analogy with a continuous function, a continuous tangential measure.

Definition 4.15. We say that a tangential measure λ is *continuous* if each λ^x is a Radon measure on $r^{-1}(x) \subset G$ and if

$$\int f(u) d\lambda^x(u)$$

is continuous in x for every continuous function f of compact support in G. This is appropriate if G is Hausdorff. If G is only locally Hausdorff we demand instead that the integral above be continuous in x when f is compactly supported inside some Hausdorff open set and is continuous there. Such a function f need not be even continuous on all of G.

As an example consider the case of a G arising from a locally compact group H acting topologically on a locally compact space X. We saw earlier in this chapter that the assignment $x \longrightarrow \lambda^x$ where λ^x is Haar measure on H carried over to $r^{-1}(x) = \{(h,h^{-1}x)\}$ by the map $h \longrightarrow (h,h^{-1}x)$ is a tangential measure. Evidently this is also a continuous tangential measure.

An obvious item of concern is to find conditions on a transverse measure ν and a tangential measure λ so that the integral $\lambda d\nu$ produces a *finite* measure on X. Rather than taking this question up in this general context we shall take it up in the more special context of primary interest when G is the graph of a foliation. We turn to that case now.

So assume that X is a locally compact foliated space with G the graph of the foliation, which we assume is Hausdorff. Then G is itself a foliated space as described in Chapter II with leaves equal to the holonomy groupoids of the leaves of the original foliation. All homomorphisms θ of G to a Lie group and in particular to $\mathbb{R}^+$ will be assumed to be tangentially smooth on the foliated space G in the sense of Chapter II.

Now suppose that ν is a transverse measure on X of modulus Δ. As suggested previously, the notion of ν being a Radon measure, to the extent that this can be defined in general, would be a condition demanding that ν be finite on some distinguished set of compact transversals. But in a foliated space there is a distinguished set of compact transversals given by the foliation structure.

Definition 4.16. Call a transversal C *open-regular* if there is an open set L in $\mathbb{R}^p$, where p is the dimension of the foliation, and an isomorphism ϕ of foliated spaces of L$\times$C onto an open subset of X, which is the identity on C. A transversal C is *regular* if it is contained in an open-regular transversal.

If U_x is one of the coordinate patches in the definition of the foliation so that $U_x \cong L_x \times N_x$, L_x open in $\mathbb{R}^p$, then any compact subset of N_x is a compact *regular* transversal. Our definition of a Radon transverse measure involves finiteness on these transversals.

Definition 4.17. A transverse measure ν on a topological groupoid is *Radon* if $\nu(C)$ is finite for every compact regular transversal.

We observe that in order to check this condition, it will suffice to check finiteness on a much smaller family of compact regular transversals. For instance, let C_i be a family of such transversals with maps ϕ_i of $L_i \times C_i$ into X, and suppose that there are relatively open subsets of C_i, $U_i \subset \bar{U}_i \subset V_i \subset C_i$ so that the open sets $\phi_i(L_i \times U_i)$ cover X.

Proposition 4.18. If $\nu(C_i)$ is finite for each i for such a family, then ν is Radon.

Proof: Let B be any compact regular transversal with a map ϕ of L×D into X with D ⊃ B. By covering argument and by shrinking L if necessary we may assume that ϕ(L×B) lies inside some $\phi_i(L_i×V_i)$ and has compact closure there. The projection mapping to the second coordinate of $L_i×V_i$ gives rise to a continuous map f of B to V_i so that b and f(b) lie in the same plaque of the coordinate neighborhood $L_i×V_i$. Using the geometry of this situation, we easily show that there is an integer n so that f^{-1}(b) has at most cardinality n. Now using the quasi-invariance properties of transverse measures, we can calculate ν(B) by the formula

$$\nu(B) = \int_{f(b)} \left[\sum_{f(b)=x} \Delta(b,x) \right] d\nu(x).$$

Since the modulus Δ is a continuous function, it is bounded and as ⊂(b,f(b)) b ∈ B⊃ is compact, the integrand is bounded. As f(B) ⊂ V_i ⊂ C_i, ν(f(B)) is finite and we are done. □

It is evident of course that a Radon transverse measure is completely determined by what it does on regular transversals. For instance, the union C = $\cup_i C_i$ in the proposition above is a complete transversal and if ν is known on C_i, it is known on the union C and then knowledge of ν on a complete transversal determines the transverse measure entirely.

Up to now transverse measures have always been positive measures. However at this point we are in a position to consider signed or even complex transverse measures. We simply take differences or complex linear combinations of (positive) *Radon* transverse measures. Such an object cannot be defined on all transversals, but clearly it can be defined on regular transverals.

Definition 4.19. A *signed or complex transverse Radon*

measure of modulus Δ is a real or complex linear combination of positive Radon transverse measures of modulus Δ defined on all finite unions of regular transversals.

By our remarks above to the effect that a positive Radon transversal measure, viewed as a measure on all transversals, is completely determined by what it is on regular transversals, the domain we have specified for signed or complex Radon transverse measures is surely large enough. They can be expanded of course to a somewhat larger class of transversals without confronting expressions like $\infty - \infty$, but not in general to all transversals. We shall make use of these objects only briefly in connection with the Riesz representation theorem (4.27) for compact foliated spaces.

On the graph of a foliation of X we can construct tangential measures of particular interest. Each set $r^{-1}(x)$ is itself a C^∞ manifold, and so has a unique equivalence class of measures, those equivalent to nonvanishing densities. As each set $r^{-1}(x)$ is a covering space of the leaf ℓ_x of x in the foliation, and as tangential measures are invariant under the deck group, giving a tangential measure λ^x (as we have already noted in Proposition 4.11) is the same as giving measures $\tilde{\lambda}^\ell$, one for each leaf ℓ.

To construct such measures, cover X by coordinate charts of the form $L_i \times N_i$ where L_i is an open ball in $\mathbb{R}^p$ and let λ_i be tangential measure on the foliated space $L_i \times N_i$ where λ_i^n is for n ∈ N_i, normalized Lebesgue measure on L_i. Now choose a partitition of unity θ_i subordinate to the covering and define $\tilde{\lambda}$ to be the sum $\Sigma \theta_i \lambda_i$. Then we lift $\tilde{\lambda}^\ell$ on each leaf ℓ to a unique measure λ^x on $r^{-1}(x)$ using counting measure on the fibres of the covering map. Proposition 4.12 implies that $\lambda = \{\lambda^x\}$ satisfies the invariance properties required and thus is a tangential measure.

Proposition 4.20. The tangential measure just constructed is a continuous tangential measure (G Hausdorff or locally Hausdorff).

Proof: We have to check the continuity of

$$x \rightarrow \int f(u)d\lambda^X(u)$$

for each f of compact support on G for G Hausdorff. Using partitions of unity we may localize the support of f so that it lies within a subset K of G on which the map (r,s) is injective into a subset of X×X contained in $\text{Supp}(\theta_i) \times \text{Supp}(\theta_j)$ for suitable (i,j) where θ_i is the original partition of unity used to define λ. One easily verifies the continuity of the integral as a function of x for such f. One proceeds similarly in the locally Hausdorff case. □

If we perform the construction using different coordinate charts, or using different partitions of unity, we obtain a tangential measure λ_1 which is equivalent to λ in the sense that λ_1^X is mutually absolutely continuous with respect to λ^X on $r^{-1}(x)$. The Radon Nikodym derivative $d\lambda_1/d\lambda$ is a continuous nonvanishing function on G which one can check is bounded from 0 and ∞ not just on compact subsets of G but also on any set $r^{-1}(C)$, C compact in X. In terms of this we may define local boundedness of a tangential measure.

Definition 4.21. A tangential measure λ' is *locally bounded* (*Lebesgue*) if $(\lambda')^X$ has a Borel density on $r^{-1}(x)$ for each x and if $d\lambda'/d\lambda$ is bounded on any set $r^{-1}(C)$, C compact, for one (and hence any) tangential measure λ of the kind constructed above by partitions of unity. If the unit space is compact, we will for simplicity call such a measure a *bounded tangential measure*.

With these definitions the desired finiteness result is quite straightforward.

Proposition 4.22. If ν is a Radon transverse measure on the graph G of a foliated space X, and if λ is a locally bounded (Lebesgue) tangential measure, then for any compact set K of X the integral $\mu(K) = \int_K \lambda d\nu$ is finite.

Proof: We consider coordinate patches U_i isomorphic as foliated spaces via ψ_i to $L_i \times N_i$ so that ψ_i extends to $L_i' \times N_i'$ where N_i' is compact and contains N_i in its interior and where L_i is a relatively compact open ball in the ball L_i'. The compact set K can be covered by a finite number of such sets U_i so it suffices to show that $\mu(U_i)$ is finite. But by the definition of μ we can evaluate $\mu(U_i)$ by the formula

$$\mu(U_i) = \int_{N_i} \left[\int_{L_i} \Delta((x_0,n), (x,n)) d\lambda^n(x) \right] d\nu_0(n)$$

where x_0 is a fixed point of L_i and ν_0 is the transverse measure ν restricted to the transversal $\psi_i(x_0,N_i)$. As this transversal is contained in a compact regular transversal, ν_0 is a finite measure. Moreover the measures λ^n on the plaques $\psi_i(L_i,n)$ have smooth densities which extend uniformly in n to a slightly larger "ball" and hence are bounded uniformly in n. The modular function Δ is continuous, and hence its values entering into the integrand are bounded. (Note that strictly speaking Δ is a function on the graph. It can be transported locally down to the equivalence relation $\mathcal{R}$ as we have done, since we are operating in coordinate patches with the plaques contractible). It now follows at once that the integral above is finite, and we are done. $\square$

We remark that one could easily obtain finiteness results for tangential measures which do not have densities by imposing similar local boundedness conditions.

Our final goal now is to relate the previous discussion, which has been mostly analytic, to more geometric and topological aspects of the foliation. We begin with a tangentially smooth homomorphism Δ of G into $\mathbb{R}^+$, such as the modular function of a transverse measure on X. Tangential smoothness is with respect to the foliation of G by the holonomy groupoids of the leaves of the original foliation. We consider $\log(\Delta)$ as a real-valued function on G and form its differential. On each set $G^x = r^{-1}(x)$, the homomorphism property of Δ implies that

$$(\log \Delta)(\gamma u) = \log(\Delta)(\gamma) + (\log \Delta)(u)$$

for γ in the holonomy group G_x^x. Hence the differential of the function $\log(\Delta)$ becomes in a natural way a differential on the quotient $G_x^x \backslash G^x$, or in other words the leaf ℓ_x of x. Again by the homomorphism property of Δ, this differential on ℓ_x is independent of x, and hence one has an intrinsically defined differential 1-form on each leaf ℓ. Moreover the tangential smoothness of Δ implies immediately that these 1-forms on the leaves fit together continuously to what we have called in Chapter III a tangentially smooth 1-form for the foliation; recall that this is a tangentially smooth section of the dual F^*X of the foliation bundle. We denote this 1-form by α (or α_Δ if there is confusion). Summarizing:

Proposition 4.23. For a tangentially smooth homomorphism Δ on G, the construction above yields a tangentially smooth 1-form α_Δ on X. The map $\Delta \longrightarrow \alpha_\Delta$ is injective.

Proof: If $U \cong L \times N$ is a coordinate patch with L a p-ball with coordinates $(\bar{x}, u) = (x_1, ..., x_p, n)$, $n \in N$, then locally Δ can be written as a function of pairs $(\bar{x}, n)$, $(\bar{y}, n)$, $\bar{x}, \bar{y} \in L$

$$\Delta((\bar{x}, n), (\bar{y}, n)) = f(\bar{x}, \bar{y}, n).$$

Then the procedure for calculating α gives

$$\alpha = \sum \frac{\partial}{\partial y_i} (\log f)(\bar{x}_0, \bar{y}, n) dy_i;$$

this expression is seen to be independent of $\bar{x}_0$. The desired properties of α follow from this explicit local formula.

To see the final statement, we observe that for a point $(x, y, [\gamma])$ in G, we can obtain the value of Δ by integrating α_Δ along a smooth version of the path γ. Recall that γ is totally on a leaf so integration of tangential 1-forms makes sense. $\square$

Now suppose that Δ is the modular function of a transverse measure ν, and suppose for simplicity that the bundle of the foliation FX is oriented. If o is the orientation, and if σ is a tangentially smooth p-form on FX (p = leaf dimension), then $\sigma_1 = o \cdot \sigma$ is a tangentially smooth volume form on FX. Then σ_1 restricted to any leaf ℓ defines a signed measure with a C^∞ density, and hence a (signed) tangential measure λ. We can write $\sigma_1 = \sigma_1^+ - \sigma_1^-$ where $\sigma_1^\pm$ have corresponding positive (negative) measures $\lambda_1^\pm$. Then assuming that ν is a Radon transverse measure, we define the integral $\mu = \int \lambda d\nu$ to be

$$\int \lambda_1 d\nu - \int \lambda_2 d\nu$$

which by Proposition 4.22 is the difference of two Radon measures on X, and is therefore a signed Radon measure defined on bounded Borel sets in X. If we further assume that the form σ has compact support in X, then evidently μ has compact support and is a Radon measure.

The integral can therefore be viewed as a linear functional C_ν on the space $\Omega_{\mathcal{T}c}^p$ of compactly-supported tangentially smooth p-forms on X, where

$$C_\nu(\sigma) = \int_X \lambda d\nu \ .$$

Such an object is what we have called a tangential p-dimensional current in Chapter III. This was first defined in Ruelle-Sullivan [RuS] and is called the *Ruelle-Sullivan current*. The point of this discussion is to determine the boundary of this current. The boundary is a p-1 dimensional current defined by

$$d_* C_\nu(\sigma) = (-1)^p C_\nu(d\sigma)$$

where d is the differential on tangential forms.

Proposition 4.24. For a compactly supported p-1 tangential form σ, we have

$$C_\nu(d\sigma) = C_\nu(\sigma \wedge \alpha)$$

where α is the tangential 1-form associated to the modulus of the transverse measure ν.

It follows that if $\alpha = 0$, or equivalently $\Delta = 1$, then C_ν is a closed current. Conversely if C_ν is closed, we can deduce that $\alpha = 0$. Thus whenever ν is an invariant transverse measure C_ν defines a tangential homology class in the tangential homology group $H_p^\tau(X, \mathbb{R})$ of Chapter III (3.31) because the map $\sigma \longrightarrow C_\nu(\sigma)$ is continuous with respect to the natural topology on $\Omega_{\tau c}^p(X)$. We denote this class by $[C_\nu]$. Summarizing, we have

Corollary 4.25. For a Radon transverse measure ν with tangentially smooth modular function Δ, the following are equivalent:

(1)   The Ruelle–Sullivan current C_ν is closed and so defines $[C_\nu] \in H_p^\tau(X, \mathbb{R})$

(2)   The 1-form $\alpha = 0$

(3)   The modular function $\Delta = 1$

(4)   The transverse measure ν is an invariant transverse measure.

The proof of Proposition 4.24 is a straightforward calculation which goes as follows: first we may assume that σ is supported inside of some coordinate patch $U \cong L \times N$ where we use coordinates $(\bar{x}, n) = (x_1, x_2 \ldots x_p, n)$. The form σ can be written as

$$\sigma = \sum a_i(\bar{x}, n) dx_1 \wedge \ldots \wedge d\check{x}_i \ldots \wedge dx_p,$$

and we can represent the modular function locally as

$$\Delta((x',u), (x,u)) = f(x',x,u).$$

The 1-form α is $\sum \frac{\partial}{\partial x_i} \log(f(x_0,x,n))dx_i$, which does not depend on x_0. Now if ν_0 is the transverse measure on the transversal given by $x = x_0$, $h(x,n)dx_1 \wedge ... \wedge dx_p$ is a p-form, and λ_p the corresponding tangential measure, the definitions yield

$$\int_U \lambda d\nu = \int_N \left[\int_L h(x,n) \, f(x_0,x,n)dx \right] d\nu_0(n).$$

Thus for our p-1 form σ,

$$C_\nu(d\sigma) = \int_N \left[\int_L (\Sigma(-1)^i \frac{\partial a_i}{\partial x_i}) f(x_0,x,n)dx \right] d\nu_0(n) .$$

On the other hand, we see that

$$C_\nu(\sigma \wedge \alpha) = \int_N \left[\int_L \Sigma(-1)^{i-p} \, n_i \frac{\partial f}{\partial x_i} (x_0,x,n)dx \right] d\nu_0(n) .$$

Integration by parts on L gives the desired result as the functions a_i vanish in a neighborhood of the boundary of L. $\square$

Corollary 4.26. If X is a compact oriented foliated space which has a non-zero invariant Radon transverse measure then $H_\tau^P(X) \neq 0$.

This is the case, for instance, when X has a closed leaf. We improve this result significantly below (4.27).

Haefliger [Hae3] has used tangential transversely smooth cohomology in connection with the question of the existence of a Riemannian metric on M for which all the leaves are minimal (in the sense of area-minimizing) submanifolds. One consequence of his work is that the group $C_\tau^\infty(N)_H$ of compactly supported functions on some complete transverse submanifold N modulo holonomy maps onto the group $H_\tau^P(M)$. The map fails to be an isomorphism (e.g., for the

Liouville–irrational flow on the torus).

Recalling the definition of a signed (or complex) Radon measure from 4.19 we see from 4.25 and 4.26 that any signed Radon invariant transverse measure will define a continuous linear functional on $H^p_\tau(X)$ or equivalently $\bar{H}^p_\tau(X)$ – the topology on these spaces was defined in connection with Proposition 3.7. In the special case of a compact space X foliated by points, so $p = 0$, $H^0_\tau(X) = \bar{H}^0_\tau(X)$ is the Banach space of continuous functions on X. An invariant Radon (signed) transverse measure is just a Radon (signed) measure on X and the Riesz representation theorem says that these measures provide all the continuous linear functionals on $\bar{H}^0_\tau(X)$. More generally, if the foliation on X arises from a fibre bundle structure on X as total space, a base B and fibre L which we take to be a p–dimensional oriented manifold, then as we have already remarked (Proposition 4.11), invariant transverse Radon measures for this foliated space can be viewed simply as Radon measures on B. On the other hand, we have seen in Chapter III that in that case $H^p_\tau(X) = \bar{H}^p_\tau(X)$ can be identified topologically as C(B), the set of continuous functions on B. Again the usual Riesz representation theorem tells us that all continuous linear functionals on $\bar{H}^p_\tau(X)$ are given by invariant transverse measures. We show that this is true in general. This result is clearly closely related to, but distinct from the Corollary contained in section 3.3 of Haefliger [Hae3] and the work of Sullivan ([Su], cf. Prop. I.8).

Let MT(X) denote the vector space of Radon invariant transverse measures on X.

Theorem 4.27 (Riesz Representation Theorem). If X is a compact oriented foliated space with leaf dimension p, then the set of continuous linear functionals on $\bar{H}^p_\tau(X)$ can be identified as the set of Radon invariant transverse measures. More precisely, the Ruelle-Sullivan map

$$MT(X) \longrightarrow \mathrm{Hom}_{\mathrm{cont}}(H^p_\tau(X), \mathbb{R})$$

is an isomorphism.

Note that $\text{Hom}_{\text{cont}}(H_{\mathcal{T}}^p(X),\mathbb{R})$ is isomorphic to $H_p^{\mathcal{T}}(X)$ for X compact by (3.32).

Given the theorem it is easy sometimes to compute the top tangential cohomology group. For example, an invariant transverse measure on the Kronecker flow on the torus corresponds by (4.10) to a measure on a transverse circle which is invariant under rotation by an irrational angle, hence a multiple of Haar measure. Hence MT(Kronecker flow) $\cong \mathbb{R}$ and so $\bar{H}_{\mathcal{T}}^1$(Kronecker flow) $\cong \mathbb{R}$. A more interesting case arises from the Reeb foliation of S^3. The only holonomy invariant measures are multiples of the counting measure associated to the unique closed leaf. Thus MT(Reeb) $\cong \mathbb{R}$ and so $\bar{H}_{\mathcal{T}}^2$(Reeb) $\cong \mathbb{R}$.

Proof. Let Φ be a continuous linear functional on $\bar{H}_{\mathcal{T}}^p(X)$. Now choose as in Definition 2.1 an open "coordinate" chart U around $x \in X$ with $U \cong B \times N$ where $B = B^p$ is an open ball in $\mathbb{R}^p$ and where N is locally compact. Then the set $N_\lambda = \mathbb{C}(\lambda,n), n \in N, \lambda$ fixed in B$\mathbb{J}$ is a transversal and if D is a compact subset of N, $D_\lambda = \mathbb{C}(\lambda,n), n \in D\mathbb{J}$ is a regular transversal in the sense of 4.12. We fix a tangentially smooth p-form σ which has compact support in B. Now if f is any compactly supported real valued function on N, $f \in C_c(N)$, the formula $f\sigma(\lambda,n) = f(n)\sigma(\lambda)$ defines a tangentially smooth p form of compact support on the foliated space $U \cong B \times N$. If we extend it by zero outside U to X, it yields a tangentially smooth p-form on X, which we also denote by $f\sigma$.

We now consider the map $\varphi: f \longrightarrow \Phi(f\sigma)$ for fixed σ. By the definition of the topology on $\Omega_{\mathcal{T}}^p(X)$, it is evident that φ is norm continuous on $C_c(N)$. By the usual Riesz representation theorem, it must be represented by a finite Radon measure μ_σ on N.

The orientation on leaves of X gives by restriction an orientation on the ball B which is an open subset of a leaf of X, and hence we may integrate the forms σ on B. By the Poincaré lemma two compactly supported forms σ_1 and σ_2 on B are cohomologous, that is, $\sigma_1 - \sigma_2 = dp$ on B if and only if their integrals are the same. It follows that $f\sigma_1$ and $f\sigma_2$ as elements of $\Omega_\tau^p(X)$ differ by a coboundary if σ_1 and σ_2 have the same integral. Now define μ on N to be μ_σ for any σ of integral one. Then clearly

$$\Phi(f\sigma) = \int_N \left[\int_B \sigma \right] f(n) d\mu(n).$$

Moreover for any λ we can identify N with N_λ by $n \longrightarrow (\lambda,n)$ and can transport μ onto N_λ, calling it μ^λ. Then we can rewrite the above as

$$\Phi(f\sigma) = \int_{N_\lambda} \left[\int_B (f\sigma)(\lambda,n) \right] d\mu^\lambda(n) .$$

Finally if σ is any tangentially smooth p-form on X with compact support inside $U \cong B \times N$, it may by a kind of Stone-Weierstrass theorem be approximated in the topology of $\Omega_\tau^p(X)$ by linear combinations of forms of the type $f\sigma$. By continuity of both sides of the formula above,

$$\Phi(\sigma) = \int_{N_\lambda} \left[\int_B \sigma(\lambda,n) \right] d\mu^\lambda(n)$$

holds for any λ.

Now each N_λ is a transversal and μ^λ is a measure on it; we have to see now that we can piece these together to construct a transverse measure. First we observe that our compact space X can be covered by a finite number of open sets of the form $U \cong B \times N$, let us say $U^1,...,U^n$ with $U^i \cong B \times N^i$. We identify each N^i with say N_b^i ($b \in B$) and then $N = \cup N^i$ is a complete transversal; we can also arrange for simplicity that the N^i are all disjoint as subsets of X. Each N^i carries a Radon signed measure denoted by μ^i from the construction above and we fit them together to give a (signed) measure on N. As we have observed before, the foliated structure on N gives rise to a countable standard equivalence relation on N in the sense of Feldman-Moore [FM]. We want to show that μ is invariant under this equivalence relation. To see this, we observe that if U^i and U^j intersect, then the projection of their intersection onto the respective transversals N^i and N^j are open subsets N^{ij} and N^{ji} respectively of N^i and N^j. Clearly for each $n \in N^{ij}$ there is a unique x' in N^{ji} which lies on the same leaf and it is evident that the map φ^{ij} taking x to x' is a homeomorphism of N^{ij} onto N^{ji}.

It is further evident that these partial homeomorphisms generate the equivalence relation on N in the obvious sense. To see that μ is invariant under this equivalence relation, it suffices by Feldman-Moore [FM] to see that each φ^{ij} is measure preserving. (The fact that here we have signed measures, while in [FM] we have positive measures is of course irrelevant.) However the formulas above for $\Phi(\sigma)$ when σ is supported in U^i or U^j in terms of μ^i and μ^j show immediately that φ^{ij} is measure preserving, for we apply these formulas to σ's which are supported in $U^i \cap U^j$.

Thus we have an invariant measure μ on the complete transversal N. To get transverse measures in the usual sense, we should first split $\mu = \mu^+ - \mu^-$ into its positive and negative parts, each of which is automatically invariant because $\mu^\pm$ are canonically defined. Then $\mu^\pm$ is extended to all transversals as in 4.10. Thus finally μ is a signed Radon transverse measure in the sense of our definition.

Let Φ_μ be the corresponding linear functional on $\bar{H}^p_\tau(X)$. Then the integral formulas above when compared to the formulas of Proposition 4.19 show that $\Phi_\mu(\sigma) = \Phi(\sigma)$ for σ supported in U^i. But then a partition of unity argument shows that these span and so $\Phi = \Phi_\mu$ and we are done. $\square$

This result identifies the dual of the topological vector space $\bar{H}^p_\tau(X)$ in an explicit fashion as the set of invariant Radon transverse measures MT(X). Then of course by duality, any $\sigma \in \bar{H}^p_\tau(X)$ defines a linear functional Fσ on MT(X) by F$\sigma(\nu) = \int \sigma d\nu$. It will be of considerable interest to us at several points to know which linear functionals F on MT(X) can be so represented. Of course there is no problem in those cases when MT(X) and $\bar{H}^p_\tau(X)$ are finite dimensional, but it is a problem in general. Following standard techniques, we introduce a "weak" topology on MT(X) with the result that those linear functionals representable as Fσ are just the ones continuous in this topology. For each open-regular transversal N (cf. 4.16) and each continuous real valued function on N of compact support, and for each $\nu \in$ MT(X), the integral $\int \text{fd}\nu_N$ is well defined, where ν_N is the transverse measure ν on the transversal N; This defines a linear functional I_f on MT(X).

Definition 4.28. The *weak topology* on MT(X) is the smallest topology making these linear functions continuous.

Proposition 4.29. The weak topology so defined coincides with the weak-* topology on MT(X) as the dual of $\bar{H}^p_\tau(X)$ and consequently a linear function F on MT(X) is representable as Fσ, $\sigma \in \bar{H}^p_\tau(X)$, if and only if it is continuous in the weak topology.

Proof. If N is an open-regular transversal, let B be a ball in $\mathbb{R}^p$; then there is a tangentially smooth homeomorphism of B $\times$ N onto an open set U in X; we shall think of U = B $\times$ N as sitting inside X. If f is a compactly supported function on N, we can easily construct a tangentially smooth p-form σ on X supported on U = B $\times$ N so that

$$\int_B \sigma(b,r) = f(n).$$

Then from our formulas for integration it is immediate that

$$I_f(\nu) = \int \text{fd}\nu_N = \int \sigma \text{d}\nu = \nu([\sigma])$$

where $[\sigma]$ is the class of σ in $\bar{H}^p_\tau(X)$. Hence the weak topology defined by the I_f is contained in the weak-* topology on MT(X) as the dual of $\bar{H}^p_\tau(X)$. Conversely we see by a partition of unity argument that a linear functional $\nu \longrightarrow \nu([\sigma])$ for any σ can be represented as a finite linear combination of I_f's. Hence the two topologies coincide and the result follows. $\square$

We note that it would suffice in defining the weak topology to restrict to any finite set of open-regular transversals N_i so that there are corresponding coordinate charts $N_i \times B$, B a ball in $\mathbb{R}^p$, which cover X.

We close Chapter IV with two examples which illustrate the Riesz representation theorem 4.27.

Suppose that T is a homeomorphism of a separable metrizable space N and that f is a positive continuous function on N. Then we may form the space X_T obtained as the quotient of the space

$$\{(t,n) \in \mathbb{R} \times N \mid 0 \leqslant t \leqslant f(n)\}$$

by the relation $(f(n),n) \sim (0,T(n))$. If $f \equiv 1$ then X_T is simply the suspension of the homeomorphism T (cf. 2.3). The space X_T has a natural oriented foliation of dimension one corresponding to the action of $\mathbb{R}$ on the first factor of $\mathbb{R} \times N$. As f changes the topological foliated conjugacy class of X_T remains the same; so in that sense at least the dependence of X_T on f is minimal. Invariant transverse measures on X_T correspond to T-invariant measures on N, denoted $M(N)^T$. Theorem 4.27 implies that $\bar{H}_\tau^1(X_T)^* \cong M(N)^T$.

Let us look at this example in more detail. The general tangential 1-form $a(t,n)dt$ is a tangential cocycle, since it is in the top degree. The function a must satisfy

$$(*) \qquad\qquad a(f(n),n) = a(0,T(n))$$

in order to be defined on X_T. If $a(t,n)dt = \dfrac{\partial b}{\partial t}(t,n)$, then b must also satisfy $(*)$. Set $b(0,n) = b_0(n)$. Then

$$b(t,n) = \int_0^t a(t,n)dt + b_0(n)$$

and hence

$$b(f(n),n) = \int_0^{f(n)} a(t,n)dt + b_0(n).$$

Now

$$\int_0^{f(n)} a(t,n)dt = b_0(T(n)) - b_0(n)$$

and so any tangential 1-coboundary must be of the form $(b_0(T(n)) - b_0(n))dt$. Thus

$$H^1_\tau(X_T) \cong \frac{C(N)}{(T-1)C(N)}$$

and

$$\bar{H}^1_\tau(X_T) \cong \frac{C(N)}{\overline{(T-1)C(N)}} .$$

It is clear then that $\bar{H}^1_\tau(X_T)^* \cong M(N)^T$, as is predicted by (4.27).

This example generalizes to the case of bundles with discrete structural group, as follows. Let B be an oriented compact manifold of dimension p with $\Gamma = \pi_1(B)$ and $\tilde{B} \longrightarrow B$ the universal cover. Suppose that Γ acts on a space F. Then the space $X = \tilde{B}x_\Gamma F$ is foliated by leaves of dimension p which are the images of $\tilde{B} \times \{x\}$ for $x \in F$ (cf. 2.2). We may regard differential forms on X as forms $\omega(b,x)$ defined on $\tilde{B} \times F$ satisfying the invariance condition

(**) $\omega(\gamma b, \gamma x) = \omega(b,x)$, $\gamma \in \Gamma$.

Fix a fundamental domain U in $\tilde{B}$. Let ω be a p-form (necessarily closed) and define $f_\omega(x) = \int_U \omega(b,x)$. If ν is an invariant transverse measure on X, then $\int f_\omega d\nu$ is independent of choice of U. If ω is a coboundary, say $\omega = d\sigma$, then

$$f_\omega(x) = \int_U \omega(b,x)$$

$$= \int_{\partial U} \sigma(b,x) \qquad \text{by Stokes' theorem}$$

and

$$\int f_\omega d\nu = \int\!\!\int_{\partial U} \sigma(b,x) d\nu$$

$$= \int_{\partial U} \left(\int \sigma(b,x)\right) d\nu$$

$$= 0$$

for any invariant transverse measure $d\nu$, since $\int \sigma(b,x)$ is a periodic function and $\int_{\partial U}$ (periodic)$d\nu = 0$.

Suppose that U is sufficiently well-behaved so that ∂U consists of 2k piecewise smooth hypersurfaces

$$\partial U = H_1^+ \cup H_1^- \cup \ldots \cup H_k^+ \cup H_k^-$$

and there are elements $\gamma_i \in \Gamma$ which reflect H_i^+ with H_i^- and generate Γ. (This sort of decomposition is quite familiar in the theory of Riemann surfaces. In general one may assume that Γ acts by isometries. Let D be an open dense PL disk in B and let $\tilde{V}$ be one component of its preimage. Then $U = \text{int}(\text{closure}(\tilde{V}))$ is an open disk in $\tilde{B}$ with PL boundary ∂U. The deck group Γ acts in a PL fashion on ∂U which decomposes into smooth hypersurfaces. However, to ensure that Γ is generated by $\{\gamma_i\}$ which act as reflections on these hypersurfaces is a very delicate (and sometimes impossible) matter. The interested reader is referred to M. W. Davis [Da] for a taste of the difficulty.) Then

$$\int_{\partial U} \sigma(b,x) = \sum_1^k \int_{H_i^+} \sigma(b,x) + \int_{-H_i^-} \sigma(b,x)$$

where $-H_i^-$ indicates H_i^- with the orientation reserved. Let $g_i(x) = \int_{H_i^+} \sigma(b,x)$. Then

$$\int_{\partial U} \sigma(b,x) = \int \sum_1^k g_i(x) - g_i(\gamma_i x)$$

so that terms cancel in pairs under integration $\int (\ ) d\nu$. We see from this analysis that the p-coboundaries correspond to the algebraic sum in $C(X)$

$$\sum_1^k (\gamma_i - 1)C(X)$$

which is also

$$(\Gamma - 1)C(X) = \sum_{\gamma \in \Gamma} (\gamma - 1)C(X).$$

Thus

$$H_\tau^p(X) \cong \frac{C(X)}{(\Gamma - 1)C(X)} \cong \frac{C(X)}{\sum_1^k (\gamma_i - 1)C(X)}$$

and

$$\bar{H}_\tau^p(X) \cong \frac{C(X)}{\overline{(\Gamma - 1)C(X)}} \cong \frac{C(X)}{\overline{\sum_1^k (\gamma_i - 1)C(X)}}$$

which is the predual of MT(X).

CHAPTER V: CHARACTERISTIC CLASSES

In this chapter we mimic as closely as possible the Milnor-Stasheff [MS] exposé of the Chern-Weil construction of characteristic classes in terms of curvature forms.

The Chern-Weil procedure begins with a vector bundle with a certain structural group G. In our situation we consider complex (tangentially smooth) bundles with structural group $GL(n,\mathbb{C})$, real vector bundles with structural group $GL(n,\mathbb{R})$, and oriented real vector bundles with structural group $SO(2n)$. Choose a tangential connection ∇, see below, that respects the structure. The associated curvature form K determines a closed tangential 2-form whose tangential cohomology class is independent of choice of the connection. Then any polynomial or formal power series P which is G-invariant determines a characteristic form. In the case $X = M$ is a manifold with $FX = TM$ then this yields the usual characteristic classes in de Rham cohomology $H^*(M)$.

We shall assume throughout that all bundles over foliated spaces are tangentially smooth and that leaf-preserving maps between foliated spaces are also tangentially smooth; this is not a real restriction, in view of our smoothing results (2.16). We use the Milnor-Stasheff [MS] sign conventions for characteristic classes.

Let $E \longrightarrow X$ be a (tangentially smooth) complex n-plane bundle over the foliated space X, and let $F^*_{\mathbb{C}} = \mathrm{Hom}_{\mathbb{R}}(F,\mathbb{C})$ be the complexified dual tangent bundle of the foliated space X.

Definition 5.1. A *tangential connection* on $E \longrightarrow X$ is a $\mathbb{C}$-linear mapping

$$\nabla: \Gamma_{\mathcal{T}}(E) \longrightarrow \Gamma_{\mathcal{T}}(F^*_{\mathbb{C}} \otimes E)$$

which satisfies the Leibnitz formula

$$\nabla(fs) = df \otimes s + f\nabla(s)$$

for every $s \in \Gamma_{\mathcal{T}}(E)$ and every $f \in C^\infty_{\mathcal{T}}(X,\mathbb{C})$. The image $\nabla(s)$ is called the tangential covariant derivative of s.

Equivalently, we may regard ∇ as a bilinear map

$$\Gamma_\tau(F_\mathbb{C}) \times \Gamma_\tau(E) \longrightarrow \Gamma_\tau(E)$$

with

$$\nabla_{fv}(s) = f\nabla_v(s) \qquad\qquad v \in F_\mathbb{C}$$

and

$$\nabla_v(gs) = v_g \cdot s + g\nabla_v(s) \qquad f, g \in C_\tau^\infty(X)$$

$$s \in \Gamma_\tau(E)$$

One may regard ∇ as a map between the Lie algebras $\Gamma_\tau(F_\mathbb{C})$ (with Lie bracket) and $\mathrm{Hom}(\Gamma_\tau(E), \Gamma_\tau(E))$ (with bracket corresponding to AB–BA for matrices). Thus

$$\nabla: \Gamma_\tau(F_\mathbb{C}) \longrightarrow \mathrm{Hom}\,(\Gamma_\tau(E), \Gamma_\tau(E)).$$

Note that ∇ is not generally a Lie algebra homomorphism.

The correspondence $s \longmapsto \nabla(s)$ *decreases supports*; that is, if the section s vanishes throughout an open subset $U \subset X$, then $\nabla(s)$ vanishes throughout U also. For given $x \in U$ we can choose a tangentially smooth function f which vanishes outside U and is identically 1 near x. The identity

$$df \otimes s + f\nabla(s) = \nabla(fs) = 0$$

evaluated at x, shows that $\nabla(s)$ vanishes at x.

Since a connection is a local operator (i.e., it decreases supports), it makes sense to talk about the restriction of ∇ to an open subset of X. If a collection of open sets U_α covers X, then a global tangential connection is uniquely determined by its restrictions to the various U_α.

If the open set is small enough so that $E\,|\,U$ is trivial, then $\Gamma_\tau(E\,|\,U)$ is a free $C_\tau^\infty(X)$-module with basis denoted, say, $s_1, \ldots, s_n$.

Tangential connections may be constructed as follows.

Proposition 5.2.   Let $[\omega_{ij}]_{1 \leqslant i, j \leqslant n}$ be an arbitrary $n \times n$ matrix of tangentially smooth complex 1-forms on U.   Then there is a unique tangential connection ∇ on the trivial bundle $E|U$ such that $\nabla(s_i) = \Sigma \omega_{ij} \otimes s_j$.

Proof: The connection ∇ is determined uniquely by the formula

$$\nabla(\sum_i f_i s_i) = \sum_i (df_i \otimes s_i + f_i \nabla(s_i)). \qquad \square$$

Henceforth, we assume that *all tangential connections are given locally as differential operators*, as in the above proposition.

There is exactly one tangential connection on a coordinate patch such that the tangential covariant derivatives of the s_i are all zero; or in other words so that the connection matrix is zero.   It is given by

$$\nabla(\sum_i f_i s_i) = \sum_i df_i \otimes s_i.$$

This particular "flat" connection depends of course on the choice of basis $\{s_i\}$.

Note that if ∇_1 and ∇_2 are tangential connections on E and g is a tangentially smooth complex-valued function on X, then the linear combination $g\nabla_1 + (1-g)\nabla_2$ is again a well-defined tangential connection on E.

Proposition 5.3.   Every tangentially smooth vector bundle $E \longrightarrow X$ with paracompact foliated base space possesses a tangential connection.

Proof: Choose open sets U_α covering X with $E|U_\alpha$ trivial, and choose a tangentially smooth partition of unity $\{r_\alpha\}$ subordinate to $\{U_\alpha\}$.   Each restriction $E|U_\alpha$ possesses a connection ∇_α by Proposition 5.2.   The linear combination $\sum_\alpha r_\alpha \nabla_\alpha$ is now a well defined global tangential connection. □

Given a tangentially smooth map $g: X' \longrightarrow X$ we can form the induced vector bundle $E' = g^* E$. Note that there is a canonical $C^\infty_\tau(X, \mathbb{C})$-linear map

$$g^*: \Gamma_\tau(E) \longrightarrow \Gamma_\tau(E').$$

Similarly, any tangentially smooth 1-form on X pulls back to a 1-form on X', so there is a canonical $C^\infty_\tau(X, \mathbb{C})$-linear mapping

$$g^*: \Gamma_\tau(F^*_\mathbb{C} \otimes E) \longrightarrow \Gamma_\tau(F'^*_\mathbb{C} \otimes E').$$

Proposition 5.4. To each tangential connection ∇ on E there corresponds one and only one tangential connection $\nabla' = g^* \nabla$ on $g^* E = E'$ so that the following diagram is commutative:

$$
\begin{array}{ccc}
\Gamma_\tau(E) & \xrightarrow{\ \nabla\ } & \Gamma_\tau(F^*_\mathbb{C} \otimes E) \\
\Big\downarrow{\scriptstyle g^*} & & \Big\downarrow{\scriptstyle g^*} \\
\Gamma_\tau(E') & \xrightarrow{\ \nabla'\ } & \Gamma_\tau(F'^*_\mathbb{C} \otimes E')
\end{array}
$$

Proof: Let $\{U_\alpha\}$ be an open cover and $\{r_\alpha\}$ a tangentially smooth partition of unity subordinate to a locally finite refinement of $\{g^{-1}(U_\alpha)\}$. On a typical set U_α, pick sections $s_1, \ldots, s_n$ with $\nabla(s_i) = \sum_j \omega_{ij} \otimes s_j$. Lift the 1-forms ω_{ij} to $\omega'_{ij} = g^* \omega_{ij}$ and lift the sections s_i to $s'_i = g^* s_i$ over $g^{-1}(U_\alpha)$. If ∇' exists then

$$\nabla'\Big|_{g^{-1}(U_\alpha)} = \sum_j \omega'_{ij} \otimes s'_j \qquad (*)$$

which shows uniqueness for ∇'. For existence, use $(*)$ to define

$$\nabla'\Big|_{g^{-1}(U_\alpha)} \text{ and then define } \nabla' \text{ globally by } \nabla' = \sum_\alpha r_\alpha \Big(\nabla'\Big|_{g^{-1}(U_\alpha)} \Big).$$

$$K = K_\nabla \colon \Gamma_\tau(F_{\mathbb{C}}) \times \Gamma_\tau(F_{\mathbb{C}}) \longrightarrow \mathrm{Hom}\ (\Gamma_\tau(E),\ \Gamma_\tau(E))$$

where K_∇ is given by the formula

$$K_\nabla = \nabla_v \nabla_w - \nabla_w \nabla_v - \nabla_{[v,w]}.$$

Thus the curvature is the obstruction to ∇ being a Lie algebra homomorphism: if the connection is flat then ∇ is a Lie algebra homomorphism and $K \equiv 0$.

Starting with the tangential curvature tensor K, we construct tangential characteristic classes as follows. Recall that $M_n(\mathbb{C})$ denotes the algebra consisting of all $n \times n$ complex matrices.

Definition 5.7. An *invariant polynomial* on $M_n(\mathbb{C})$ is a function

$$P \colon M_n(\mathbb{C}) \longrightarrow \mathbb{C}$$

which may be expressed as a complex polynomial in the entries of the matrix and satisfies

$$P(XY) = P(YX)$$

for all matrices X, Y, or equivalently

$$P(TXT^{-1}) = P(X)$$

for all X and for all non-singular matrices T. (The structural group is, of course, $GL(n, \mathbb{C})$.)

The trace and determinant functions are well-known examples of invariant polynomials on $M_n(\mathbb{C})$.

If P is an invariant polynomial, then an exterior form $P(K) \in \Gamma_\tau(\Lambda^* F_{\mathbb{C}}^*) = \oplus_m \Gamma_\tau(\Lambda^m F_{\mathbb{C}}^*)$ is defined as follows. Choose a local basis

143

$s_1,...,s_n$ for the sections in a neighborhood U of x, so that $K(s_i)$ = $\sum_j \Omega_{ij} \otimes s_j$. The matrix $\Omega = [\Omega_{ij}]$ has entries in the commutative algebra over $\mathbb{C}$ consisting of all exterior forms of even degree. It makes good sense to form $P(\Omega)$. This lies a priori in $\Omega_\tau^*(U)$ but patches together to form $P(K) \in \Omega_\tau^*(X)$, since a change of basis will replace Ω by a matrix $T\Omega T^{-1}$ and $P(T\Omega T^{-1}) = P(\Omega)$.

If P is a homogeneous polynomial of degree r then $P(K) \in \Omega_\tau^{2r}(X)$. If P is an invariant formal power series of the form

$$P = P_0 + P_1 + \dots$$

where each P_r is an invariant homogeneous polynomial of degree r, then $P(K)$ is still well-defined since $P_r(K) = 0$ for $2r > p$ (the leaf dimension).

Fundamental Lemma 5.8. For any invariant polynomial (or invariant formal power series) P, the exterior form $P(K)$ is *closed*; that is, $dP(K) = 0$. Thus $P(K)$ represents an element $[P(K)]$ in the tangential de Rham cohomology group $H_\tau^*(X;\mathbb{C})$.

Proof: We summarize the proof found in Milnor-Stasheff [MS, p. 296-8]. Given any invariant polynomial or formal power series $P(A) = P([A_{ij}])$ form the matrix $[\partial P/\partial A_{ij}]$ of formal first derivatives and let $P'(A)$ denote the transpose of this matrix. Let $\Omega = [\Omega_{ij}]$ be the curvature matrix with respect to some basis for the restriction of the bundle to U. Then

$$dP(\Omega) = \sum(\partial P/\partial \Omega_{ij})d\Omega_{ij} = \text{Trace}(P'(\Omega)d\Omega).$$

Since $\Omega = d\omega - \omega \wedge \omega$, taking exterior derivatives yields the Bianchi identity

$$d\Omega = \omega \wedge \Omega - \Omega \wedge \omega.$$

The matrix $P'(A)$ commutes with A, and hence

$$\Omega \wedge P'(\Omega) = P'(\Omega) \wedge \Omega.$$

Now

$$dP(\Omega) = \text{trace}((P'(\Omega) \wedge \omega) \wedge \Omega - \Omega \wedge (P'(\Omega) \wedge \omega))$$

$$= \sum (P'(\Omega) \wedge \omega)_{ij} \wedge \Omega_{ji} - \Omega_{ji} \wedge (P'(\Omega) \wedge \omega)_{ij}.$$

Since each $(P'(\Omega) \wedge \omega)_{ij}$ commutes with the 2-form Ω_{ji}, this sum is zero, which proves the lemma. $\square$

Corollary 5.9. The cohomology class $[P(K)] \in H_{\tau}^{*}(X)$ is independent of the choice of tangential connection ∇.

Proof: Let ∇_0 and ∇_1 be two different tangential connections on E. Map $X \times \mathbb{R}$ to X by the projection $(x,t) \longmapsto x$ and form the induced bundle E' over $X \times \mathbb{R}$, the induced tangential connections ∇_0' and ∇_1' and the linear combination

$$\nabla = t\nabla_1' + (1-t)\nabla_0'.$$

Thus $P(K_\nabla)$ is a tangential de Rham cocycle on $X \times \mathbb{R}$ (foliated of dimension p+1).

Consider the map $i_\epsilon: x \longmapsto (x,\epsilon)$ from X to $X \times \mathbb{R}$, where ϵ equals 0 or 1. Evidently the induced tangential connection $(i_\epsilon)^* \nabla$ on $(i_\epsilon^*)E'$ may be identified with the tangential connections ∇_ϵ on E. Therefore

$$(i_\epsilon^*)(P(K_\nabla)) = (P(K_{\nabla_\epsilon})).$$

But the mappings i_0 and i_1 are homotopic and hence

$$[P(K_{\nabla_0})] = [P(K_{\nabla_1})]. \qquad \square$$

The polynomial P determines a tangential characteristic

cohomology class in $H^*_\mathcal{T}(X;\mathbb{C})$ which depends only upon the isomorphism class of the vector bundle E. If a tangentially smooth map g: $X' \longrightarrow X$ induces a bundle $E' = g^*E$ with induced tangential connection ∇', then clearly

$$P(K_{\nabla'}) = g^* P(K_\nabla).$$

Thus these characteristic classes are well behaved with respect to induced bundles.

The entire treatment may be repeated for real vector bundles, and one obtains characteristic cohomology classes $[P(K)] \in H^*_\mathcal{T}(X;\mathbb{R})$ for any $GL(n,\mathbb{R})$-invariant polynomial P on $M_n(\mathbb{R})$.

For any square matrix A, let $\sigma_k(A)$ denote the k-th elementary symmetric function of the eigenvalues of A, so that

$$\det(I + tA) = 1 + t\sigma_1(A) + \dots + t^n\sigma_n(A).$$

It is well known (Milnor-Stasheff [MS, p. 299]) that any invariant polynomial on $M_n(\mathbb{C})$ can be expressed as a polynomial function of $\sigma_1, \dots, \sigma_n$.

Definition 5.10. Let E be a tangentially smooth complex vector bundle with tangential connection ∇. The *tangential Chern classes* $c^\mathcal{T}_m(E)$ are defined for m = 1,2,... by

$$c^\mathcal{T}_m(E) = \frac{1}{(2\pi i)^m} [\sigma_m(K_\nabla)] \in H^{2m}_\mathcal{T}(X;\mathbb{C}).$$

The tangential Chern classes do not depend on the choice of tangential connection, by Corollary 5.9. The fact that any invariant polynomial on $M_n(\mathbb{C})$ can be expressed as a polynomial function of $\sigma_1, \dots, \sigma_n$ implies that any characteristic class $c = [Q(K)]$ can be expressed as a polynomial in the Chern classes. If g: $X' \longrightarrow X$ is a tangentially smooth map then

$$g^* c^\mathcal{T}_m(E) = c^\mathcal{T}_m(g^*E)$$

146

by Proposition 5.4. If E has a flat tangential connection then all characteristic classes vanish and in particular $c_m^T(E) = 0$.

If $X = M$ is a compact smooth manifold foliated by one leaf then a tangential connection is a connection, tangential curvature is curvature, and $c_m^T(E) = c_m(E) \in H^{2m}(M;\mathbb{C})$; the tangential Chern classes are Chern classes. In general, however, this cannot be the case. If X is a compact foliated space with leaves of dimension p then $H_T^m(X;\mathbb{C}) = 0$ for $m > p$, so $c_m^T(E) = 0$ for $2m > p$. On the other hand, the ordinary Chern classes $c_m(E)$ (defined topologically, since we do not assume that X is a manifold) need not vanish. The following proposition explains the relation between the c_m and the c_m^T.

Proposition 5.10. Let E be a tangentially smooth complex vector bundle over a compact foliated space X. Then

$$c_m^T(E) = r_* c_m(E)$$

where r_*: $H^*(X;\mathbb{C}) \longrightarrow H_T^*(X;\mathbb{C})$ is the canonical map.

Proof: Since X is compact there is a compact Grassmann manifold $G_k(\mathbb{C}^{n+k})$ with universal n-plane bundle E^n and a continuous map g: X $\longrightarrow G_k(\mathbb{C}^{n+k})$ (which we may assume to be tangentially smooth) such that $E = g^* E^n$. Let ∇ be a connection on E^n, so that $c_m(E^n) = \frac{1}{(2\pi i)^m}[\sigma_m(K_\nabla)] \in H^{2m}(G_k(\mathbb{C}^{n+k}); \mathbb{C})$. If $\nabla' = g^*\nabla$ is the induced tangential connection on X, then

$$c_m(E) = g^* c_m(E^n)$$

$$= \frac{1}{(2\pi i)^m} g^* [\sigma_m(K_\nabla)].$$

To complete the proof, then, we need only show that the diagram

$$H^{2m}(G_k(\mathbb{C}^{n+k});\mathbb{C}) \xrightarrow[\cong]{r_*} H^{2m}_{\mathcal{T}}(G_k(\mathbb{C}^{n+k});\mathbb{C})$$

$$\downarrow g^* \qquad\qquad\qquad\qquad \downarrow g^*$$

$$H^{2m}(X;\mathbb{C}) \xrightarrow{r_*} H^{2m}_{\mathcal{T}}(X;\mathbb{C})$$

commutes; this follows from the naturality of r. $\square$

Corollary 5.11. The tangential Chern classes satisfy the following properties:

1) If $g: X' \longrightarrow X$ is tangentially smooth then
$$g^* c^{\mathcal{T}}_m(E) = c^{\mathcal{T}}_m(g^* E).$$

2) $c^{\mathcal{T}}_m(E \oplus E') = \sum\limits_{i=0}^{m} c^{\mathcal{T}}_i(E)\, c^{\mathcal{T}}_{m-i}(E').$

3) If E is a line bundle then $c^{\mathcal{T}}_0(E) = 1$ and
$c^{\mathcal{T}}_m(E) = 0$ for $m > 1$.

4) If E is of dimension n then $c^{\mathcal{T}}_m(E) = 0$ for $m > n$.

Proof: We have established 1) previously. The rest of the corollary follows from the analogous properties of Chern classes and the fact that $r_*: H^*(X;\mathbb{C}) \longrightarrow H^*_{\mathcal{T}}(X;\mathbb{C})$ is a ring map. $\square$

There are several important combinations of Chern classes. Here are two of them. The *tangential Chern character* $ch^{\mathcal{T}}(E) \in \oplus_m H^{2m}_{\mathcal{T}}(X;\mathbb{C})$ is the characteristic class associated to the invariant formal power series

(5.12) $$ch^{\mathcal{T}}(A) = \text{trace }(e^{A/2\pi i}).$$

The *tangential total Chern class* of E is the formal

sum

(5.13) $$c^T(E) = 1 + c_1^T(E) + c_2^T(E) + \ldots$$

which lies in $\oplus H_T^{2m}(X;\mathbb{C})$, and satisfies $c^T(E\oplus E') = c^T(E)c^T(E')$. It corresponds to the invariant polynomial det $(I + A/2\pi i)$.

Let σ_i be the elementary symmetric polynomials and let s_i be the universal polynomials determined inductively by Newton's formula

$$s_n - \sigma_1 s_{n-1} + \sigma_2 s_{n-2} - \ldots \mp \sigma_{n-1}s_1 \pm n\sigma_n = 0.$$

For example,

$$s_1(\sigma_1) = \sigma_1$$

$$s_2(\sigma_1,\sigma_2) = \sigma_1^2 - 2\sigma_2$$

$$s_3(\sigma_1,\sigma_2,\sigma_3) = \sigma_1^3 - 3\sigma_1\sigma_2 + 3\sigma_3.$$

Proposition 5.14. The tangential Chern character has the following properties:

1) $ch^T(E) = n + \displaystyle\sum_{k=1}^{\infty} s_k(c^T(E))/k!$
where $n = \dim E$. In particular, if E is a line bundle, then

$$ch^T(E) = \sum_{k=0}^{\infty} c_1^T(E)^k/k! = \exp(c_1^T(E)).$$

2) $ch^T(E\oplus E') = ch^T(E) + ch^T(E').$

3) $ch^T(E\otimes E') = ch^T(E)ch^T(E').$

4) $ch^T: K^0(X) \longrightarrow \underset{m}{\oplus} H_T^{2m}(X;\mathbb{C})$ is a ring map. $\square$

As an example we compute the Chern classes of the canonical bundles of $\mathbb{C}P^n$ (regarded as a foliated space with one leaf). Let E be the canonical complex line bundle over $\mathbb{C}P^n$ and D its orthogonal complement, so that $E \oplus D = \mathbb{C}P^n \times \mathbb{C}^{n+1}$ (the trivial complex (n+1)-plane bundle.). A geometric argument, which we omit, implies that the tangent bundle of $\mathbb{C}P^n$ (a complex n-plane bundle, as $\mathbb{C}P^n$ is a complex manifold) satisfies $T(\mathbb{C}P^n) \cong \text{Hom } (E,D)$. Then

$$T(\mathbb{C}P^n) \oplus (\mathbb{C}P^n \times \mathbb{C}) \cong \text{Hom } (E, D \oplus (\mathbb{C}P^n \times \mathbb{C}))$$

$$\cong \text{Hom } (E, \mathbb{C}P^n \times \mathbb{C}^{n+1})$$

$$\cong \bar{E} \oplus ... \oplus \bar{E}$$

where $\bar{E}$ is the conjugate bundle of E. In general we have $c_k(\bar{E}) = (-1)^k c_k(E)$ for any bundle E. Thus

$$c(T(\mathbb{C}P^n)) = c(T(\mathbb{C}P^n) \oplus (\mathbb{C}P^n \times \mathbb{C}))$$

$$= c(\bar{E} \oplus ... \oplus \bar{E}) \qquad \text{(n+1 times)}$$

$$= c(\bar{E})^{n+1}$$

$$= [1 + c_1(\bar{E})]^{n+1}$$

$$= [1 - c_1(E)]^{n+1}.$$

For example, if n = 1 so that $\mathbb{C}P^n = \mathbb{C}P^1 = S^2$ then

$$c(T(S^2)) = [1 - c_1(E)]^2$$

$$= 1 - 2c_1(E)).$$

(The classes $c_1(E)^k$ vanish for k > 1 since $H^{2k}(S^2) = 0$ for k > 1.) Thus $c_1(T(S^2)) = -2c_1(E)$. Similarly,

$$c(T(\mathbb{C}P^2)) = 1 - 3c_1(E) + 3c_1(E)^2.$$

For real vector bundles we may use real tangential connections or else complexify. If E is a real tangentially smooth n-plane vector bundle with complexification $E_{\mathbb{C}}$, then $c_i(E_{\mathbb{C}}) = 0$ in $H^{2i}(X;\mathbb{C})$ for i odd (Milnor-Stasheff [MS], p. 174), and hence $c_i^{\mathcal{T}}(E_{\mathbb{C}}) = 0$ for i odd. Define the *tangential Pontrjagin classes* $p_i^{\mathcal{T}}(E)$ by

(5.15) $$p_i^{\mathcal{T}}(E) = (-1)^i c_{2i}^{\mathcal{T}}(E_{\mathbb{C}}).$$

Define the *tangential total Pontrjagin class* to be the unit

(5.16) $$p^{\mathcal{T}}(E) = 1 + p_1^{\mathcal{T}}(E) + p_2^{\mathcal{T}}(E) + \dots$$

Note that $p_i^{\mathcal{T}}(E) = 0$ for $i > n/2$. We may regard $p_i^{\mathcal{T}}(E) \in H_{\mathcal{T}}^{4i}(X;\mathbb{R})$ as $(-1)^i c_{2i}(E) \in H^{4i}(X;\mathbb{Z})$ via the topological definition. The following properties of the tangential Pontrjagin classes follow immediately from the corresponding properties of tangential Chern classes.

Proposition 5.17. For each tangentially smooth real vector bundle E over a compact foliated space X there are tangential Pontrjagin classes $p_i^{\mathcal{T}}(E) \in H_{\mathcal{T}}^{4i}(X;\mathbb{R})$ satisfying the following properties:

1) If $g: X' \longrightarrow X$ is tangentially smooth then

$$g^* p_i^{\mathcal{T}}(E) = p_i^{\mathcal{T}}(g^* E)$$

2) $p_m^{\mathcal{T}}(E \oplus E') = \sum_{i=0}^{m} p_i^{\mathcal{T}}(E) \, p_{m-i}^{\mathcal{T}}(E')$ $\quad (p_0^{\mathcal{T}} = 1)$.

3) If E is of real dimension n then $p_i^{\mathcal{T}}(E) = 0$ for $i > n/2$.

4) The total tangential Pontrjagin class

$$p^{\mathcal{T}}(E) = 1 + p_1^{\mathcal{T}}(E) + p_2^{\mathcal{T}}(E) + \dots$$

corresponds to the invariant polynomial $\det(I + \frac{A}{2\pi})$.

151

To complete our discussion of characteristic classes it remains to define the tangential Euler class. For this we need to assume that each leaf of X is given a Riemannian structure which varies continuously in the transverse direction.

Proposition 5.18. The dual tangent bundle F^* possesses one and only one symmetric tangential connection which is compatible with its metric.

This preferred tangential connection ∇ is called the *Riemannian or Levi-Civita tangential connection*. A tangential connection on F^* is *symmetric* if the composition

$$\Gamma_\tau(F^*) \xrightarrow{\nabla} \Gamma_\tau(F^* \otimes F^*) \xrightarrow{\wedge} \Gamma_\tau(\Lambda^2 F^*)$$

is equal to the exterior derivative d.

Proof: Let $s_1..,s_n$ be an orthonormal basis for $\Gamma_\tau(F^*|_U)$. There is one and only one skew-symmetric matrix $[\omega_{kj}]$ of 1-forms such that

$$ds_k = \sum_j \omega_{kj} \wedge s_j.$$

(See Milnor Stasheff [MS, p.302-3]). Define the tangential connection ∇ over U by

$$\nabla(s_k) = \sum_j \omega_{kj} \otimes s_j$$

and extend by partitions of unity to all of X. $\square$

Let E be an *oriented* tangentially smooth real 2n-plane bundle with a tangentially smooth Euclidean metric. Choose an oriented orthonormal basis for the sections $\Gamma_\tau(E|_U)$ for some coordinate patch U. Then the tangential curvature matrix Ω obtained from a symmetric tangentially smooth connection is skew-symmetric. There is a unique polynomial with integer coefficients on skew–symmetric matrices called the *Pfaffian* and written Pf with the property that

$$Pf(A)^2 = det(A)$$

and

$$Pf(diag(S,S,...,S)) = 1.$$

The Pfaffian satisfies the invariance condition

$$Pf(BAB^t) = Pf(A)det(B)$$

and is hence SO(2n)–invariant. (For the linear algebra we omit, see Milnor–Stasheff [MS, p.309–310]). Thus $Pf(\Omega) \in \Omega_\tau^{2n}(U)$ makes sense. Choosing a different oriented orthonormal basis for the sections over U, this exterior form will be replaced by $Pf(B\Omega B^t)$ where the matrix B is orthogonal $(B^{-1} = B^t)$ and orientation–preserving $(det(B) = 1)$. Thus these local forms coalesce to create a global 2n–form

$$Pf(K) \in \Omega_\tau^{2n}(X).$$

As before, this class is a cocycle and hence represents a tangential characteristic cohomology class. It is convenient to normalize. Define the *tangential Euler class* $e^\tau(E) \in H_\tau^{2n}(X;\mathbb{R})$ by

(5.19) $$e^\tau(E) = [Pf(K/2\pi)].$$

The tangential Euler class is well–defined and independent of choice of symmetric tangential connection. Here are its elementary properties.

Proposition 5.20. To each 2n–dimensional oriented tangentially smooth real vector bundle E with a Euclidean metric over a compact foliated space X there is associated a tangential Euler class

$$e^{\mathcal{T}}(E) \in H_{\mathcal{T}}^{2n}(X;\mathbb{R})$$

which is of the form $e^{\mathcal{T}}(E) = [Pf(K)/2\pi]$ and is independent of choice of symmetric connection. Further,

1) If $g: X' \longrightarrow X$ is tangentially smooth then
$g^{*}e^{\mathcal{T}}(E) = e^{\mathcal{T}}(g^{*}E)$.

2) $e^{\mathcal{T}}(E{\oplus}E') = e^{\mathcal{T}}(E)e^{\mathcal{T}}(E')$.

3) If E has a nowhere zero tangentially smooth section then $e^{\mathcal{T}}(E) = 0$.

4) the tangential Pontrjagin class $p_{n}^{\mathcal{T}}(E)$ is equal to the square of the tangential Euler class $e^{\mathcal{T}}(E)$:
$p_{n}^{\mathcal{T}}(E) = e^{\mathcal{T}}(E)^2$.

5) If E is classified by $f: X \longrightarrow G_{k}(\mathbb{R}^{n+k})$ then in $H_{\mathcal{T}}^{2n}(X;\mathbb{R})$

$$e^{\mathcal{T}}(E) = rf^{*}e(E^{n}) = re(E).$$

Note: In topological treatments of characteristic classes matters are somewhat different. Classes take values in *integral* cohomology and may very well be torsion classes. The resulting formulas are more complicated. (Our classes are the images of those under $H^{*}(;\mathbb{Z}) \longrightarrow H^{*}(;\mathbb{R})$). (For example, in integral cohomology, formula 5.17(2) holds only mod 2.) In the Chern-Weil approach the classes take values in cohomology with *real* or *complex* coefficients, so torsion has been destroyed. There is apparently no way known of showing directly from the Chern-Weil approach that Chern classes are integral cohomology classes; proofs known to us rely on the topological construction.

Recall that for any compact foliated space X we have defined the tangential Chern character

$$\mathrm{ch}^{\mathcal{T}} : K^0(X) \longrightarrow \underset{m}{\oplus} H_{\mathcal{T}}^{2m}(X;\mathbb{C})$$

to be that map which, in the Chern–Weil setting, arises from the invariant polynomial $\mathrm{trace}(e^{A/2\pi i})$. It is not hard to show that $\mathrm{ch}^{\mathcal{T}}$ is a ring map and that it extends to a natural transformation of $\mathbb{Z}/2$-graded functors

$$\mathrm{ch}^{\mathcal{T}} : K^*(X) \longrightarrow H_{\mathcal{T}}^{**}(X;\mathbb{C})$$

where $H_{\mathcal{T}}^{**}$ is the $\mathbb{Z}/2$-graded functor $(\underset{m}{\oplus} H_{\mathcal{T}}^{2m}) \oplus (\underset{m}{\oplus} H_{\mathcal{T}}^{2m+1})$.

Proposition 5.21. If M is a compact smooth manifold then

$$\mathrm{ch} \otimes 1 : K^*(M) \otimes \mathbb{C} \longrightarrow H^{**}(M;\mathbb{C})$$

is an isomorphism.

We omit the proof of this proposition. The actual situation is the following:

1)   ch extends to $K^*(X) \longrightarrow \overset{\vee}{H}{}^{**}(X;\mathbb{C})$ for any locally compact Hausdorff space X. (Here $K^*(X)$ refers to K-theory with compact supports: $K^0(X) \equiv \overline{K}^0(X^+)$.)

2)   The map

$$\mathrm{ch} \otimes 1 : K^*(X) \otimes \mathbb{C} \longrightarrow \overset{\vee}{H}{}^{**}(X;\mathbb{C})$$

is an isomorphism on those spaces.

3)   In fact, there is an isomorphism

$$\mathrm{ch} \otimes 1 : K^*(X) \otimes Q \longrightarrow \overset{\vee}{H}{}^{**}(X;Q).$$

To prove 3) one checks first that $\mathrm{ch} \otimes 1$ is an isomorphism for X a sphere. Induction (or a spectral sequence argument) implies that $ch \otimes 1$ is an isomorphism for all finite complexes (and in particular for all compact smooth manifolds). Since K^* and $\check{H}^{**}$ respect inverse limits, $ch \otimes 1$ is an isomorphism for $X = \varprojlim X_j$ the inverse limit of finite complexes. Any compact metric space arises in this manner (cf. Eilenberg-Steenrod [ES]), so (2) holds for compact metric spaces. Finally, a one-point compactification argument implies the full result.

Lest the reader fall into an obvious trap, we note that the natural map

$$ch^{\tau} \otimes 1 \; : \; K^*(X) \otimes \mathbb{C} \longrightarrow H^{**}_{\tau}(X;\mathbb{C})$$

is not an isomorphism in general for foliated spaces, or even for foliated manifolds. In the diagram

only $ch \otimes 1$ is an isomorphism. Any bundle E for which $c_i(E) = 0$ $i = 1,...,p$ will be in the kernel of $ch^{\tau} \otimes 1$, even though $[E] \neq 0$ in general in $K^0(X) \otimes \mathbb{C}$. On the other hand, $H^{**}_{\tau}(X;\mathbb{C})$ is infinitely generated in some cases. So $ch^{\tau} \otimes 1$ is neither injective nor surjective in general.

Next we consider Thom isomorphisms. Recall (from 3.29) that if X is a compact foliated space and if $E \longrightarrow X$ is a tangentially smooth oriented real n-plane bundle then there is a unique Thom class $u_E \in H^n_{\tau c}(E)$ and a Thom isomorphism

$$\Phi_{\tau} : H^k_{\tau}(X) \xrightarrow{\ \cong\ } H^{k+n}_{\tau c}(E)$$

given by

$$\Phi_\tau(\omega) = u_E \omega.$$

The same result holds in ordinary cohomology. Precisely, there is a Thom class $\tilde{u}_E \in H_c^n(E)$ and a Thom isomorphism

(5.22)
$$\Phi: H^k(X) \xrightarrow{\ \cong\ } H_c^{k+n}(E)$$

given by

$$\Phi(\omega) = \tilde{u}_E \omega.$$

The proof of this fact is essentially identical to the proof of the tangential Thom isomorphism (cf. Bott [Bo], §§6,7). Further, the restriction map $r_*: H_c^*(E) \longrightarrow H_{\tau c}^*(E)$ respects Thom classes:

$$r_*(\tilde{u}_E) = u_E$$

and hence there is a commutative diagram

$$
\begin{array}{ccc}
H^k(X) & \xrightarrow{\ r_*\ } & H_\tau^k(X) \\
\Big\downarrow{\scriptstyle \Phi} & & \Big\downarrow{\scriptstyle \Phi_\tau} \\
H_c^{k+n}(E) & \xrightarrow{\ r_*\ } & H_{\tau c}^{k+n}(E).
\end{array}
$$

There is also a Thom isomorphism in K-theory. To obtain it, however, it is necessary to assume that the structural group of the bundle reduces to the group $spin^c$. (This is slightly more than orientability.) For instance, it suffices to assume that $E \longrightarrow X$ is a complex vector bundle (which is all we shall require).

If $E \longrightarrow X$ is indeed a $spin^c$-bundle (of even real dimension for convenience) then there is a K-theory Thom class $u_E^K \in K^0(E)$. (This means K-theory with compact supports: $K^0(E) = \tilde{K}^0(E^+)$.) Further, multiplication by this class induces an isomorphism

(5.23) $$\Phi_K: K^0(X) \longrightarrow K^0(E)$$

given by

$$\Phi_K(x) = u_E^K x.$$

(For the proof of this theorem the reader may consult Atiyah [At1] and Karoubi [Kar].) All three Thom isomorphisms extend to the case X locally compact – see Karoubi [Kar].

It would be natural to suppose that Thom isomorphisms commute with the Chern character, i.e. that the diagram

$$
\begin{array}{ccc}
K^0(X) & \xrightarrow{\;ch_\tau\;} & H_{\tau\,c}^{**}(X;\mathbb{R}) \\
\Big\downarrow{\Phi_K} & & \Big\downarrow{\Phi_\tau} \\
K^0(E) & \xrightarrow{\;ch_\tau\;} & H_{\tau\,c}^{**}(E;\mathbb{R})
\end{array}
$$

would commute. Let $1 \in K^0(X)$ denote the class of the complex one-dimensional trivial bundle over X, which is the identity of $K^0(X)$. Then

$$ch_\tau \Phi_K(1) = ch_\tau(u_E^K)$$

$$\Phi_\tau ch_\tau(1) = \Phi_\tau(1) = u_E$$

Thus commutativity boils down to the relation between the cohomology Thom class u_E and the Chern character of the K-theory Thom class $ch_\tau(u_E^K)$. Generally these classes are not equal. Define the *tangential Todd class*

$$Td_\tau(E) \in H_\tau^{**}(X;\mathbb{R})$$

by the formula

(5.23) $\qquad$ $Td_\tau(E) = [\Phi_\tau^{-1}(ch_\tau(u_E^K))]^{-1}.$

Thus

$$Td_\tau(E) = 1 \iff u_E = ch_\tau(u_E^K).$$

Our definition is of course modeled after the classical Todd class which is defined by

$$Td(E) = [\Phi^{-1}(ch(u_E^K))]^{-1} \in H^{**}(X;\mathbb{R})$$

so that

$$Td(E) = 1 \iff \tilde{u}_E = ch(u_E^K).$$

We list the elementary properties of the tangential Todd class.

Proposition 5.24. The tangential Todd class has the following properties:

1) $r_*Td(E) = Td_\tau(E)$, where $r_*: H^*(X;\mathbb{R}) \longrightarrow H_\tau^*(X;\mathbb{R})$ is the restriction map.

2) $Td_\tau(E\oplus E') = Td_\tau(E)Td_\tau(E').$

3) If f: $X \longrightarrow Y$ is a tangentially smooth map and $E \longrightarrow Y$ is a tangentially smooth bundle then

$$Td_\tau(f^*E) = f^*Td(E).$$

4) Td_τ is the tangental characteristic class associated with the invariant power series $\dfrac{A}{1-e^{-A}}$.

Proof: The first property follows from the fact that $r_*\tilde{u}_E = u_E$. The remaining properties may be proved directly or deduced from the analogous properties of the classical Todd class as in Karoubi

159

[Kar], p. 285. □

Note that Td(E) is a unit in the ring $H^{**}(X;\mathbb{R})$. Let E' be a real bundle such that E⊕E' is a trivial bundle. Then

$$1 = Td(E⊕E') \qquad \text{since Td(1) = 1}$$

$$= Td(E)Td(E') \qquad \text{by 1).}$$

It is customary to define the *Todd genus* of a smooth manifold M by

$$Td(M) = Td(TM⊗\mathbb{C})$$

and following custom we define the *tangential Todd genus* of a foliated space X by

(5.25) $\qquad\qquad Td_\tau(X) = Td_\tau(FX⊗\mathbb{C}).$

We emphasize that we may regard classes such as $Td_\tau(X)$ as tangential *forms* in $\Omega_\tau^*(X)$ given by certain universal polynomials in the tangential curvature form K_E. Given a tangential connection, these forms are uniquely defined (not just up to cohomology class.) Changing the tangential connection changes the form but preserves the cohomology class of the form.

We may see the Todd class very explicitly. The power series $\dfrac{x}{1-e^{-x}}$ expands as

$$1 + \frac{x}{2} + \sum_{s=1}^{\infty} (-1)^{s-1} \frac{B_s}{(2s)!} x^{2s}$$

where $B_s \in Q$ is the s-th Bernoulli number (cf. the appendix of Milnor-Stasheff [MS]):

$$B_1 = 1/6 \qquad\qquad B_5 = 5/66$$

$$B_2 = 1/30 \qquad\qquad B_6 = 691/2730$$

$$B_3 = 1/42 \qquad\qquad B_7 = 7/6$$

$$B_4 = 1/30 \qquad\qquad B_8 = 3617/510.$$

For example, if X is a foliated space with leaf dimension $p \leqslant 8$ then $H_{\mathcal{T}}^k(X) = 0$ for $k > 8$ and the polynomial has the form

$$1 + \frac{x}{2} + \frac{B_1}{2}x^2 - \frac{B_2}{24}x^4 =$$

$$= 1 + \frac{x}{2} + \frac{1}{12}x^2 - \frac{1}{720}x^4$$

Thus if E is a complex line bundle over X then

$$Td_{\mathcal{T}}(E) = 1 + \frac{1}{2}c_1^{\mathcal{T}}(E) + \frac{1}{12}c_1^{\mathcal{T}}(E)^2 - \frac{1}{720}c_1^{\mathcal{T}}(E)^4$$

Note that this is a non-homogeneous class sitting as is usual in the group

$$H_{\mathcal{T}}^{ev}(X;\mathbb{R}) = \underset{m}{\oplus} H_{\mathcal{T}}^{2m}(X;\mathbb{R}).$$

For instance, if $X = M = \mathbb{C}P^2$ with canonical complex line bundle E^1 then write $\omega = \dfrac{K}{2\pi i}$. Then

$$c(T\mathbb{C}P^2) = 1 - 3\omega + 3\omega^2.$$

The Chern character and Todd class are given at the form level by

$$ch(T\mathbb{C}P^2) = 2 - 3\omega + \frac{3}{2}\omega^2$$

and

$$Td(T\mathbb{C}P^2) = 1 - \frac{3}{2}\omega + \omega^2.$$

We follow the usual convention in interpreting expressions of

161

the type $\int \omega d\nu$ where ω is a non–homogeneous form; take the part of ω which lies in Ω^p and ignore the rest. For instance,

$$\int \mathrm{Td}_{\tau}(E)d\nu$$

is to be understood as follows: write $\mathrm{Td}_{\tau}(E) = \Sigma \ \mathrm{Td}_{\tau}^m(E)$ where $\mathrm{Td}_{\tau}^m(E) \in H_{\tau}^m(X)$, and define

$$\int \mathrm{Td}_{\tau}(E)d\nu \equiv \int \mathrm{Td}_{\tau}^p(E)d\nu.$$

Finally, note that if M is a foliated manifold with tangent bundle TM and foliation bundle FM then the normal bundle to the foliation $NM = TM/FM$ has a flat connection in the leaf direction and so its relevant characteristic classes vanish. Thus if ω is any tangential cohomology class, then

$$\int \omega \mathrm{Td}(M)d\nu \equiv \int \omega \mathrm{Td}(TM)d\nu = \int \omega \mathrm{Td}(FM)d\nu$$

and so we may use $\mathrm{Td}(M)$ and $\mathrm{Td}(FM)$ interchangeably in index formulas.

CHAPTER VI: OPERATOR ALGEBRAS

We turn now to the discussion of operator algebras that can be associated with groupoids and in particular to the groupoid of a foliated space. For this discussion we start with a locally compact second countable topological groupoid G and we assume given a continuous tangential measure λ (see Chapter IV for the definition). Thus for each x in the unit space X of G we have a measure λ^x on $G^x = r^{-1}(x)$ with certain invariance and continuity properties as described in Chapter IV. For the moment we do not need to assume that the groupoid has discrete holonomy groups as in Chapter IV, but all the examples and all the applications will satisfy this condition. If in addition the support of the measure λ^x is equal to $r^{-1}(x)$, as is usual in our examples, then λ is called a Haar system.

In this chapter we construct the C^*-algebra of the groupoid and we determine this algebra in several important special cases. We describe the Hilsum–Skandalis stability theorem. Assuming a transverse measure, we construct the associated von Neumann algebra and develop its basic properties and important subalgebras. This leads us to the construction of the weight associated to the transverse measure; it is a trace if and only if the transverse measure is invariant. Finally, we introduce the K-theory index group $K_0(C^*_r(G))$ and construct a partial Chern character

$$c: K_0(C^*_r(G(X))) \longrightarrow MT(X)^* \cong \bar{H}^p_\tau(X)$$

which is given explicitly as follows. If $[u] \in K_0(C^*_r(G(X)))$ and ν is an invariant transverse measure with associated trace ϕ_ν, then

$$c([u])(\nu) = \phi^n_\nu(e - f)$$

where e and f are suitably chosen projections in $C^*_r(G(X))^+ \otimes M_n$ whose difference represents u and $\phi^n_\nu = \phi_\nu \otimes Tr$. The partial Chern character applied to the symbol of a tangential, tangentially elliptic operator D yields the cohomology analytic index class

$[\iota_D] \in \bar{H}^p_7(X)$.

We shall first review the construction of a C^* algebra associated to the pair (G,λ) (cf. Connes [Co3], Renault [Ren1]). Suppose first that G is Hausdorff. On the space $C_c(G)$ of continuous compactly supported functions on G one defines a multiplication and an involution

(6.1)
$$(f\ast g)(u) = \int f(v)g(v^{-1}u)d\lambda^{r(u)}(v)$$

$$f^*(u) = \overline{f(u^{-1})}.$$

That these define an associative algebra with involution on $C_c(G)$ is a straightforward calculation paralleling the case when G is a locally compact group, (cf. Pedersen [Ped], p. 233). The invariance property of tangential measures allows one to rewrite the convolution as

$$(f\ast g)(u) = \int f(uv)g(v^{-1})d\lambda^{s(u)}(v).$$

In the case when G is an equivalence relation R on a space X, then a tangential measure is simply a measure λ^x for each x such that $\lambda^x = \lambda^y$ if $x \sim y$. Functions on $G = R$ are viewed as partially defined functions of two variables, and the formulas become

$$(f\ast g)(x,z) = \int f(x,y)g(y,z)d\lambda^x(y)$$

$$f^*(x,y) = \overline{f(y,x)}$$

where in the first formula λ^x could be λ^y or λ^z as x, y, and z in the formula are all in the same equivalence class.

There are two ways of norming the involutive algebra $C_c(G)$. For the first way, there is for each $x \in X$ a natural homomorphism π_x of $C_c(G)$ into the algebra of bounded operators on the Hilbert space $L^2(G^x, \lambda^x)$ defined essentially by convolution:

(6.2)
$$(\pi_x(f)\varphi)(u) = \int f(u^{-1}v)\varphi(v)d\lambda^x(v).$$

The integral is clearly well defined for $\varphi \in L^2(G^x, \lambda^x)$, $f \in C_c(G)$ and it yields a bounded operator $\pi_x(f)$ on $L^2(G^x, \lambda^x)$. That π_x defines a *-homomorphism is likewise easily checked. Note that the formula (6.2) above displays $\pi_x(f)$ in effect as right convolution by f' $(= \bar{f}^*)$ where $f'(u) = f(u^{-1})$.

One norms $C_c(G)$ by $|f| = \sup_x |\pi_x(f)|$; the completion of $C_c(G)$ under this norm is virtually by construction a C^* algebra, for we obtain it by embedding $C_c(G)$ into bounded operators on a Hilbert space (the sum of the $L^2(G^x, \lambda^x)$) and closing up the image.

Definition 6.3. The *reduced C^*-algebra of the groupoid* G is the completion of $C_c(G)$ with respect to the norm $|f|$ above; it is denoted $C_r^*(G)$.

This construction is analogous to the construction of the reduced C^* algebra of a locally compact group by closing up the image of the regular representation. Therefore it is sensible to call the C^* algebra above the reduced C^* algebra of the groupoid, $C_r^*(G)$. Connes and his students write this algebra as $C^*(V,F)$ when G is the graph of a foliated manifold (V,F).

The second way of norming $C_c(G)$ corresponds to the full C^*-algebra of a group. Namely we first put a kind of an L_1 norm on $C_c(G)$

$$|f|_1 = \max(\sup_x \int |f(u)| \, d\lambda^x(u), \quad \sup_x \int |f(u^{-1})| \, d\lambda^x(u))$$

so that it becomes a normed *-algebra. Then we form the C^*-completion of this algebra using all bounded * representations. This is denoted $C^*(G)$ and is called the *full C^* algebra* of the groupoid. As the representations π_x are among all bounded representations, it is evident that the reduced C^* algebra $C_r^*(G)$ is a quotient of $C^*(G)$.

As our concern here will be with analysis and differential operators on foliated spaces where the representations π_x play the central role, it is evident that it is the reduced C^* algebra $C_r^*(G)$ rather than the full C^* algebra that will be the focus of attention. Of course the construction of both $C_r^*(G)$ and $C^*(G)$ presupposes a tangential measure or a Haar system on the groupoid G. These algebras can depend on this choice, and if one is being absolutely precise, the underlying Haar system should be included in the notation. The reader is referred to Renault [Ren1] for a more extended discussion on this point. This will not be an issue for us because in the first place if λ is a tangential measure (resp. Haar system) and if $\lambda' = f\lambda$ where f is a continuous everywhere positive function on G that is constant on the fibres of the map $u \longrightarrow (r(u),s(u))$ of G into $X \times X$, then λ' is also a tangential measure (resp. Haar system). In this case it is easy to check that the C^* algebras $C_r^*(G)$ and $C^*(G)$ do not depend on whether one uses λ or λ'. Secondly, in the case of the graph of a foliated space, there is as we have already pointed out in Chapter IV (4.20 and remarks following) a choice of a class of Haar systems (each λ^x should have a continuous, or even tangentially C^∞, density on the leaves in local coordinates), any two of which differ like λ and λ' above. If the total space of the foliation is compact one can also ensure that the function relating λ and λ' is bounded above and below. At all events when we speak about $C_r^*(G)$ or $C^*(G)$ in the context of the graphs of foliations, we shall always understand that standard choice of Haar system. It is evident that the hypothesis that G be second countable makes the C^* algebra $C_r^*(G)$ separable.

All of this discussion has assumed that the groupoid G is Hausdorff, but we know that the groupoid of a foliation need not be Hausdorff. At all events the groupoid is locally Hausdorff so it may be covered by a family of open sets each one of which is Hausdorff. We still assume that the space is second countable. Then as one can take this family to be a countable family U_i, it is straightforward using standard techniques to see that G is at least a standard Borel groupoid. (A set E is Borel if and only if $E \cap U_i$ is Borel for each i.) This will be useful later when we introduce von Neumann algebras associated with these groupoids.

It is still quite straightforward to introduce the C^* algebra in the locally Hausdorff case. We gave a definition in (4.15) of what was meant by a continuous tangential measure $\lambda = \{\lambda^\ell\}$ in this situation. Now instead of considering all continuous compactly supported functions on G, let us consider instead the set of finite linear combinations of functions $f = \Sigma f_i$ where each f_i is a compactly supported function continuous on some open Hausdorff subset U_i of G extended to be zero on the rest of G. (Note that while f_i is continuous on U_i its extension to G is not in general a continuous function on G.) These functions can be convolved using the same formulas as in the Hausdorff case to give an *-algebra. Then one follows the same recipe for norming it and constructing a separable C^* algebra $C_r^*(G)$.

The C^* algebra associated to the groupoid of a foliated space X plays a key role in the analysis and geometry of X as we shall see. In particular its K-theory group $K_0(C_r^*(G))$ is the natural place where indices for operators live (cf. Connes-Skandalis [CS1,CS2]). One may also think of it as a non-commutative replacement for the algebra of functions on the quotient space X/R where X is the unit space of G and R is the equivalence relation on X defined by R. For a foliated space this is the space of leaves. Indeed when X/R is a "good" space such as when G is the groupoid of a foliated space which is a fibration, then $C_r^*(G)$ looks very much like C(X/R), as we shall see presently; they are in fact stably isomorphic. The interpretation of $C_r^*(G)$ as functions on the leaf space is enhanced by the following result of Fack and Skandalis [FS].

Theorem 6.4. If G is the groupoid of a foliated space, then $C_r^*(G)$ is simple as a C^* algebra if and only if every leaf of the foliated space is dense.

We shall not prove this result except to remark that if ℓ is a proper closed leaf, then the representation π_x above for any x in ℓ has a non-trivial kernel so that $C_r^*(G)$ is not simple.

One of the most important classes of examples of topological groupoids comes from group actions. Suppose that a locally compact group H acts as a topological transformation group on a space X with $h \in H$ acting on a point $x \in X$ denoted $h \cdot x$. The product space

$G = X \times H$ becomes in a natural way a groupoid where the product in G, $(x,g) \cdot (y,h)$, is defined if and only if $g \cdot y = x$ and then

$$(x,g) \cdot (y,h) = (x,gh).$$

X is the space of units and the range and source maps are given by $r(x,h) = x$, $s(x,h) = h^{-1} \cdot x$. Then $G^x = \{(h,x),\ x$ fixed, h arbitrary in $H\} \simeq H$. If $u = (g,y)$ is an element of G which can multiply G^x on the left, that is $s(u) = s(y,g) = g^{-1}y = x$, then left multiplication by u maps G^x onto G^y, $y = g \cdot x$ and the map is $(x,h) \longrightarrow (y,gh)$. Now giving a tangential measure or a Haar system on the groupoid G is giving a measure λ^x on each G^x which is invariant under these left multiplications. There is a natural choice in this case, namely take for λ^x on $G^x \simeq H$, a fixed left Haar measure on H. Evidently this tangential measure is continuous in the sense of 4.15. This example is of course the reason one calls such objects Haar systems. Of course, for the groupoid just defined to fall strictly within the context of Chapter IV where we assume discrete holonomy groups, the various isotropy groups of the action of H on X, $H_x = \{h\colon h \cdot x = x\}$ must be discrete. If for instance H is a Lie group acting on a manifold X then the action of H will give rise to a foliation of X only in this case.

In the case of a group H acting on a space X as above, one may form (cf. Pedersen [Ped]) a C^* algebra, the *reduced crossed product* algebra of C(X) by H, written $C(X)_r \rtimes H$. It is clear that the general construction for groupoids should and does yield the reduced crossed product construction in this case.

Proposition 6.5. If $G = X \times H$ for an action of H on X (no assumptions on isotropy) then

$$C_r^*(G) \simeq C(X)_r \rtimes H.$$

Proof. One checks that there are dense subalgebras of both sides that are algebraically identical. The algebra $C_r^*(G)$ is obtained by completing in the norm defined by the family of representations π_x. for $x \in X$. Each of these comes from a covariant representation of the pair (C(X).G) and one has to prove that these give the same norm as the subfamily of all covariant representations of the pair used to define the reduced crossed product. We omit the details. □

Another, somewhat trivial case, is of interest; let X be locally compact and let $G = X \times X$ be the equivalence relation (principal groupoid) with all points equivalent. If X is a manifold foliated by one leaf, G is its groupoid. A Haar system is simply a measure λ on X whose support is all of X. Evidently elements of the dense subalgebra of the definition can be realized as integral kernel operators on $L^2(X,\lambda)$ with compactly supported kernels. The completion is obviously all compact operators $\mathcal{K}$ on $L^2(X,\lambda)$.

Proposition 6.6. In this case $C_r^*(G) \cong \mathcal{K}$. □

An important theme in Feldman–Moore [FM], Feldman–Hahn–Moore [FHM], and Ramsay [Ra] is that for measured groupoids or equivalence relations, the special case when the orbits are discrete is much easier to handle and that in some sense the general case could be reduced to this special case. We want to see that the same is true in this context. First of all we need a definition.

Definition 6.7. The (locally Hausdorff) topological groupoid G has *discrete orbits* if the range and source maps are local homeomorphisms.

It follows that each equivalence class (or leaf) of the associated equivalence relation is countable and discrete in the relative topology from G (although not in the relative topology from X). In this case there is a natural choice of tangential measure, namely the counting measure on each leaf. It follows from the definition of discreteness that this is a continuous tangential measure, and so we can define the C^* algebra $C_r^*(G)$.

It is useful to point out that in the principal case, where G is simply an equivalence relation, that G can be covered by sets of a very simple kind. Let O be an open set in X and let b be a homeomorphism onto an open subset of X; let $U(b,O) = \{(x,b(x)), x \in O\}$.

Proposition 6.8. If G is discrete and principal, then the $U(b,O)$ are open sets and form a cover of G. □

The dense subalgebra A of compactly supported functions (or its substitute in the non-Hausdorff case) can be thought of as generalized matrices especially if there is no holonomy so that G is a principal groupoid, i.e. an equivalence relation R. Then as we have already seen the formulas simplify and the product of two functions on $R \subset X \times X$ is given by $(f*g)(x,z) = \Sigma f(x,y)g(y,z)$ where the sum is extended over all y which lie in the same class as x and z. The condition that f and g be compactly supported implies that the sum is finite. Written this way the product really does look like matrix multiplication. When there is holonomy, multiplication is still given by a sum rather than an integral, but the sum must include summation (convolution) on the discrete holonomy groups.

For general groupoids, the process of completing the dense subalgebra A of functions to obtain $C_r^*(G)$ leads to elements in the C^* algebra which cannot be represented as functions on G. One of the nice features of discrete groupoids is that an element of the C^* algebra can be represented by a continuous function on G, at least if G is Hausdorff.

Proposition 6.9. For $f \in A$ and $u \in G$,

$$|f(u)| \leqslant |f|$$

where $|f| = \sup_x |\pi_x(f)|$ is the C^*-norm.

Proof. The representation π_x of A in $L^2(G^x, \lambda^x)$ in the definition of $C_r^*(G)$ is a representation by matrices since G^x is a countable discrete set and λ^X is counting measure.   It is evident moreover that $f(u)$ is just one of the matrix coefficients of $\pi_x(f)$, and so the inequality is obvious.   □

   Thus if f_n is a sequence in A with a limit in $C_r^*(G)$, f_n must converge uniformly as continuous functions on G to a limit f and we have the desired result for Hausdorff G.   Moreover multiplication of elements of $C_r^*(G)$ is given by the same "matrix multiplication" formulas for the functions which represent them.   The sums are no longer finite but are absolutely convergent as is easily seen using the argument of Proposition 6.9.   The situation for non-Hausdorff G can be handed by localization to Hausdorff subsets.

   Another feature of the discrete case is that the set of units X is an open subset of G because r and s are local homeomorphisms. Hence the set of compactly supported functions $C_c(X)$ is a subset of the algebra A used to define $C^*(G)$.   Moreover it is a subalgebra; elements of $C_c(X)$ correspond to diagonal matrices in the description above of A as generalized matrices.   Hence the C^* algebra $C_o(X)$ becomes a subalgebra of $C_r^*(G)$.   If X is compact, as it will be in most cases, then $C_o(X) = C(X)$ has a unit which is also a unit for $C_r^*(G)$.

   Finally, suppose that G is discrete and principal--that is, an equivalence relation and Hausdorff (note that it would be Hausdorff automatically if X itself is Hausdorff since G can be mapped continuously into X × X).   In this case the subalgebra $C_o(X)$ of $C_r^*(G)$ has a very special property--namely it is a diagonal subalgebra of $C_r^*(G)$ in the language of Kumjian [Kum] (and a Cartan subalgebra in the language of Renault [Ren1]).   We are inclined to change terminology and call Kumjian's diagonal subalgebras Cartan subalgebras.

Definition 6.10 (Kumjian [Kum]). A *Cartan subalgebra* B of a unital C^* algebra A is an abelian subalgebra (that contains the unit of A) with a faithful conditional expectation P: A ⟶ B with the property that the kernel of P is spanned by all elements a of A such that

(i) $aBa^* \subset B$,

(ii) $a^{*}Ba \subset B$.

(iii) $a^2 = 0$.

(Kumjian calls these free normalizers of B.) If A is not unital, a Cartan subalgebra of A is a subalgebra B such that B^+ is a Cartan subalgebra of A^+ where $(\ )^+$ is the operation of appending a unit.

To see (Kumjian [Kum]) that $C_o(X)$ is a Cartan subalgebra of $C_r^*(G)$, one has to define first a conditional expectation P. If $m \in C_r^*(G)$, it is represented by a function on G by Proposition 6.9 and then one restricts the function to the diagonal to get an element in $C_o(X)$. Next note that free centralizers can be obtained by taking a function a supported on a set of the form U(f,O) in G where f has no fixed points and such that $a(x,f(x))a(f(x),f^2(x)) = 0$. An easy localization argument shows that any compactly supported function on $G-\Delta X$ can be written as a finite sum of such functions, and hence that there are enough normalizers to span the kernel of P. Conversely it is evident from the condition $a^2 = 0$ that any such a viewed as a function on G must vanish on the diagonal and so is in the kernel of P.

Kumjian proves, complementing an earlier result of Renault [Ren1], a powerful converse to this exercise. Roughly stated it says that every pair (A,B) where B is a Cartan subalgebra (diagonal subalgebra in the language of Kumjian [Kum]) arises uniquely from a discrete equivalence relation but with a "twist" coming from a kind of two cocycle as in Feldman-Moore [FM] and Renault [Ren1]. In fact the topological objects which classify the pairs (A,B) are called twists. We shall not pursue this topic further here as it would take us afield.

If G_1 and G_2 are topological groupoids then their product $G_1 \times G_2 = G$ is also. If λ_i is a Haar system on G_i, i = 1,2 then it is evident that we can define a Haar system $\lambda_1 \times \lambda_2$ on $G = G_1 \times G_2$ for if X_i is the unit space of G_i, $X = X_1 \times X_2$ is the unit space of G and the range map r of G is $r_1 \times r_2$. Hence $r^{-1}(x_1,x_2) = r_1^{-1}(x_1) \times r_2^{-1}(x_2)$ and $(\lambda_1 \times \lambda_2)_1^{(x_1,x_2)}$ is defined to be the product measure. The following is then straightforward.

Proposition 6.11. $C_r^*(G_1 \times G_2) \cong C_r^*(G_1) \otimes C_r^*(G_2)$ where $\otimes$ denotes the minimal or spatial tensor product.

Proof.   If $A_i.A$ are the algebras of functions on $G_i.G$ used to define these C^* algebras, it is evident that the algebraic tensor product $A_1 \otimes A_2$ can be identified as a dense subalgebra of A.   The completions $C_r^*(G_i)$ are defined by a family of $*$ representations $\pi_{x_{ij}}$ ($x_{ij} \in X_i$) and it is clear that the completion $C_r^*(G)$ can be defined exactly by the family of tensor products $\pi_{x_{1,j}} \otimes \pi_{x_{2,j}}$.   The result follows. $\square$

As a corollary of this, suppose that G_1 is a groupoid and that $G_2 = X_2 \times X_2$ is a groupoid of the type in Proposition 6.6, for instance the groupoid of a manifold X_2 foliated by a single leaf. Then form $G = G_1 \times G_2$.   If for instance G_1 is the groupoid of a foliated space X_1 and X_2 is a fixed manifold foliated as a single leaf, then G is the groupoid of the foliated space $X = X_1 \times X_2$ where the leaves of X are $\ell \times X_2$ where ℓ is a leaf in X_1.   In other words we have fattened up the leaves of X_1 by crossing with a fixed manifold.   The foliated spaces X_1 and X have the same transversal structure.   As a consequence of 6.6 and 6.11 we have for any G_1 and any X_2 the following:

Proposition 6.12. $C_r^*(G) \cong C_r^*(G_1) \otimes K$ where K is the algebra of compact operators.

As a further example let us consider the groupoid arising from a fibration $p: X \longrightarrow B$ with standard fibre F.   We let G be the equivalence relation on X where $x \sim y$ if $p(x) = p(y)$.   If the fibration is locally trivial (cf. Steenrod [St]) and the standard fibre is a manifold, then X is a foliated space with leaves equal to the fibre of the fibration.   If $U \subset B$ is an open set over which the fibration is trivial, then U defines an open subfoliated space which is the product of U foliated by points with F foliated by one leaf.   Hence "locally" the C^* algebra is a product $C^*(U) \otimes K$ by above. But this works globally at the algebra level, at least if B is finite dimensional. In order to avoid degenerate cases we assume in the following that the standard fibre is not a finite set. The algebra is formed with respect to any given continuous Haar system.

Proposition 6.13. If G is the groupoid of a locally trivial fibration with base B, then $C_r^*(G)$ is Morita equivalent to $C(B) \otimes K$. If B is finite-dimensional then

$$C_r^*(G) \cong C(B) \otimes K.$$

Proof. For each $b \in B$, let $p^{-1}(b)$ be the fibre over b. A Haar system is simply the assignment in a "smooth" fashion of a measure λ^b on $p^{-1}(b) \cong F$, where smoothness means that in each local trivialization of $p^{-1}(U) \cong U \times F$, the λ^b for b in U viewed as measures on F vary continuously. The Hilbert spaces $L^2(p^{-1}(b), \lambda^b)$ then form a continuous field of Hilbert spaces over B (cf. Dixmier [Di2]) and it is evident from Proposition 6.12 that $C_r^*(G)$ consists of the sections of the corresponding field of operator algebras $K(L^2(p^{-1}(b), \lambda^b))$. The Dixmier-Douady invariant [DD] is trivial and hence $C_r^*(G)$ is Morita equivalent to $C(B) \otimes K$. If B is finite-dimensional then the field of Hilbert spaces is trivial and so $C_r^*(G) \cong C(B) \otimes K$. □

If in this example, the fibration $X \longrightarrow B$ has a cross section s, then s(B) is a complete transversal homeomorphic to B. Then s(B) is a groupoid of a trivial sort--equivalence classes are points. Thus $C_r^*(s(B)) = C(B)$ and so $C_r^*(G) \cong C_r^*(s(B)) \otimes K$. This is in fact quite a general phenomenon at least for groupoids of foliated spaces as is shown by Hilsum and Skandalis [HS]. We describe this result, which will be of considerable use to us, in some more detail.

In Chapter IV we discussed regular transversals for foliated spaces. These were locally compact subsets N of the foliated space X so that $N \subset N'$ with $\bar{N} \subset N'$ and $\bar{N}$ compact, and such that there exists an open ball B in $\mathbb{R}^p$, p the leaf dimension, with a homeomorphism of N' × B onto an open subset U of X with the map an isomorphism of foliated spaces. For this discussion, we shall also assume that there is a larger ball B' containing B with an extension of the homeomorphism of N' × B to N' × B' onto some U'; we shall also assume that N is open in N' so that N × B corresponds to an open set. To simplify notation, let us take the N × B to be subsets of X. If X is compact one can clearly find a finite number of such N_i so that the union is a complete transversal. If X is locally compact, then as in Fack–Skandalis [FS] one can find a locally finite such family and can also arrange that the N_i × B are disjoint from each other. At all events if N = $\cup N_i$, finite or infinite, then there is a ball B in $\mathbb{R}^p$ so that N × B = U is an open subset of X. We can also arrange that U^c contains a set of exactly the same form N × B using the fact that for the original N_i we had a N_i × B' $\supset$ N_i × B.

Let G be the groupoid of the foliated space (Hausdorff or not) and let G_N^N be the groupoid relativized to N; $G_N^N = \{u \in G: r(u), s(u) \in N\}$. Then H = G_N^N is a topological groupoid in its own right and it has discrete orbits. Then $C_r^*(H)$ is an algebra of the kind discussed earlier in the chapter. If we form U = N × B then G_U^U is an open subgroupoid of G and is clearly the product

$$G_U^U \simeq G_N^N \times (B \times B)$$

where B × B is the principal groupoid (equivalence relation) with unit space B and with all points equivalent. It follows from Proposition 6.12 that

$$C_r^*(G_U^U) \simeq C_r^*(G_N^N) \otimes \mathcal{K}$$

where $\mathcal{K}$ is the algebra of compact operators.

Further as G_U^U is an open subgroupoid of G, we can extend functions in the dense subalgebra defining $C_r^*(G_U^U)$ to functions on G. Moreover the choice of Haar system for defining $C_r^*(G_U^U)$ and $C_r^*(G)$ are compatible. It follows now that the natural injection of the dense algebra of compactly supported functions on G_U^U into functions on G produces a map i on the C^* algebra level of $C_r^*(G_U^U)$ into $C_r^*(G)$. The result of Hilsum and Skandalis [HS], in a slightly strengthened version is the following, which shows in some sense that $C_r^*(G)$ is no more complicated than $C_r^*(G_U^U)$ and also that the complete transversal N controls the structure of these algebras.

Theorem 6.14 **(Hilsum–Skandalis [HS]).** The algebra $C_r^*(G)$ is isomorphic to the algebra $M_2(C_r^*(G_U^U))$ of 2×2 matrices over $C_r^*(G_U^U)$ and the injection map i above corresponds to the natural inclusion of $C_r^*(G_U^U)$ into 2×2 matrices $a \longrightarrow \begin{bmatrix} a & 0 \\ 0 & 0 \end{bmatrix}$. Hence $C_r^*(G)$ is also isomorphic to $C_r^*(G_N^N) \otimes \mathcal{K}$ where $\mathcal{K}$ is the algebra of compact operators.

We shall not reproduce the details of the argument but will note some highlights. Fixing the complete transversal N as above, they define for each open set V of the unit space X, a C^*-module (cf. Kasparov [Kas2]) H(V) over $C_r^*(G_N^N)$. These add for disjoint U's and H(V) = H(W) if V =W - F where F is closed in X and meets each leaf in a null set. They also establish that the algebra of "compact operators" (in Kasparov's terminology) on H(V) is $C_r^*(G_V^V)$. For the particular choice of U = N X B, the "tube" around the transversal N which was constructed above, it is easy to see that $H(U) = H_\infty \otimes C_r^*(G_N^N)$ where H_∞ is an infinite dimensional Hilbert space. Since by construction we can find another transversal N' which looks just like N and a "tube" around it, N' X B inside of $\bar{U}^C$, one may use the above to argue that H(X) = H(U) $\oplus$ H($\bar{U}^C$) and that H($\bar{U}^C$) contains a submodule H(U') isomorphic to H(U). The Kasparov stablization theorem of [Kas2] says that H($\bar{U}^C$) is isomorphic to H(U) which establishes the result. $\square$

We note that there is another interesting approach to the isomorphism (6.14) due to Haefliger (cf. Theorem A5.1 in appendix A), Renault [Ren2] and Muhly-Renault-Williams [MRW]. One shows that G is equivalent to G_N^N as topological groupoids. Then one shows that under mild hypotheses, the C^*-algebras associated to equivalent groupoids are strongly Morita equivalent and hence (by

Brown-Green-Rieffel [BGR]) stably isomorphic.

We now turn our attention to another and closely related operator algebra that one can construct. Let G be a locally compact groupoid with discrete holonomy groups together with a given Haar system λ. We also assume that the underlying Borel groupoid has a complete transversal--a condition that is *a fortiori* satisfied for the groupoid of a foliated space. We also assume given on this Borel groupoid a positive transverse measure ν, not necessarily invariant--cf. Chapter IV. Indeed for the coming discussion we can and shall neglect any topological structure and simply work as in Chapter IV with a standard Borel groupoid with a complete transversal, a fixed tangential measure and with countable holonomy groups G^x_x.

The integration process of Chapter IV where we integrate the tangential measure λ with respect to the transverse measure ν produces a measure $\mu = \int \lambda d\nu$ on the unit space X. Then as noted in Chapter IV one may turn G into a *measured groupoid* (Mackey [Ma5], Ramsay [Ra2]) by defining a measure $\tilde{\mu}$ on G by

$$\tilde{\mu}(E) = \int_X \lambda^x(E \cap G^x) d\mu(x)$$

(i.e., $\tilde{\mu} = \mu_r$ in the terminology of Chapter IV). We note that conversely if G is a standard measured groupoid, then by a result of Peter Hahn [Hap1] there is a Haar measure λ (= tangential measure $\sim$ Haar system) on G and a measure ν on X so that the original measure on G is given by the formula above.

Now any measured groupoid has a regular representation (Hahn [Hap1], Connes-Takesaki [CT]) which in form looks just like the construction defining $C^*_r(G)$. We form the Hilbert space $\tilde{H} = L^2(G, \tilde{\mu})$ which is decomposed as a direct integral

$$\tilde{H} = \int H^x d\mu(x)$$

where $H^x = L^2(G^x, \lambda^x)$. If $\varphi = (\varphi^x)$ is an element of H and if f is a suitable function (see below) on G one defines just as in (6.2):

$$(\pi(f)\varphi)(u) = \int f(u^{-1}v)\varphi(v)d\lambda^x(v).$$

In order for this to define a bounded operator f has to satisfy conditions as in Hahn [Hap1].

Definition 6.15. A measurable function f on G equipped with a Haar system λ and a transverse measure ν is *left integrable* with respect to the Haar system λ if

$$\text{ess} \sup_u \int |f(u^{-1}v)| \, d\lambda^{r(u)}(v) < \infty,$$

the essential sup being taken with respect to $\tilde{\mu}$; f is *right integrable* if f^* is left integrable, $f^*(u) = \overline{f(u^{-1})}$, and *integrable* if left and right integrable. A function is *left (right, two sided) square integrable* if $|f|^2$ is left (right, two sided) integrable.

It is not hard to see that the integrable functions form a *-algebra under the same operations (6.1) we used to define $C_r^*(G)$. Note that the integrability conditions are not the same as f being in $L^2(G,\tilde{\mu})$, and that the condition depends only on the equivalence class, i.e. the null sets, of the transverse measure ν, and not on ν itself.

We observe that if f is integrable with respect to λ the operator $\pi_x(f)$ in $L^2(G^x,\lambda^x)$ is given by a kernel function which, when we unravel the respective definitions, is integrable in the sense of Proposition 1.14. Hence $\pi_x(f)$ defines a bounded operator with a norm that is essentially bounded in x by Proposition 1.14 and so defines a bounded operator $\pi(f)$ on $L^2(G,\tilde{\mu})$. Further, $f \longrightarrow \pi(f)$ is a *-homomorphism. We note parenthetically that if $f \in C_0(G)$, the continuous compactly supported functions on a locally compact topological groupoid with Haar system λ (or the replacement for $C_0(G)$ in the non-Hausdorff case), then the integrability conditions are satisfied with ordinary suprema instead of essential suprema.

Definition 6.16. The *von Neumann algebra* $W^*(G,\tilde{\mu})$ associated to the measured groupoid $(G,\tilde{\mu})$ is the weak closure of the *-algebra generated by the operators $\pi(f)$, for all integrable functions f.

This algebra is realized on the space $\tilde{H}$ and quite evidently commutes with the abelian algebra A_r of multiplication operators generated by all bounded measurable functions on G which depend only on the range r(u) of a point u in G.

The fact that elements of $W^*(G,\tilde{\mu})$ commute with A_r means by direct integral theory (cf. Takesaki [Tak2], IV, §8) that any $m \in W^*(G,\tilde{\mu})$ may be decomposed as a direct integral. Specifically the abelian algebra A_r on $\tilde{H}$ decomposes $\tilde{H}$ as a direct integral

$$\tilde{H} = \int_X H^x d\nu$$

where evidently $H^x = L^2(G^x, \lambda^x)$. Then any operator m commuting with A_r, and in particular any m in $W^*(G,\tilde{\mu})$ has a direct integral decomposition

$$m = \int m^x d\nu$$

where m^x is an operator on $L^2(G^x, \lambda^x)$. Conversely every bounded Borel field of operators $x \longrightarrow m^x$ on the Borel field of Hilbert spaces $L^2(G^x, \lambda^x)$ defines an operator that commutes with A_r.

Moreover for each $u \in G$, left translation L_u by u defines a bijection from $G^{s(u)}$ to $G^{r(u)}$ which maps $\lambda^{s(u)}$ to $\lambda^{r(u)}$. (This is the definition of invariance for λ.) Consequently left multiplication by u^{-1} gives rise to a unitary operator U_u which is a unitary equivalence of $H^{s(u)}$ to $H^{r(u)}$, and these evidently satisfy $U_u U_v = U_{uv}$. It is further easily verified that the convolution operators $\pi(f)$ which are dense in $W^*(G,\tilde{\mu})$ have the further property that their disintegration products above $\pi(f)^x$ satisfy

$$U_u \pi(f)^{s(u)} = \pi(f)^{r(u)} U_u \text{ for almost all u.}$$

Consequently the same holds for any $m \in W^*(G,\tilde{\mu})$, namely

(*) $$U_u m^{s(u)} = m^{r(u)} U_u \text{ for almost all u.}$$

The intuitive reason for this is that the m's are a kind of right convolution operator on the groupoid and so a sum of right translations. The U_u are left translation operators and right translations always commute with left translations. Finally it is true that $W^*(G,\tilde{\mu})$ is exactly the set of operators which commute with A_r and whose disintegration products satisfy the above relations (*).

There are some important subalgebras of $W^*(G,\tilde{\mu})$ that will occur. First of all if φ is a bounded measurable function on $(G,\tilde{\mu})$ with the property that $\varphi(u)$ depends only on the source $s(u)$ of u (so $\varphi(u) = \varphi'(s(u))$), then m_φ, multiplication by φ on $L^2(G)$ defines a bounded operator which evidently commutes with A_r and whose direct integral disintegration products m_φ^x on $L^2(G^x,\lambda^x)$ are multiplication operators by the function $\varphi'(s(v))$ $v \in G^x$. This field of operators evidently satisfies (*) because left translation by u^{-1} maps G_y^x to G_y^z where $u^{-1} \in G_z^x$. Hence m_φ defines an element of $W^*(G,\tilde{\mu})$. The set of such is evidently a von Neumann subalgebra of $W^*(G,\tilde{\mu})$, denoted A_s and isomorphic to $L^\infty(X)$. In case G is principal--that is, an equivalence relation--this is the usual diagonal subalgebra, and if the equivalence relation is countable, it is a Cartan subalgebra (Feldman-Moore [FM]) that plays a key role in the structure of $W^*(G,\tilde{\mu})$.

Another slightly larger subalgebra of $W^*(G,\tilde{\mu})$ which takes account of the holonomy is also useful. Let $E = \{u \in G, r(u) = s(u)\}$. Then E can be viewed as the union $\cup_y G_y^y$ of the (discrete) holonomy groups. If f is any Borel function on E which is not only bounded, but for which $\Sigma|f(u)|$ $(u \in G_y^y)$ is bounded in y, then right convolution by f restricted to G_y^y defines an operator $R(f)_y^x$ on $L^2(G_y^x)$ for any x because G_y^x is a principal homogeneous space on the right for G_y^y. Then as $L^2(G^x)$ can be regarded as the direct integral

$$L^2(G^x) = \int L^2(G_y^x) d\lambda^x(y)$$

$R(f)_y^x$ integrates to give an operator $R(f)^x$ on $L^2(G^x)$. For exactly the same reasons as above, this field satisfies (*) and so defines an element R(f) of $W^*(G,\tilde{\mu})$. Suppose that f happens to be supported on the subset of $E = \cup G_y^y$ consisting of the identity elements of the (discrete) groups G_y^y. Then f is in effect a function on the unit space X, and $\varphi(u) = f(s(u))$ defines a function on $\tilde{G}$ depending only on s(u) that in turn defines an element m_φ of the algebra A_s. A moment's thought shows that $m_\varphi = R(f)$. The closure D_s of the set of operators R(f), which is evidently a von Neumann subalgebra of $W^*(G,\tilde{\mu})$ contains A_s. This generalized diagonal subalgebra D_s has a readily apparent structure.

Proposition 6.17. The algebra D_s is a direct integral

$$D_s = \int_X R^x d\nu(x)$$

of the right group von Neumann algebras R^x of the discrete groups G_x^x with $A_s \cong L^\infty(X)$ the obvious subalgebra. □

The algebra $W^*(G,\tilde{\mu})$ on the Hilbert space $\tilde{H}$ is in standard form (cf. Takesaki [Tak1]) in the sense that there exists a conjugate linear isometry J of $\tilde{H}$ onto $\tilde{H}$ such that $J^2 = id$, and

$$JW^*(G,\tilde{\mu})J = W^*(G,\tilde{\mu})'$$

where N' denotes the commutant of N in $\mathcal{B}(\tilde{H})$. In fact the J that works is quite easy to write down. Recall from Chapter IV that the transverse measure ν, which we started with here, has a modular function, or modulus, Δ which is a positive function on the groupoid G satisfying $\Delta(u)\Delta(v) = \Delta(uv)$ whenever uv is defined. This function measures the extent that ν is not an invariant transverse measure. This modular function has the further property that if $i(v) = v^{-1}$ is the inversion map on G, then i transforms the measure $\tilde{\mu}$ on G into a measure $i_*(\tilde{\mu})$ $(i_*(\tilde{\mu}) = \mu_s$ in the language of Chapter IV) which is equivalent to $\tilde{\mu}$ with Radon–Nikodym derivative given by

$$\frac{d\tilde{\mu}}{d(i_*(\mu))}(v) = \Delta(v).$$

It follows then that for $\varphi \in \widetilde{H} = L^2(G, \widetilde{\mu})$,

$$(J\varphi)(u) = \bar{\varphi}(u^{-1})\Delta(u^{-1})^{1/2}$$

defines a conjugate linear involutive isometry. The following summarizes results that can be found in Hahn [Hap1], Connes-Takesaki [CT], but also see Takesaki [Tak1].

Theorem 6.18. With the above notation

$$JW^*(G, \widetilde{\mu})J = W^*(G, \widetilde{\mu})'$$

and $W^*(G, \widetilde{\mu})$ is the algebra of all operators m commuting with A_r so that the corresponding disintegration products m^x satisfy

$$U^u m^{s(u)} = m^{r(u)} U^u$$

for almost all $u \in G$. Moreover $JA_s J = A_r$.

We shall not go into the somewhat tedious details of the proof; the idea is that what works for groups works for groupoids. The algebra $W^*(G, \widetilde{\mu})$ consists of right convolutions and conjugation by J makes them into left convolutions which at the von Neumann algebra level are each others commutants. As to the second part, A_r is clearly in the commutant of $W^*(G, \widetilde{\mu})$ and A_r together with "smoothed" versions of the U^u suffice to generate this commutant. □

It is evident from the definitions that once we fix a Haar system λ, the C^* algebra $C_r^*(G)$ has a natural representation into $W^*(G, \widetilde{\mu})$ for any choice of transverse measure ν and corresponding $\widetilde{\mu}$. Recall that $C_r^*(G)$ is defined by representations π_x of a dense subalgebra A. These representations π_x take place on $H^x = L^2(G^x, \lambda^x)$ and so the direct integral π of the π_x gives a representation of $C_r^*(G)$ into $W^*(G, \widetilde{\mu})$. The following is clear.

Proposition 6.19. The image of $C_r^*(G)$ is dense in $W^*(G, \widetilde{\mu})$ and the representation is faithful if the support of the transverse measure ν

is all of X in the sense that the measure $\int \lambda d\nu$ has support equal to X. □

It is well to reflect for a moment on the geometric meaning of these algebras for a topological groupoid G. First of all each orbit or leaf ℓ of the equivalence relation on X associated to G has a "holonomy covering" $\tilde{\ell}$ which we can take to be G^x for any $x \in \ell$. The left translations L_u which map $G^{s(u)}$ to $G^{r(u)}$ provide canonical identifications between these models of $\tilde{\ell}$ when $x = s(u)$, and $y = r(u)$ are points of ℓ. In addition the "holonomy group" G^x_x operates by left translation freely on $G^x \simeq \tilde{\ell}$, and the quotient $G^x_x \backslash \tilde{\ell}$ is exactly the original leaf ℓ. Each leaf ℓ and its covering $\tilde{\ell}$ come equipped with a measure so we have Hilbert spaces $L^2(\tilde{\ell})$ which are just $L^2(G^x, \lambda^x)$ for any $x \in \ell$.

Then elements of $W^*(G, \tilde{\mu})$ can be thought of as providing for almost all holonomy coverings $\tilde{\ell}$ an operator $m(\tilde{\ell})$ on $L^2(\tilde{\ell})$. These are supposed to be bounded and to vary in a Borel way with $\tilde{\ell}$. The exact meaning of the last statement is that when we identify $L^2(\tilde{\ell})$ with $L^2(G^x, \lambda^x)$ for any $x \in \ell$ and get a field of operators m^x, then the m^x are Borel sections of the Borel field of Hilbert spaces $L^2(G^x, \lambda^x)$ over X. The commuting relations in the second part of Theorem 6.18 say in part that whether we identify $L^2(\tilde{\ell})$ with $L^2(G^x, \lambda^x)$ or with $L^2(G^y, \lambda^y)$ with $x, y \in \ell$, we get the same operator on $L^2(\tilde{\ell})$. Finally $m(\tilde{\ell})$ is not an arbitrary operator on $L^2(\tilde{\ell})$ but the commuting relations in 6.18 say also that $m(\tilde{\ell})$ must commute with left translation by G^x_x, and that these are the only restrictions.

Elements of the C^*-algebra $C^*_r(G)$ have a very similar interpretation. Each m in this algebra defines an operator $m(\tilde{\ell})$ on all (not almost all) holonomy coverings of the leaves which commutes with left translation by G^x_x and which is further restricted to be a uniform limit of such operators that can be defined by convolution with suitable continuous kernel functions. Finally the $m(\tilde{\ell})$ have to vary continuously as $\tilde{\ell}$ varies in a manner that is fairly clear heuristically.

It is evident that the von Neumann algebra $W^*(G, \tilde{\mu})$ depends only on the equivalence class, in the sense of absolute continuity, of the measure μ on the unit space X of G because of its definition in terms of fields of operators. In turn the equivalence class of μ depends only on the equivalence class of the transverse measure ν from which it is constructed (regarding of course the Haar system $\{\lambda^x\}$ as fixed once and for all). The Hilbert space $\tilde{H}$ upon which we have realized $W^*(G,\tilde{\mu})$ of course depends on μ itself but for two equivalent μ's there is a natural unitary equivalence of the two spatial realizations of the algebra. For simplicity we sometimes write $W^*(G)$ where the Haar system and the equivalence class of transverse measures entering into the definition are understood.

If N is a complete transversal for the equivalence relation on X defined by the groupoid G, then as in Chapter IV, the transverse measure ν defines a measure on N, and the part of G over N, G_N^N becomes a measured groupoid $(G_N^N, \tilde{\mu}_N)$ whose orbits are countable. There should be a close relation between $W^*(G,\tilde{\mu})$ and $W^*(G_N^N, \tilde{\mu}_N)$ paralleling Theorem 6.14 and indeed there is. For convenience we assume that the tangential measure $\{\lambda^x\}$ on G^x that we are given at the very beginning of the discussion has the property that all (or almost all) the measures λ^x have no atoms. This will surely be the case for the groupoid of a foliated space with the usual choice of tangential measures. In this case, the arguments of Theorem 5.6 of Feldman-Hahn-Moore [FHM], trivially modified to cover the case of non-principal groupoids, shows that as a measured groupoid $(G,\tilde{\mu})$ is isomorphic to $(G_N^N, \mu_N) \times \mathcal{Q}$ where $\mathcal{Q}$ is the principal groupoid (equivalence relation) with unit space the interval $I = [0,1]$ with all points equivalent and with the measure μ on I Lebesgue measure, the measure on each leaf also Lebesgue measure. With this structural result for G the following is clear.

Proposition 6.20. Under the conditions above there is an isomorphism

$$W^*(G,\tilde{\mu}) \cong W^*(G_N^N, \tilde{\mu}_N) \otimes \mathcal{B}(L^2(I)).$$

The importance and usefulness of this result is that it allows most questions about $W^*(G,\tilde{\mu})$ to be reduced to questions about $W^*(G_N^N, \tilde{\mu}_N)$. Since G_N^N has countable orbits, the structure and properties of the algebra built on it is far more understandable and transparent, and there are far fewer technical details to wrestle with. In particular an operator in the algebra is represented by a "matrix" over G_N^N (that is, formally it is of the form $\pi(f)$ for a function f on G_N^N), and the study of unbounded weights on $W^*(G, \tilde{\mu})$ will often reduce to the study of (bounded) states on $W^*(G_N^N, \tilde{\mu}_N)$.

It is evident that the abelian and diagonal subalgebras A_s and D_s of $W^*(G, \tilde{\mu})$ introduced above decompose naturally with respect to the tensor product decomposition. Let A_s^N and D_s^N be the corresponding abelian and diagonal subalgebras of $W^*(G_N^N, \tilde{\mu}_N)$.

Proposition 6.21. In the decomposition of Proposition 6.20 we have isomorphisms

$$A_s \simeq A_s^N \otimes L^\infty(I)$$
$$D_s \simeq D_s^N \otimes L^\infty(I)$$

where $L^\infty(I)$ is the subalgebra of $\mathcal{B}(L^2(I))$ consisting of multiplications by bounded measurable functions. $\square$

One example of the usefulness of the reduction to a cross section is the following which of course could be established directly but less transparently.

Proposition 6.22. The relative commutant of A_s in $W^*(G, \tilde{\mu})$ is D_s, and the relative commutant of D_s is the center of D_s, which in the direct integral decomposition of Proposition 6.17 is the direct integral of the centers Z^X of the right group von Neumann algebras R^X of the holonomy groups G_x^X. In particular if almost all of the holonomy groups are infinite conjugacy class (i.c.c.) groups then the relative centralizer of D_s is A_s.

Proof. By the previous proposition, the question is reduced to A_s^N and D_s^N. All operators are given by "matrices" as in [FM]; then easy computation in this discrete case does the trick. As to the final statement, recall that a discrete group H is i.c.c. (all non-trivial conjugacy classes are infinite) if and only if the center of the group von Neumann algebra is trivial. $\square$

The next step begins with the crucial observation that the algebra $W^*(G)$ comes with a natural family of normal semi-finite weights. Indeed each (positive) transverse measure ν in the fixed equivalence class will define in a natural way a weight ϕ_ν on $W^*(G)$; this weight will be a trace if and only if ν is an *invariant* transverse measure. There are several different ways to define these weights; one way starts by utilizing the natural Hilbert algebra structure that is implicit in the construction of $W^*(G,\tilde{\mu})$ and uses the basic Tomita-Takesaki construction of weights from a Hilbert algebra (cf. Takesaki [Tak1]). We will rather approach the matter through the ideas developed in Chapter I of locally traceable operators; we can give a very simple and direct definition as follows.

Suppose given a transverse measure ν and associated von Neumann algebra $W^*(G,\tilde{\mu})$. We wish to define ϕ_ν on the positive part $W^*(G,\tilde{\mu})^+$ and taking values in $[0,\infty]$. Here is a rough idea of the construction of the weight. To each $m \in W^*(G,\tilde{\mu})^+$ we shall associate a tangential measure λ_m which has the property that if one decomposes m to a field of operators m^x on $L^2(G^x, \lambda^x)$, then the local trace of m^x determines the measure λ_m^x on $G_x^x \backslash G^x \equiv \ell(x)$ uniquely. Then the weight ϕ_ν corresponding to the transverse measure ν is given by

$$\phi_\nu(m) = \int \lambda_m(\ell) d\nu(\ell)$$

where the integral is taken in the sense of Chapter IV. Now here are the details.

Any $m \in W^*(G,\tilde{\mu})^+$ corresponds to a field of positive operators $m(\tilde{\ell})$, one for almost all holonomy coverings $\tilde{\ell}$ or equivalently a field m^x of positive operators on $L^2(G^x, \lambda^x)$ for almost all x. Then since m^x is positive we can define its local trace as a positive measure $Tr(m^x)$ on G^x. This measure may be identically plus infinity. At all events it is defined even in this degenerate sense and recall that our definition of m^x being locally traceable was that this measure should be σ-finite (or Radon if G^x comes with a locally compact topology). These measures are always absolutely continuous with respect to λ^x by their definition. The invariance properties satisfied by the m^x as stated in 6.18 imply by the analysis in Chapter I that left translation L_u which maps $G^{s(u)}$ to $G^{r(u)}$ must transform $\text{Tr}(m^{s(u)})$ into $\text{Tr}(m^{r(u)})$ for almost all $u \in G$. Thus for almost all pairs x,y with x ~ y, $\text{Tr}(m^x)$ on G^x is the same as $\text{Tr}(m^y)$ on G^y after identifying G^x and G^y. Moreover the countable group G^x_x acts by left translation on G^x and hence on $L^2(G^x, \lambda^x)$ and m^x commutes with these translations. Again by Chapter I, $\text{Tr}(m^x)$ is invariant under G^x_x and hence $\text{Tr}(m^x)$ uniquely determines a measure $\text{Tr}'(m^x)$ on $G^x_x \backslash G^x$. But this quotient space is just the equivalence class $\ell(x)$ of x. Hence for each leaf ℓ, and each $x \in \ell$ we obtain a positive measure $\text{Tr}'(m^x)$ on ℓ. The invariance properties cited above tell us that this measure does not depend on which x we choose and depends only on the leaf ℓ; we denote it by $\lambda_m(\ell)$.

This description is simpler if there is no holonomy so that G is an equivalence relation. Then G^x is the equivalence class or leaf of x, and the local trace of m^x gives a measure $\text{Tr}(m^x)$ on G^x; invariance properties say that $\text{Tr}(m^x) = \text{Tr}(m^y)$ and so there is a measure $\lambda_m(\ell)$ depending only on the leaf ℓ; this can be thought of as the local trace of $m(\ell)$ for all or almost all ℓ. But now $\lambda_m(\ell)$ is what we called a tangential measure and it is the sort of object that can be integrated against a transverse measure to give a number.

Proposition 6.23. For every $m \in W^*(G, \widetilde{\mu})^+$, the above prescription yields a tangential measure $\lambda_m(\ell)$ (perhaps not σ-finite). The integral in the sense of Chapter IV

$$\phi_\nu(m) = \int \lambda_m(\ell) d\nu(\ell)$$

(finite or not) defines a semi-finite normal weight on $W^*(G, \widetilde{\mu})$.

Proof. For the assignment of a measure $\lambda_m(\ell)$ to each leaf to be a tangential measure, it must satisfy some smoothness conditions transversally. From Chapter IV we see that these amount to the requirement that the field of measures $\lambda^x_m = \text{Tr}(m^x)$ on G^x should be Borel viewed as measures on G in that

$$\int f(u) d\lambda^x$$

should be a Borel function of x for any non-negative Borel function f on G. This is clearly satisfied by the local traces of the Borel field of operators m^x. We note once more that m^x is not assumed to be locally traceable in the sense that λ^x is a σ-finite measure. The integral we write down in the statement still always makes sense as everything is non-negative. It is clear that ϕ_ν as defined is additive and positively homogeneous. That it is normal is clear from the properties of the local trace and the integration process of Chapter IV. Equivalently it is not hard to produce a family of vectors ξ_i in the Hilbert space $\tilde{H}$ such that $\phi_\nu(m) = \Sigma(m\xi_i, \xi_i)$, which is an equivalent definition of normality. Finally the dense subalgebra used in Hahn [Hap] to define the algebra $W^*(G, \tilde{\mu})$ synthetically contains a weakly dense set of operators where ϕ_ν is evidently finite so that ϕ_ν is semi-finite. □

One of the features of this definition is that it is clear for which positive operators ϕ_ν is finite.

Corollary 6.24.   Let $m \in W^*(G, \tilde{\mu})^+$.   Then $\phi_\nu(m) < \infty$ if and only if m^x is locally traceable on almost all G^x in the sense that $\lambda^x_m = \mathrm{Tr}(m^x)$ is a σ-finite measure; if so, then the integral

$$\int \lambda_m(\ell) d\nu(\ell)$$

is finite.   □

If an operator $a \in W^*(G, \tilde{\mu})$ is given by a kernel function f so that $a = \pi(f)$ and

$$(\pi(f)\psi)(u) = \int f(u^{-1}v)\psi(v) d\lambda^x(v) \qquad u \in G^x, \ \psi \in L^2(G^x, \lambda^x)$$

with f integrable in the sense of Chapter IV, then we can give an alternate formula for $\phi_\nu(a)$. If $b = a^*a$ then b is given as $\pi(g)$, where

$$g(u) = f * f(u) = \int \bar{f}(v^{-1}) f(v^{-1}u) d\lambda^{r(u)}(v)$$

according to formula 6.1. If $x \in X$, the unit space of G, then x can be thought of as an element of G and to keep matters straight let us call this element e(x). (In case G is an equivalence relation on X, $e(x) = (x,x)$ is a diagonal element.) Now although f and g above are measurable functions on G defined only almost everywhere and as the units e(X) form a null set in G, the restriction of g to e(X) appears to have no sense. However if $u = e(x)$ is a unit, then

$$g(e(x)) = \int | f(v^{-1}) |^2 d\lambda^x(v)$$

has a well defined meaning for almost all x. When we write g(e(x)) for a g of the form $f^* f$, it is this function that we shall understand. The following shows that as one expects, traces of integral operators are obtained by integrating the kernel on the diagonal.

Proposition 6.25. For a transverse measure ν, let $\mu = \int \lambda^\ell d\nu$ be the integral of the tangential measure λ with respect to ν, the result viewed as a measure on the unit space X. For an operator $b = \pi(g) \in W^*(G, \tilde{\mu})$ with $g = f^* f$, then $\phi_\nu(b) = \int g(e(x)) d\mu(x)$ where g(e(x)) is as defined above. Equivalently $g(e(x)) \cdot \lambda$ defines a new tangential measure λ' whose derivative with respect to λ is g(e(x)). Then

$$\phi_\nu(b) = \int_X \lambda' d\nu$$

(the integral of λ' with respect to ν).

Proof. This is simply a matter of identifying the tangential measure λ' (or rather $(\lambda')^X$ as a measure on G^X for each x) as the local trace of the operator $\pi_x(f^* f)$ on $L^2(G^X, \lambda^X)$; this is self evident as $\pi_x(f^* f) = \pi_x(f)^* \pi_x(f)$ where $\pi_x(f)$ is given by a kernel defined by the function f. Then the result follows. □

189

By the general Tomita-Takesaki theory (cf. Takesaki [Tak1]), any semi-finite normal faithful weight φ on a von Neumann algebra has associated to it a one parameter group of automorphisms of the algebra, the so called *modular automorphism group*, $\sigma_\varphi(t)$. The standard construction of this group via unbounded operators can be exploited easily to construct this group explicitly for the weights ϕ_ν above. (These weights will always be normal, faithful and semi-finite as the transverse measure ν was restricted to lie in the same equivalence class that defines the von Neumann algebra itself.) This is worked out in Feldman-Moore [FM], Hahn [Hap], Connes-Takesaki [CT].

Proposition 6.26. Let Δ be the modular function of the transverse measure ν (cf. Definition 4.9). Then the modular automorphism group σ_ν associated to the weight ϕ_ν of $W^*(G,\tilde{\mu})$ is spatially implemented by the one parameter group of unitary operators $U_\nu(t)$ on $L^2(G,\tilde{\mu})$ defined by multiplication by the functions Δ^{it} on G. Thus

$$\sigma_\nu(t)m = U_\nu(t)mU_\nu(-t) \text{ for } m \in W^*(G,\tilde{\mu}).$$

Moreover for operators of the form $\pi(f)$ in $W^*(G,\tilde{\mu})$ (cf. Definition 6.15)

$$\sigma_\nu(t)\pi(f) = \pi(f\Delta^{it})$$

where $f\Delta^{it}$ is pointwise multiplication of f and Δ^{it}.

Proof. The operators $\pi(f)$ form a Hilbert algebra with the * operator given very concretely by $f^*(u) = \overline{f(u^{-1})}$. One then easily computes the polar decomposition of the unbounded conjugate linear operator $f \longrightarrow f^*$ and following the standard recipe in Takesaki [Tak1], one finds the result. The final formula is a simple calculation. □

Recall that the centralizer of a weight ϕ on a von Neumann algebra R is equivalently the von Neumann subalgebra generated by those unitaries u in the algebra such that $\phi(uxu^*) = \phi(x)$, or equivalently it is the fixed point algebra of the modular automorphism group (Pedersen [Ped], Lemma 8.14.6). A weight is a trace if and only if its centralizer is the entire algebra. As the modular automorphism group σ_ν of the weight ϕ_ν on $W^*(G,\tilde{\mu})$ is given explicitly and clearly fixes the diagonal subalgebra D_s of $W^*(G,\tilde{\mu})$ (Proposition 6.22), the first half of the following is immediate.

Proposition 6.27. The centralizer of ϕ_ν contains the diagonal subalgebra D_s. Conversely if almost all of the holonomy groups are i.c.c. (cf. Proposition 6.22) then any faithful normal semi-finite weight whose centralizer contains D_s is of the form ϕ_ω for some transverse measure ω.

Proof. For the second part we fix a weight ϕ_ν and let ψ be any other faithful normal semi-finite weight with centralizer containing D_s. Then compute the Radon-Nikodym derivative $(\psi:\phi_\nu)_t$ (Connes [Co1], or cf. Takesaki [Tak1], p. 23). This is a one parameter family of unitary operators in $W^*(G,\tilde{\mu})$ satisfying a certain cocycle condition. Since D_s centralizes both ψ and ϕ_ν, it follows that $(\psi:\phi_\nu)_t$ must commute with D_s for each t. But under the condition on G_x^x, the relative commutant of D_s is by Proposition 6.22 the abelian subalgebra A_s. Because of commutation properties, the derivative $(\psi:\phi_\nu)_t$ is actually a one parameter unitary group in A_s and so has the form $\exp[ith(x)]$ where h is a measurable function on X, which by positivity properties of ψ and ϕ_ν is positive. Then $w = h\nu$ is another transverse measure, and it is evident that $\psi = \phi_{h\nu}$. $\square$

The argument just given provides an answer in general to the question of finding all weights whose centralizer contains D_s, but one has to introduce an extended class of weights. As we will not need this, we sketch this only briefly. Suppose that in addition to a transverse measure ν on X one is given for each $x \in X$, a semi-finite normal faithful trace τ^x on R^x, the group von Neumann algebra of G_x^x. Then one can construct in an obvious way a trace τ

on the diagonal algebra D_s because D_s is given as a direct integral of the algebras R^X. The transverse measure ν itself also defines a weight τ_ν on D_s. The Radon-Nikodym derivative $(\tau:\tau_\nu)$ computed in D_s can then be used to define a weight ϕ on $W^*(G,\tilde{\mu})$ by the condition $(\tau:\tau_\nu) = (\phi:\phi_\nu)$ (computed in $W^*(G,\tilde{\mu})$).

The weights ϕ constructed in this fashion are, we claim, the most general weights on $W^*(G,\tilde{\mu})$ with centralizer containing D_s. The data entering into ϕ, namely a transverse measure ν and a family of traces τ^X on R^X are not independent for we can multiply each τ^X by a positive scalar $c(x)$, replace ν by the transverse measure ν' with $d\nu'/d\nu = c(x)^{-1}$, and the resulting weight will be the same. When the traces τ^X are finite, then they can be normalized so $\tau^X(1) = 1$ and then the transverse measure ν is determined. Of course when τ^X is taken to be the Plancherel trace, then the resulting weight is ϕ_ν that we constructed previously. It is evident that values of the more general weights discussed in this paragraph can be given by integral formulas analogous to those in Propositions 6.23 and 6.25. In addition it is not difficult to compute the modular automorphism group of these weights because there is a simple formula for the Radon-Nikodym derivative of these with respect to a ϕ_ν where we already know the modular automorphism group.

Returning to the ϕ_ν, we see that we have determined when ϕ_ν is a trace because this is true if and only if the modular automorphism group is trivial.

Corollary 6.28. The weight ϕ_ν is a trace if and only if ν is an invariant transverse measure, that is, its modular function Δ is identically one almost everywhere. □

It is not so easy to tell when the more general weights defined by fields of traces τ^X together with a ν are traces because in general it is hard to determine what the center of $W^*(G,\tilde{\mu})$ is.

To conclude this chapter let us return to the topological and geometric context of a locally compact topological groupoid G, or in particular the holonomy groupoid of a foliated space. As before G is assumed to come equipped with a fixed continuous tangential measure.

Then for any transverse measure ν, the reduced C^* algebra $C_r^*(G)$ has a natural representation into $W^*(G,\tilde{\mu})$ as described in Proposition 6.19. The weight ϕ_ν may be restricted then to the image $C_r^*(G)$ to produce a weight on this C^*-algebra, which we denote by the same symbol. If the transverse measure ν is *finite* relative to the tangential measure λ in the sense that $\mu = \int \lambda d\nu$ is a finite measure on the unit space X of G (and in particular if it is a Radon transverse measure on the groupoid of a foliation in the sense of 4.17) then the restriction of ϕ_ν to $C_r^*(G)$ enjoys finiteness properties. In particular for any g in $C_c(G)$, the norm dense subalgebra of compactly supported functions on G used in the definition of $C_r^*(G)$, the positive element $f = g^*g$ satisfies $\phi_\nu(f) < \infty$ in view of Proposition 6.25 or Corollary 6.24. This finiteness property plus the known continuity properties of ϕ_ν on $W^*(G,\tilde{\mu})$ assure that ϕ_ν as a weight on $C_r^*(G)$ is densely defined and lower semi-continuous (Pedersen [Ped], 5.6.7). Quite evidently we can recapture the von Neumann algebra $W^*(G,\tilde{u})$ from $C_r^*(G)$ and ϕ_ν via the GNS construction as the image of $C_r^*(G)$ is dense in $W^*(G,\tilde{\mu})$ by Proposition 6.19.

If ν is an invariant transverse measure, then ϕ_ν is of course a trace on $C_r^*(G)$, and as $C_r^*(G)$ is dense in $W^*(G,\mu)$, the converse is true. Thus Corollary 6.28 and Corollary 4.25 combine to yield the following corollary in the setting of foliated spaces.

Corollary 6.29. For a Radon transverse measure ν on a compact foliated space X with continuous tangentially smooth modular function Δ, the following are equivalent:

(1)   The Ruelle-Sullivan current C_ν is closed and so defines $[C_\nu] \in H_p^7(X;\mathbb{R})$.

(2)   The 1-form $\sigma = 0$.

(3)   The modular function $\Delta \equiv 1$.

(4)   The transverse measure ν is an invariant transverse measure.

(5)   The weight ϕ_ν on $W^*(G(X),\tilde{\mu})$ is a trace.

In general, $C_r^*(G)$ will have traces other than ϕ_ν; for instance if G is the holonomy groupoid of the Reeb foliation, the closed leaf and its holonomy produces a quotient isomorphic to $C^*(\mathbf{Z}^2) \otimes \mathcal{K}$ where $C^*(\mathbf{Z}^2)$ is the group C^* algebra of $\mathbf{Z}^2$ and $\mathcal{K}$ is the compact operators. The only ϕ_ν which factors through this quotient comes by taking ν to be the transverse measure corresponding to the closed leaf; then ϕ_ν is $P \otimes Tr$ where P is the Plancherel trace on $C^*(\mathbf{Z}^2)$.

However in the absence of holonomy, traces are always given, as one suspects, by transverse measures.

Theorem 6.30.   Let G be the groupoid of a compact foliated space X and assume there is no holonomy (so that G is the equivalence relation). If ϕ is any densely defined lower semi-continuous trace on the C^* algebra $C_r^*(G)$, then there is a unique invariant transverse Radon measure ν on X with $\phi = \phi_\nu$.

Proof.   We pick a complete transversal N and an open neighborhood U of it as in the discussion preceding Theorem 6.14. We make use of the structural fact that $C_r^*(G) \cong C_r^*(G_N^N) \otimes \mathcal{K}$, and we recall that any densely defined lower semi-continuous trace is finite on the Pedersen ideal--the unique minimal dense two-sided ideal (cf. Pederson [Ped], Theorems 5.6.1, 5.6.7). As this ideal intersects any subalgebra in a dense ideal, it follows that ϕ is densely defined on the subalgebra $C_r^*(G_N^N) \otimes e \simeq C_r^*(G_N^N)$ where e is a minimal projection in $\mathcal{K}$. Finally since the equivalence relation when restricted to N is discrete, $C_r^*(G_N^N)$ contains a Cartan subalgebra $C_0(N)$ by the remarks following Definition 6.10. For the same reasons as above, ϕ is densely defined on $C_0(N)$ and so is given by a Radon measure ν on N. Moreover by the construction of N, there is a larger transversal N' containing N with the closure $\bar{N}$ of N in N' compact. As the measure ν is by the same reasoning the restriction of a Radon measure ν' on N' it follows that ν is a finite measure on N. Since $C_0(N)$ contains an approximate identity for $C_r^*(G_N^N)$ and ϕ remains bounded on this approximate identity, it follows that ϕ is a finite trace on $C_r^*(G_N^N)$.

Since there is no holonomy, G_N^N is an equivalence relation on N. We know by Proposition 6.8 that G_N^N has a covering by open sets of the form $U(f,O) = \{(x,f(x)), x \in O\}$ where O is an open set in N and f is a homeomorphism of O onto an open subset of N with $f(x) \sim x$ where $\sim$ is the equivalence relation on N. As the diagonal ΔN of N in G_N^N is open and closed, its complement may be covered by sets of the form $U(f,O)$ where f has no fixed points. If a is any compactly supported function on $U(f,O)$ and b any compactly supported function on $\Delta N = U(id,N)$, then viewed as elements in $C_r^*(G_N^N)$ their convolution products in both orders are again compactly supported on open sets $U(f,O)$ and

$$(a^*b - b^*a)(x,f(x)) = a(x,f(x)) \ \{b(f(x),f(x)) - b(x,x)\}$$

Since $\phi(c)$, for c compactly supported in $U(f,O)$ can be expressed as

$$\phi(c) = \int c d\lambda$$

for a Radon (signed) measure on $U(f,O)$, the equality $\phi(a^*b - b^*a) = 0$ plus the fact that b has no fixed points tells us that λ is zero. As any compactly supported function on $G_N^N - \Delta N$ can be written as a finite sum of functions supported on open sets $U(f,O)$, it follows that

$$\phi(a) = \int a(x,x) d\nu(x)$$

for every compactly supported function on G_N^N where ν is the measure on N constructed above.

An argument similar to the one above shows that ν as a measure on N is invariant under the equivalence relation; that is, its modular function on N is trivial. Then, as in Chapter IV, ν can be extended to all Borel transversals to give an invariant Radon transverse measure, which we denote by ν. Then clearly

$\phi = \phi_\nu$ on $C_r^*(G_N^N)$ and hence on $C_r^*(G)$. □

We have seen in Chapter IV that von Neumann factors of type II_∞ and of type III_λ for all λ occur as the von Neumann algebras of foliated spaces. Proposition 6.20 shows that with minimal assumptions on tangential measures, the von Neumann algebra has the form $\widetilde{W} \otimes B(H)$ for an infinite-dimensional Hilbert space H. We also have seen that the von Neumann algebra comes equipped with a family of semi-finite normal faithful weights, with corresponding modular automorphism groups. Given this much structure, it is natural to wonder just which von Neumann algebras can occur as the von Neumann algebra of a foliated space. Here is the answer.

Theorem 6.31. Any purely infinite approximately finite von Neumann algebra A is isomorphic to $W^*(X,\mu)$ for some compact foliated space X and transverse measure μ.

Proof. According to the classification of such algebras (Connes [Co1], Haagerup [Ha2], Kreiger [Kr]) one may find a Borel space Y, an automorphism ϕ of Y (so that there is an associated action of $\mathbb{Z}$ on Y), and a transverse measure μ_0 so that the group measure construction associated to these data produces a von Neumann algebra A_0 so that

$$A \cong A_0 \otimes B(H).$$

Equivalently, if G is the measure groupoid generated by (Y,ϕ,μ), then A_0 is the von Neumann algebra of this measure groupoid as defined in Chapter VI.

Now according to Theorem 3.2 of Varadarajan [Var] we may assume without loss of generality that Y is a compact metric space and that the map ϕ is a homeomorphism. Form the associated compact foliated space X obtained by suspending (Y,ϕ), and let μ be the associated transverse measure on X constructed from μ_0 as in Chapter IV. Then the von Neumann algebra of (X,μ) is A as desired. □

We note that we have proved more than we stated, for the foliated space produced is always of leaf dimension 1. There are obvious questions which arise in this connection. May Y be chosen to be zero-dimensional? May Y be chosen to be a smooth manifold and ϕ a diffeomorphism, so that X is a smooth manifold? We do not know the answers to these questions.

We conclude this chapter with a brief discussion of some aspects of the K-theory of operator algebras in the context of the C^*-algebras of groupoids. In the following chapter, K-theory will enter in a more extended fashion. We assume that the reader is familiar with the basics of the K-theory of operator algebras (cf. Karoubi [Kar], Atiyah-Singer [ASI] and especially Blackadar [Bl]). Recall that for a unital C^* algebra A, one looks at all projections in $\cup_n M_n(A)$ ($M_n(A)$ is the n × n matrices over A) and subjects them to the natural equivalence relation that e ~ f if there are $u,v \in \cup_n M_n(A)$ with uv = e, vu = f. These classes form a semi-group, and one forms the associated Grothendieck group which is denoted $K_0(A)$. One may think of it as classes of formal differences of projections. If A does not have a unit, append one to obtain A^+, compute $K_0(A^+)$ as above, and note that the natural homomorphism e: $A^+ \longrightarrow \mathbb{C}$ induces a homomorphism $e_*: K_0(A^+) \longrightarrow K_0(\mathbb{C})$ where the latter group is easily seen to be isomorphic to the integers. Then define $K_0(A)$ to be the kernel of e_*. For a compact space X, $K_0(C(X))$ is the usual topological K-theory of compact spaces $K^0(X)$. For X locally compact, $K_0(C_0(X))$ is the usual K-theory of the space X with compact supports (cf. Atiyah-Singer [ASI], Karoubi [Kar]). We define $K_1(A) = K_0(SA)$ where $SA = C_0((0,1),A)$. Then Bott periodicity asserts that $K_j(A) \cong K_j(S^2 A)$.

We recall two further properties of K-theory. First, $K_*(A)$ is homotopy-invariant; that is, if $f^t: A \longrightarrow A'$ is a 1-parameter family of *-homomorphisms (continuous in the sense that the associated map A $\longrightarrow C([0,1],A')$ is a *-homomorphism) then

$$f_*^0 = f_*^1: K_*(A) \longrightarrow K_*(A').$$

Second, if J is a closed ideal of A then there is a natural long exact sequence

$$
\begin{array}{ccccc}
K_0(J) & \longrightarrow & K_0(A) & \longrightarrow & K_0(A/J) \\
\uparrow & & & & \downarrow \\
K_1(A/J) & \longleftarrow & K_1(A) & \longleftarrow & K_1(J) .
\end{array}
$$

The group $K_0(C_r^*(G))$ is going to be a central player in index theory and will be the group where the index lives. If G is the groupoid of a compact manifold foliated by a single leaf, $C_r^*(G) = \mathcal{K}$ is the compact operators and it is well known and easily seen that $K_0(\mathcal{K}) = \mathbb{Z}$, and the usual index of an elliptic operator is interpreted as an element of this group.

Let $\mathcal{P}$ denote the C^*-algebra of norm limits of pseudodifferential operators of order ≤ 0 (say, with matrix coefficients) on a compact manifold M. There is a natural sequence of C^*-algebras

$$
0 \longrightarrow \mathcal{K} \longrightarrow \mathcal{P} \overset{\pi}{\longrightarrow} C(S^*M) \otimes M_n \longrightarrow 0
$$

where S^*M is the cosphere bundle. If $P \in \mathcal{P}$ is elliptic with principal symbol σ, then $\pi(P) = \sigma$ and $[\sigma] \in K_1(C(S^*M) \otimes M_n) \cong K^{-1}(S^*M)$. The boundary map

$$
\partial : K_1(C(S^*M) \otimes M_n) \longrightarrow K_0(\mathcal{K})
$$

corresponds to the Fredholm index map

$$
\partial[\sigma] = \text{index } (P)
$$

as may be seen easily by a naturality argument involving

$$
\begin{array}{ccccccccc}
0 & \longrightarrow & \mathcal{K} & \longrightarrow & \mathcal{P} & \longrightarrow & C(S^*M) \otimes M_n & \longrightarrow & 0 . \\
& & \| & & \downarrow & & \downarrow & & \\
0 & \longrightarrow & \mathcal{K} & \longrightarrow & \mathcal{B}(H) & \longrightarrow & \mathcal{B}(H)/\mathcal{K} & \longrightarrow & 0
\end{array}
$$

198

If G is the groupoid of a foliation coming from a fibration with base B, $C_r^*(G)$ is, as we have seen, isomorphic to $C(B) \otimes \mathcal{K}$, and so by stability,

$$K_0(C_r^*(G)) \cong K_0(C(B)) \cong K^0(B).$$

Recall that the Atiyah-Singer index for families of elliptic operators [ASIV] with a parameter space B is an element of $K^0(B)$.

What we want to describe here is a kind of Chern character on $K_0(C_r^*(G(X)))$, G the groupoid of a foliation, or more properly a *partial* Chern character. This Chern character will take values in the reduced tangential cohomology group $\bar{H}_\tau^p(X)$ in top degree p (the leaf dimension) as defined in Chapter III. We shall assume without further notice that the foliation is tangentially oriented and that the groupoid of the foliation is Hausdorff. This partial Chern character sees only part of the structure of $K_0(C_r^*(G(X)))$, specifically the part that transverse measures can see. The "full" Chern character is conjecturally a homomorphism from $K_0(C_r^*(G))$ into the cyclic homology $H_*^\lambda(A^0)$ of a suitable dense subalgebra A^0 of $C_r^*(G)$, (see Connes-Skandalis [CS2] and, for cyclic theory, Connes [Co8], [Co9]). While the outline of this is clear and specific cases are known, there do remain some details. The "partial" Chern character that we will define directly would be obtained in general by composing the full Chern character with a natural homomorphism from $H_*^\lambda(A^0)$ to $\bar{H}_\tau^p(X)$.

For the definition of our Chern character, we start with a typical element of K-theory, $u = [e] - [f]$, where e and f are projections in $M_n(C_r^*(G)^+)$ with the same images in $K_0(\mathbb{C})$, where $G = G(X)$. Then we can assume without loss of generality that the images of e and f in $M_n(\mathbb{C})$ are exactly the same. Let ν be a positive Radon invariant transverse measure on X and form the corresponding trace ϕ_ν on $C_r^*(G)$. Extend ϕ_ν to $\phi_\nu^n = \phi_\nu \otimes \mathrm{Tr}$ on $M_n(C_r^*(G))$.

Theorem 6.31.

(a)   The element e-f may be chosen to be in the ideal of definition of ϕ_{ν}^{n}.

(b)   The map $\nu \longrightarrow \phi_{\nu}(e-f)$ extends to a linear functional on the set of all Radon signed transverse measures MT(X) which depends only on the K-theory class $u = [e] - [f] \in K_0(C_r^*(G))$. Denote it by c'(u).

(c)   The map c' takes values in the weak * continuous functionals on MT(X) (viewed the dual space of $\bar{H}_{\tau}^{p}(X)$ as in (4.27), (4.29)) and hence yields uniquely a map

$$c: K_0(C_r^*(G)) \longrightarrow \bar{H}_{\tau}^{p}(X)$$

which we call the *partial Chern character* c(u) of u.

Before turning to the proof, we offer some observations. Note that the partial Chern character is given very explicitly as follows. If $[u] \in K_0(C_r^*(G))$ is represented by $[e]-[f]$, where $e, f \in M_n(C_r^*(G)^+)$ with common images in $M_n(\mathbb{C})$ and if e and f are in the domain of ϕ_{ν}^{n}, then $c[u]$ is the cohomology class of the tangentially smooth p-form ω_u which (after identifying p-currents with Radon invariant transverse measures) is given by

$$\omega_u(\nu) = \phi_{\nu}^{n}(e-f)$$

where ϕ_{ν}^{n} is the trace $\phi_{\nu} \otimes Tr$ on $C_r^*(G) \otimes M_n$ associated to the invariant transverse measure ν.

Suppose that [u] is the index class of a tangential, tangentially elliptic operator D on X. One might try to construct ω_u as follows. The restriction of D to a leaf ℓ is locally traceable and has an associated p-form $(\rho_u)_\ell \in \Omega^p(\ell)$. One is tempted, then, to try to amalgamate the p-forms $(\rho_u)_\ell$ to a p-form $\rho_u \in \Omega_{\tau}^{p}(X)$. Unfortunately the forms $(\rho_u)_\ell$ do not vary continuously in the transverse direction and it is not at all clear that it is possible to alter the $(\rho_u)_\ell$ in some direct fashion to obtain a global class. We avoid this difficulty by regularizing at the C^*-algebra level with respect to MT(X).

Proof of 6.31. In view of the Hilsum–Skandalis result, Theorem 6.14, and the fact that $K_0(A \otimes \mathcal{K}) \cong K_0(A)$, we may assume that $u \in K_0(C_r^*(G_N^N))$ for a transversal N of the type described in (6.14) and consequently that e and f are in $M_n(C_r^*(G_N^N)^+)$.

We need to look more carefully at how $C_r^*(G_N^N)$ sits inside $C_r^*(G)$. As before we may arrange matters so that there is a larger transversal N' containing N and the closure $\bar{N}$, which we may assume is compact. Moreover we may arrange that a neighborhood U' of N' has the form $U' = N' \times \mathbb{R}^p$, p the leaf dimension, so that the second coordinates describe the leaves locally. Then $G_U^U = G_{N'}^{N'} \times \mathbb{R}^p \times \mathbb{R}^p$. Suppose that the graph is Hausdorff. Then elements of $G_{N'}^{N'}$ and G_N^N can be represented by Proposition 6.9 as continuous functions vanishing at ∞ on these spaces. We then pick a fixed compactly supported function φ on $\mathbb{R}^p$ and extend a function ψ on $G_{N'}^{N'}$ to one on G_U^U by the formula $\psi_U(g,x,y) = \psi(g)\varphi(x)\varphi(y)$ and one extends ψ_U to ψ_G on all of G by making it zero on the complement of G_U^U. In particular if ψ is supported on G_N^N and represents an element of $C_r^*(G_N^N)$, then ψ_U and ψ_G have compact support and ψ_G represents an element of the dense subalgebra A of functions used to define $C_r^*(G)$. It may be checked that this map $\psi \longrightarrow \psi_G$ gives an embedding i of $C_r^*(G_N^N)$ into $C_r^*(G)$. In the non–Hausdorff case the same argument works after localizing to open Hausdorff subsets. It follows from the discussion here and in Theorem 6.14 that the isomorphism θ of $C_r^*(G_N^N) \otimes \mathcal{K}$ with $C_r^*(G)$ can be arranged so that $\theta(x \otimes e_1) = i(x)$ where e_1 is a one dimensional projection.

In particular any finite matrix of elements in $C_r^*(G_N^N)$ is always represented in $C_r^*(G_N^N)$ by a kernel operator where the kernel is continuous and has compact support. Further, the kernel is tangentially smooth (Chapter III) for the natural foliation of G. Finally as every element of $C_r^*(G_N^N)$ can be written as a linear combination of elements of the form a*a, it follows by Proposition 6.25

that any element b of $C_r^*(G)$ represented as a finite matrix of elements of $C_r^*(G_N^N)$ via the isomorphism θ is in the ideal of definition of any weight ϕ_ν for any (positive) Radon transverse measure, and that $\phi_\nu(b)$ is given by integrating the kernel of b on the unit space. Specifically if ν is a Radon transverse measure, and λ is the fixed smooth tangential measure, then the integral of λ with respect to ν

$$\mu = \int \lambda d\nu$$

is a measure on X. If k_b is the kernel function on G for b, then

$$\phi_\nu(b) = \int k_b(e(x))d\mu(x)$$

where e is the function embedding X as the set of units in G. As k_b is tangentially smooth on G, $k_b(e(x))$ is tangentially smooth on X. Finally as the foliation is oriented, we may view the tangential measure λ as a tangentially smooth p-form, and then $\omega_b = k_b(e(\cdot))\lambda$ is also a tangentially smooth p-form. Recasting the formula above, we see that

$$\phi_\nu(b) = \int \omega_b d\nu$$

is given by integrating the tangentially smooth p-form ω_b against the transverse measure ν.

The proposition is now obvious for we can arrange the two projections e and f defining the K-theory element u = [e] – [f] to be in $M_n(C_r^*(G_N^N)^+)$ and their difference e-f to be a finite matrix over $C_r^*(G_N^N)$ to which the above analysis applies. For an invariant (positive) Radon transverse measure, $\phi_\nu(e-f)$ can be given by integrating a fixed tangentially smooth p-form ω_{e-f} against ν. This in fact constructs the value of the partial Chern character c(u) in $\bar{H}_\tau^p(X)$; namely it is the class of the form ω_{e-f}. That it is well defined and independent of the choice of e and f results from the fact that ϕ_ν is a trace and the duality result, Proposition 4.29. □

As we will be working with projections that often do not lie in the C^*-algebras under consideration, but rather in a von Neumann algebra containing the C^*-algebra of interest, we shall add a few words about K-theory for von Neumann algebras. As any von Neumann algebra W is a C^*-algebra one could just define $K_0(W)$ using the C^* definition. However, the presence of infinite projections leads to bad behavior; for instance, $K_0(\mathcal{B}(H)) = 0$ for an infinite dimensional Hilbert space H. We want to stick to finite projections. Recall the definition, which is reminiscent of Dedekind's definition of a finite set.

Definition 6.32. A projection e in a von Neumann **algebra** W is *finite* if e is not equivalent in the sense above to a proper projection of itself.

Equivalently, one may first define a von Neumann algebra W to be finite if given $w \in W^+$ there is a finite normal trace ϕ on W^+ with $\phi(w) \neq 0$; then define $e \in W$ to be a finite projection if eWe is a finite von Neumann algebra. For this approach, cf. Dixmier [Di1].

One then forms the semi-group of classes of finite projections in $\bigcup_n M_n(W)$ and then the corresponding Grothendieck group to obtain a group we denote $K_0^f(W)$. Evidently $K_0^f(\mathcal{B}(H)) = \mathbb{Z}$ while $K_0^f(W) = \mathbb{R}$ if W is a factor of type II, and $K_0^f(W) = 0$ if W is a factor of type III. From this one can readily compute $K_0^f(W)$ for any W.

Now if we start with a C^* algebra A and a representation π of A into a von Neumann algebra W, we would like to define, at least under some conditions, a homomorphism

$$\pi_*: K_0(A) \longrightarrow K_0^f(W).$$

At the very least we would want this map to exist when $A = C_r^*(G(X))$ and $W = W^*(G(X), \tilde{\mu})$ with $\tilde{\mu}$ arising from a Radon invariant transverse measure for the foliation.

Proposition 6.33. Let A is a C^*-algebra of the form $A = B \otimes \mathcal{K}$ with a representation π into a von Neumann algebra W, such that

203

$\pi(A)$ is dense and when we write $\pi(B)'' = eWe$ for a projection e in W, then e is a finite projection. Then there is a well defined homomorphism

$$\pi_*: K_0(A) \longrightarrow K_0^f(W).$$

We omit the obvious proof that is based on the equality $K_0(B) \cong K_0(A)$ and observe that the conditions are satisfied in the case at hand since $A = C_r^*(G(X)) = C_r^*(G_N^N) \otimes \mathcal{K}$, and since the trace ϕ_ν has been shown to be finite on $C_r^*(G_N^N)$ in Theorem 6.31 it follows that $\phi_\nu(e)$ is finite where e is the projection in the statement of the Proposition. Finally since ϕ_ν is a faithful trace on $W^*(G,\tilde\mu)$ it follows that e is a finite projection. $\square$

Corollary 6.34. For any finite Radon invariant transverse measure ν there is a natural homomorphism

$$\pi_\nu: K_0(C_r^*(G)) \longrightarrow K_0^f(W^*(G,\tilde\mu))$$

and the associated trace ϕ_ν on $C_r^*(G)$ and on $W^*(G,\tilde\mu)$ extends to yield a commuting diagram

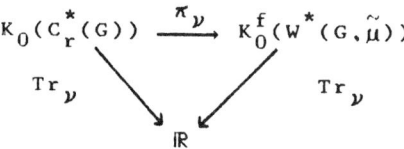

We note that if $W^*(G,\tilde\mu)$ is a factor then it is of Type II_∞ and $Tr_\nu: K_0^f(W^*(G,\tilde\mu)) \longrightarrow \mathbb{R}$ is an isomorphism.

Looking ahead more explicitly to the next chapter, we consider the following situation: we have an exact sequence

$$0 \longrightarrow A \longrightarrow P \longrightarrow C \longrightarrow 0$$

of C^* algebras. We assume that P has a representation π into a von Neumann algebra W such that $\pi(A)$ is weakly dense. As above we assume that $A = B \otimes \mathcal{K}$ where $\pi(B)'' = eWe$ with e a finite projection. We suppose that P and hence C are unital. Now let $d \in P$ and suppose that $\hat{d}$, its image in C is invertible (that is, to adumbrate the following chapter, d is elliptic). We regard $\hat{d}$ as an element of $K_1(C)$ and then according to the exact sequence of K-theory, the index of $\hat{d}$ is a well defined element ind(d) in $K_0(A)$.

On the other hand, we can view $\pi(d)$ as an element of the von Neumann algebra W and then its kernel, $\ker(\pi(d))$ and $\ker(\pi(d^*))$ are projections in W. What one hopes, using the map of Proposition 6.33 is indeed the case, as follows from Breuer's theory of Fredholm operators [Bre] in von Neumann algebras. The following summarizes the result and will play a crucial role in Chapter VII.

Proposition 6.35. Under the assumptions above, $\ker(\pi(d))$ and $\ker(\pi(d^*))$ are finite projections in W. Moreover the difference $[\ker(\pi(d))] - [\ker(\pi(d^*))]$ (the analytic index of d) is an element of $K_0^f(W)$ and if π_* is the map from $K_0(A)$ to $K_0^f(W)$ of Proposition 6.33, then

$$\pi_*(\mathrm{ind}(d)) = [\ker(\pi(d))] - [\ker(\pi(d^*))].$$

Proof. If m is the smallest norm closed ideal in W containing the finite projections, then evidently $m \supset \pi(B)$ by hypothesis, and hence $m \supset \pi(B \otimes \mathcal{K}) = \pi(A)$. Since $\hat{d}$ is invertible in C, it follows that the image of $\pi(d)$ in W/m is invertible and hence that $\pi(d)$ is Fredholm by Theorem 1 of Breuer [Bre]. It follows from Breuer that $\ker(\pi(d))$ and $\ker(\pi(d^*))$ are finite projections in W. Finally a close examination of the definition of the index map $K_1(C) \longrightarrow K_0(A)$ as for instance given in Blackadar [Bl] Definition 8.7 shows directly that that $\pi_*(\mathrm{ind}(\hat{d})) = [\ker(\pi(d))] - [\ker \pi(d^*)]$. □

The point here is that the naively defined "spatial" analytic index, $[\ker \pi(d)] - [\ker \pi(d^*)]$ of d in the von Neumann algebra W, a very measure theoretic type of object, is always the image via π_*

of an element in $K_0(A)$, an element which in turn has important topological invariance properties. This will be applied to the extension

$$0 \longrightarrow C_r^*(G) \longrightarrow \bar{\wp}^0 \longrightarrow \Gamma(S^*F, \text{End}(E)) \longrightarrow 0$$

of pseudodifferential operators of a foliated space. The von Neumann algebra W will be $W^*(G, \tilde{\mu})$ constructed from a Radon invariant transverse measure ν.

CHAPTER VII: PSEUDODIFFERENTIAL OPERATORS

This chapter is devoted to the study of tangential pseudodifferential operators and their index theory. The chapter has four topics, treated in turn. They are

A) the general theory of pseudodifferential operators on foliated spaces (7.1 - 7.19);

B) differential operators and finite propagation (7.20 - 7.27);

C) Dirac operators and the McKean–Singer formula (7.28 - 7.39);

D) Superoperators and the asymptotic expansion of the heat kernel (7.40 - 7.51).

A.   **Pseudodifferential operators.** We begin the chapter by introducing the machinery of tangential differential operators, smoothing operators, and pseudodifferential operators, first in a local setting and then globally. We demonstrate that a tangentially elliptic pseudodifferential operator has an inverse modulo compactly smoothing operators. Letting $\bar{\wp}$ denote the closure of the *-algebra of pseudodifferential operators of order 0 on a bundle E, there is a short exact sequence

$$0 \longrightarrow C_r^*(G(X)) \longrightarrow \bar{\wp} \longrightarrow \Gamma(S^*F, \text{End}(E)) \longrightarrow 0$$

where S^*F is the cotangent sphere bundle of the foliated space. This leads to formulas which relate the abstract index class $\text{ind}(P) \in K_0(C_r^*(G(X)))$, the Connes index $\text{ind}_\nu(P)$, and the Type II von Neumann index. In general, the index of a tangential, tangentially elliptic operator may be regarded as a class in $K_0(C_r^*(G(X)))$ or in

$K_0^f(W^*(G(X),\tilde{\mu}))$. The natural map

$$\beta_*: K_0(C_r^*(G(X))) \longrightarrow K_0^f(W^*(G(X),\tilde{\mu}))$$

commutes with the homomorphisms from these groups to $\mathbb{R}$ induced by an invariant transverse measure ν, so

$$\text{ind}_\nu(P) = \phi_\nu(\ker[\beta P] - \ker[\beta P^*])$$

where ϕ_ν is the trace associated to the invariant transverse measure ν. These results imply that the index of the operator depends only upon the homotopy class of the tangential principal symbol of the operator.

B.   Differential operators and finite propagation. Turning next to tangential differential operators, we introduce bounded geometry and finite propagation conditions.   We show that a tangential differential operator D on a compact foliated space has a unique (leafwise) closure, so that the Hilbert fields Ker(D) and Ker(D*) are well-defined. It then makes sense to form the index measure ι_D and then to define the index by

$$\text{ind}_\nu(D) = \int \iota_D \, d\nu.$$

This is formally the same as the definition for operators of order zero, of course, but some further work is required to make the connection between the two more concrete and transparent.

C.   Dirac operators and the McKean-Singer formula. The key differential operators for the purposes of index theory are the tangential Dirac operators.   Having introduced these operators in an abstract context and having verified that the general machinery of Section B applies to these operators, we establish the McKean-Singer formula: for $t > 0$,

$$\mathrm{ind}_{\nu}(D) = \phi_{\nu}([e^{-tD^*D}] - [e^{-tDD^*}]).$$

D. **Superoperators and the asymptotic expansion.** We introduce superoperators and restate the McKean–Singer formula in the form

$$\mathrm{ind}_{\nu}(D) = \phi_{\nu}^{s}(e^{-t\hat{D}})$$

where $\hat{D}$ is the superoperator

$$\hat{D} = \begin{bmatrix} 0 & D \\ D^* & 0 \end{bmatrix}^2$$

and ϕ_{ν}^{s} is the supertrace. Next we introduce complex symbols and prove that as $t \longrightarrow 0$ there is an asymptotic expansion

$$\phi_{\nu}^{s}(e^{-t\hat{D}}) \sim \sum_{j \geq -p} t^{j/2p} \int_{X} \lambda_{j}(\hat{D})d\nu$$

where each $\lambda_{j}(\hat{D})$ is a signed tangential measure independent of t. As $\mathrm{ind}_{\nu}(D)$ is independent of t, an easy argument shows that

$$\mathrm{ind}_{\nu}(D) = \int \omega_{D}(g,E)d\nu = <[\omega_{D}(g,E)],[C_{\nu}]>$$

where

$$\omega_{D}(g,E) = \lambda_{0}(\hat{D})|d\lambda| = (\lambda_{0}(D) - \lambda_{0}(D^*))|d\lambda|$$

is the associated tangentially smooth p-form and $[C_{\nu}]$ is the homology class of the Ruelle–Sullivan current associated to ν. The identification of ω_{D} for twisted signature operators and the completion of the proof is left to Chapter VIII.

We are deeply grateful to Steve Hurder, Peter Gilkey, Jerome Kaminker, John Roe, and Michael Taylor for their enormously helpful assistance in the preparation of this chapter.

A. PSEUDODIFFERENTIAL OPERATORS

Fix a tangential Riemannian metric on X and corresponding tangential Riemannian metric on G(X). This determines a volume form on each leaf of X and on each leaf of G(X). There is a corresponding tangential measure λ_X on X and a tangential measure $\lambda_{G(X)}$ on G(X). Recall that

$$\lambda_{G(X)} = \{\lambda^x_{G(x)}\}$$

where $\lambda^x_{G(x)}$ is a measure on G^x. The measure $\lambda_{G(X)}$ is invariant under the left action of the holonomy groupoid. Precisely, if $u \in G^y_x$ and if f is a non-negative Borel function on G, then

$$\int f(uu')d\lambda^x_{G(x)}(u') = \int f(u')d\lambda^y_{G(x)}(u').$$

We fix once and for all a transverse measure $d\nu$. Note that in view of the results of Chapter IV, $d\nu$ may be regarded as a measure on the transversals of G(X) or equivalently as a measure on the transversals of X. [For most of this chapter there would be no harm in letting ν have a non-trivial modular function, but our applications require that ν be an _invariant_ transverse measure, so we assume that as needed.] This determines measures $\mu = \lambda_X d\nu$ on X and $\lambda_{G(X)}d\nu$ on G(X) by the procedure of Chapter IV.

Let U be an open subset of $\mathbb{R}^p \times N$ with the induced foliated structure. Define

$$d^\alpha_x = (\partial/\partial x_1)^{\alpha_1}...(\partial/\partial x_p)^{\alpha_p}$$

and

$$D^\alpha_x = (-i)^{|\alpha|}d^\alpha_x.$$

Recall that $C^\infty_\mathcal{T}(U)$ denotes the continuous, tangentially smooth functions on U, and $C^\infty_{\mathcal{T}c}(U)$ denotes those which are compactly supported. We topologize these by insisting that convergence means convergence on compact subsets of a function and its tangential derivatives.

Definition 7.1. Let X be a foliated space with foliation bundle F.

The *bundle of densities of order* α *on* X (a complex line bundle) is defined by

$$|F|_{\alpha,x} = \{\phi: \Lambda^p F_x - \{0\} \longrightarrow \mathbb{C} \mid \phi(\lambda w) = |\lambda|^\alpha \phi(w)$$

$$\text{for all } \lambda \in \mathbb{R} - \{0\}, \; w \in \Lambda^p F_x - \{0\}\}.$$

Define $|F| = |F|_1$. Densities of order 1 on a leaf are measures on that leaf, so it makes sense to define

$$C_o^\infty(|\ell|) = \Gamma_o(\ell, |T\ell|)$$

and then *distributions* on the leaf ℓ by

$$\mathscr{D}'(\ell) = (C_o^\infty(|\ell|))^*.$$

Similarly, *compactly supported distributions* $\mathcal{E}'(\ell)$ are defined on the leaf ℓ as dual to $\Gamma(\ell, |T\ell|)$.

These are examples of an assignment to each leaf ℓ of a topological vector space $E(\ell)$, and we shall informally speak of such an assignment as a field of topological vector spaces, leaving undefined what kind of transverse measurability is required. Further examples include

$$\mathscr{B}_\mathcal{T}^\infty(X) = \{C^\infty(\ell)\}$$

and

$$\mathscr{B}_{\mathcal{T}c}^\infty(X) = \{C_c^\infty(\ell)\}.$$

One particular case of importance is when these spaces are Hilbert spaces.

Definition 7.2. A *Borel field of Hilbert spaces* E over a foliated space X is an assignment of a (separable) Hilbert space E_x to each x in X which is Borel in the sense of direct integral theory (cf. Chapter VI, p. 183 and [Tak2], IV, §8) together with a map $u \longrightarrow u_*$ from G(X) into unitary operators from $E_{s(u)}$ to $E_{r(u)}$ in the language of Chapter IV such that $(uv)_* = u_* v_*$ and $(u^{-1})_* = (u^{-1})_*$ and so that u_* is a Borel function of u. In this case we say that u defines a representation of G(X) on the field E. A *bounded operator* P: E $\longrightarrow$ E' of Borel fields of Hilbert spaces is a Borel family (cf. p. 183) of operators P_x: $E_x \longrightarrow E'_x$ with uniformly bounded norms which is invariant under the left action of each G^x_x.

If one has a tangential measure λ^x on X, one may form $E_x = L^2(\ell^x, \lambda^x)$ as in Chapter IV. This field is clearly a Borel field of Hilbert spaces where u_* for any u is defined as the identity map from $E_{s(u)}$ to $E_{r(u)}$. This is called the *regular representation* of the groupoid G(X) with the tangential measure λ.

Let E and E' be finite-dimensional tangentially smooth complex bundles over X. A *tangential operator* from E to E' is a family P = $\{P_x$: x∈X$\}$ where, for each x, P_x is a linear map

(7.3) $$P_x: C^\infty_c(G^x, s^*(E)) \longrightarrow C^\infty(G^x, s^*(E'))$$

which is invariant under the left action of each G^x_x. Left invariance implies that there exists a vector-valued distribution on G such that for each x $\in$ X the distributional kernel associated to P_x (on G^x) is $K(\gamma, \gamma') = K(\gamma^{-1}\gamma')$ so that

(7.4) $$(P_x \xi)(\gamma) = \int K(\gamma^{-1}\gamma') \xi(\gamma') d\lambda^x_{G(X)}(\gamma')$$

for all $\xi \in C^\infty_c$. Note that in this generality, the operators $\{P_x\}$ vary measurably but not necessarily continuously in the transverse direction. To obtain continuous control transversely one must assume that the distribution kernel varies continuously transversely.

If we assume that G(X) is Hausdorff then the distributions $K_x = K(\gamma, \cdot)$ x = r(γ) $\in$ X corresponding to the operator P_x fit together to form a distribution P on G(X) because G(X) is a fibre space over X with G^x as the fibre over x. One defines

$$K(\varphi) = \int K_x(\varphi_x) d\mu(x)$$

where φ_x is the restriction of a compactly supported test function φ on $G(X)$ to G^x and μ is the measure on X obtained by integrating the tangential measure λ with respect to the fixed transverse measure ν as in Chapter IV. The distribution K is called the *distribution kernel* of P.

The usual constructions for operators on manifolds may be conducted leaf by leaf. For instance, if T is a tangential operator then a *formal adjoint* T^t on C_c^∞ is defined leafwise by

$$\langle \psi, T^t \phi \rangle = \langle T\psi, \phi \rangle.$$

__Definition__ 7.5. A *tangential differential operator*

$$D: \Gamma_\tau(E) \longrightarrow \Gamma_\tau(E')$$

is a continuous linear operator which, locally, is given by a linear combination of partial differential operators along the leaves. We extend D to tangential distributional sections ψ by

$$\langle D\psi, \phi \rangle = \int \psi(x) D^t \phi(x) dx.$$

where D^t is the formal adjoint of D. A tangential differential operator D has a local expansion on a coordinate patch of the form

$$D = \sum_{|\alpha| \leq m} a_\alpha(x) D^\alpha$$

where the a_α vary continuously in x and vary smoothly on each leaf. The maximal global value for m is the *order* of D. A tangential differential operator from E to E' induces an operator

$$D: \mathcal{B}_\tau^\infty(E) \longrightarrow \mathcal{B}_\tau^\infty(E')$$

by restriction. This operator varies continuously as one moves transversely. More generally, one sometimes wishes to consider operators $D = \{D_\ell\}$ where the transverse variation is only measurable.

The *Hodge–Laplace operator* provides a key example of a tangential differential operator. Suppose given a foliated space with a tangential Riemannian connection. Recall that the de Rham operator is a map $d = \{d_x\}$ where, for $x \in \ell$, $d: \Omega^k(\ell) \longrightarrow \Omega^{k+1}(\ell)$. The orientation on F determines the Hodge *-operator

$$*: \Omega^j(\ell) \longrightarrow \Omega^{p-j}(\ell).$$

Define

$$\delta = (-1)^{pk+p-1}*d* : \Omega^{k+1}(\ell) \longrightarrow \Omega^k(\ell)$$

and

$$\Delta_k = d\delta + \delta d : \Omega^k(\ell) \longrightarrow \Omega^k(\ell).$$

This determines the tangential Hodge–Laplace operator

$$\Delta_k : \Omega^k_\tau(X) \longrightarrow \Omega^k_\tau(X)$$

on forms over X and similarly on forms over G(X). Each Δ_k is a second order tangential differential operator. In flat space, $\Delta_0 = -\Sigma(\partial^2/\partial x_i^2)$ is the classical Laplacian.

Given a tangential differential operator D, define the *tangential (total) symbol* of D by

$$\sigma(x, \xi) = \sum_{|\alpha| \leqslant m} a_\alpha(x)\xi^\alpha.$$

and define the *tangential principal symbol* of D by

$$\sigma_m(x, \xi) = \sum_{|\alpha| = m} a_\alpha(x)\xi^\alpha.$$

The tangential total symbol is a purely local notion; it depends on the choice of coordinate system. In contrast, the local tangential principal symbols patch together to yield the global tangential principal symbol $\sigma_m(D)$ on the cosphere bundle S^*F of F. If $\sigma_m(D)$ is invertible then D is said to be *tangentially elliptic*. For example, $\sigma_2(\Delta_\ell) = -\Sigma\xi_i^2$ which is invertible on unit vectors, so the tangential Hodge–Laplace operator Δ is a tangentially elliptic operator.

Next recall a bit of the classical theory of pseudodifferential operators – see [Tay], [AS2], [Gi3], for more detail. Suppose first that U is an open subset of $\mathbb{R}^p$. If

$$P(x,D) = \sum_{|\alpha|\leq m} a_\alpha(x)D^\alpha$$

is a differential operator with smooth coefficients one can write for u $\in C_c^\infty(U)$, extended to $\mathbb{R}^p$, ($\hat{u}$ the Fourier transform,)

(*) $$P(x,D)u(x) = \sum_{|\alpha|\leq m} a_\alpha(x)\int \xi^\alpha e^{2i\pi<x,\xi>}\hat{u}(\xi)d\xi$$

so that, with the symbol of P given by

$$\sigma(x,\xi) = p(x,\xi) = \sum_{|\alpha|\leq m} a_\alpha(x)\xi^\alpha \in C^\infty(U\times\mathbb{R}^p),$$

one has

(**) $$P(x,D)u(x) = \int p(x,\xi)e^{2i\pi<x,\xi>}\hat{u}(\xi)d\xi.$$

Sometimes one writes

$$P(x,D) = OP(p(x,\xi)).$$

The class of differential operators is not large enough to include, for instance, the parametrix of a differential operator of positive order, since such an operator would have negative order. The general idea then is to admit a larger class of symbols and then use (**) to define a larger class of operators. We define two such classes, $S^m(U)$ and $S^m_o(U)$, as follows. For any integer m, let $S^m(U)$ be the set of all smooth functions $p(x,\xi)$ on $U \times \mathbb{R}^p$ which satisfy the following condition: for each compact subset K of U and for all multi-indices α, β,

$$|D^\beta_x D^\alpha_\xi p(x,\xi)| \leq c_{\alpha,\beta,K}(1 + |\xi|)^{m-|\alpha|}$$

$$\text{for } x \in K, \xi \in \mathbb{R}^p.$$

For instance, polynomials in ξ of degree m with smooth coefficients lie in $S^m(U)$. More generally, if ϕ is some smooth function, let

$$p(x,\xi) = \phi(x)(1 + |\xi|^2)^{m/2}I.$$

This is an elliptic symbol of order m whenever $\phi \neq 0$. For $p \in S^m(U)$, define

$$P = OP(p): C^\infty_c(U) \longrightarrow C^\infty(U)$$

by

$$Pu(x) = (2\pi)^{-p} \int p(x,\xi) e^{i\langle x,\xi\rangle} \hat{u}(\xi) d\xi.$$

The class $S^m_o(U)$ consists of those symbols $p \in S^m(U)$ which satisfy the following additional condition: for each non-zero value of ξ, the limit

$$\sigma_m(p)(x,\xi) = \lim_{\eta \to \infty} p(x,\eta\xi)/\eta^m$$

exists. Then $\sigma_m(p)$ is a C^∞ function on $U \times (\mathbb{R}^p - 0)$ and it is homogeneous of degree m in ξ.

Finally, a *pseudodifferential operator* is an operator $P: C^\infty_c(U) \longrightarrow C^\infty(U)$ such that for each $f \in C^\infty_c(U)$ the associated operator Pf is a pseudodifferential operator in local coordinates; i.e., it is of the form $OP(p_f)$ for some $p_f \in S^m_o$. The set of such operators is denoted $\mathcal{P}^m(U)$. There is an obvious extension to

216

matrix-valued functions.

Lemma 7.6. Let $r(x, \xi, y)$ be a matrix-valued symbol which is smooth in each variable. We suppose that r has compact x-support inside U (an open set in $\mathbb{R}^p$ with compact closure) and that there are estimates

$$|D_x^\alpha D_\xi^\beta D_y^\gamma r| \leq C_{\alpha, \beta, \gamma}(1 + |\xi|)^{m - |\beta|}$$

for all multiindices (α, β, γ), where $m < -p-k$. so that the associated operator

$$OP(r)f(x) = \iint e^{i(x-y)\xi} r(x, \xi, y) f(y) dy d\xi$$

is a pseudodifferential operator of order m. The *distribution kernel* $K(x,y)$ is given by

$$K(x,y) = \int e^{i(x-y)\xi} r(x, \xi, y) d\xi.$$

Then K is C^k in (x,y) and

$$OP(r)f(x) = \int K(x,y)f(y) dy.$$

Proof. See Gilkey [Gi3] page 19, Lemma 1.2.5. □

Lemma 7.7. Let $K(x,y)$ be a smooth kernel with compact x,y support in U (an open set in $\mathbb{R}^p$ with compact closure). Let P be the operator defined by K. If k is a non-negative integer, then

$$|K|_{\infty, k} \leq C(k) |P|_{-k, k}.$$

Proof. See Gilkey [Gi3], page 21, Lemma 1.2.9. □

The *principal symbol* $\sigma_m(P)$ of a pseudodifferential operator P is defined by

$$\sigma_m(P)(x, \xi) = \sigma_m(p_f)(x, \xi)$$

where f is any function equal to 1 near x. The algebra $\wp^m(U)$ is invariant under diffeomorphisms of U and hence determines uniquely a corresponding class of operators $\wp^m(M)$ for a (paracompact) manifold M, and, more generally, for $\wp^m(E,E')$, where E and E' are smooth bundles over M. The principal symbol yields a map

$$\sigma_m(P):\ S^*M\ \longrightarrow\ \mathrm{Hom}(\pi^*(E),\ \pi^*(E'))$$

where π is the canonical projection of the cotangent sphere bundle of M to M. Give $\wp^m(E,E')$ the natural Fréchet topology using coordinate neighborhoods. Then

$$\sigma_m:\ \wp^m(E,E')\ \longrightarrow\ \Gamma(S^*M,\ \mathrm{Hom}(\pi^*(E),\pi^*(E'))).$$

If M is compact then

$$\wp^m(M)\ \longrightarrow\ \mathcal{L}(W^s(M),W^{s-m}(M))$$

is continuous for each m and s, so bounded families of symbols yield bounded families of operators.

A pseudodifferential operator P from E to E' on a compact manifold M is *smoothing* if for all s,t, P induces bounded maps

$$P:\ W^s(E)\ \longrightarrow\ W^{s+t}(E'),$$

where $W^s(E)$ denotes the (classical) Sobolev space associated to the smooth sections over the (compact) manifold M. Equivalently, P is smoothing if

$$P:\ \mathcal{E}'(E)\ \longrightarrow\ C^\infty(E').$$

The conditions are equivalent since

$$\bigcup_s\ W^s(E) = \mathcal{E}'(E)\qquad\text{and}\qquad\bigcap_s\ W^s(E) = C^\infty(E)$$

by the Sobolev lemma. A smoothing operator has a smooth distributional kernel.

Let us return to the realm of foliated spaces. Let X be a compact foliated space with leaves of dimension p equipped with a tangential Riemannian metric and let G = G(X) be its holonomy groupoid, which we assume to be Hausdorff. Let E and E' be finite-dimensional tangentially smooth complex bundles over X.

Definition 7.8. Fix a real number s. The *tangential Sobolev field* $W_\tau^s = \{W_x^s\}$ is defined as follows: W_x^s is the completion of $Dom(1 + \Delta_x)^{s/2}$ with respect to the norm

$$\| \xi \|_{s,x} = \| (1 + \Delta_x)^{s/2} \xi \|_{L^2(G^x)}.$$

The representation of G(X) on the Hilbert field $W_\tau^s(G(X))$ by left translation is by construction equivalent to the regular representation of G(X), (cf. 7.2). Note that up to equivalence the field $W_\tau^s(G(X))$ is independent of choice of tangential Riemannian metric.

Definition 7.9. A tangential operator P is *smoothing* if P induces a bounded operator

$$P\colon W_\tau^s(G(X)) \longrightarrow W_\tau^{s+t}(G(X))$$

for all s,t. The distribution kernel which determines P is in fact a smooth function on each leaf, though it may be only measurable transversely. The kernel dies off in a complicated way on each leaf; it is not necessarily compactly supported on G(X). A tangential operator P is *compactly smoothing* if P is smoothing and if the distribution kernel of P is compactly supported on G(X).

If C is a compact subset of G(X) then the *support* of P is in C if the distribution vanishes off C, i.e.,

$$\mathrm{Supp}(P_x \xi) \subset \mathrm{Supp}\ \xi \circ C^{-1}$$

for all ξ. Then

$$\text{Supp}(P_1 \circ P_2) \subset \text{Supp}(P_1)\text{Supp}(P_2)$$

for P_1 and P_2 compactly supported. A tangential operator P is *pseudolocal* if for all neighborhoods S of G^0 there is a compactly smoothing operator R with $\text{Supp}(P+R) \subset S$. Say $P_1 \sim P_2$ if $P_1 - P_2$ is compactly smoothing.

Suppose that $\Omega \cong L \times N$ is a distinguished coordinate patch of the holonomy groupoid G(X) with L open and connected in $\mathbb{R}^{2p}$. A tangential operator P from E to E' over Ω corresponds by invariance to a measurable family $P = \{P_n : n \in N\}$ where

$$P_n : C^\infty(L \times \{n\}, E) \longrightarrow C^\infty(L \times \{n\}, E').$$

To make P a tangential pseudodifferential operator one naturally requires that each P_n be a classical pseudodifferential operator and that these operators vary continuously in n. The invariance condition on the family of operators translates into the condition that the distribution kernel $K(\gamma, \gamma', n)$ is really a function of $\gamma^{-1}\gamma'$, so write

$$K(\gamma, \gamma', n) = K(\gamma^{-1}\gamma', n).$$

Thus K may be regarded as being defined on an open set of G(X) itself. On the question of what support for K should be allowed, one has some choice. We insist that K has compact support on G(X). The set of such P of order $\leqslant$ m is denoted $\mathcal{P}_c^m(\Omega, E, E')$. Each element of $C^\infty_{\tau c}(G(\Omega))$ determines a compactly smoothing operator.

If $P \in \mathcal{P}_c^m(\Omega, E, E')$ with distribution kernel K, then K extends naturally to all of G(X) (by setting it equal to zero outside of Ω). It is then the distribution kernel for a unique tangential operator on G, denoted P'. This operator decomposes as $P' = \{P'_x\}$ where P'_x has support contained in $G^x \cap s^{-1}(\Omega)$. Finally,

Definition 7.10. A *tangential pseudodifferential operator on X* is a finite linear combination of compactly smoothing operators with transversely continuous distribution kernels and operators of the form P' above. By construction, each such operator is pseudolocal and has a continuous compactly supported distribution kernel. Transverse continuity implies that the tangential principal symbol of such an operator is continuous.

Let $\wp^m(G(X), E, E')$ be the linear space of tangential pseudodifferential operators of order $\leqslant$ m from E to E'; that is, finite linear combinations of operators arising on the various $\wp^m(\Omega, E, E')$ and compactly smoothing operators. The linear space

$$\wp^{-\infty}(E, E') = \bigcap_m \wp^m(E, E')$$

consists of the compactly smoothing operators with transversely continuous tangentially smooth kernels, which is precisely the image of $C^\infty_{\tau c}(G(X))$. When the context is appropriate we abbreviate to $\wp(E, E')$ or to $\wp$.

All of this has been for G(X) Hausdorff. If G(X) is only locally Hausdorff then we modify as in the construction of $C^*_r(G(X))$. Cover the space G(X) by open Hausdorff sets Ω, for which $\wp^m(E, E')$ does make sense, and then define $\wp^m(G(X), E, E')$ to be the algebra of linear combinations of these local pseudodifferential operators and compactly smoothing operators.

The following proposition is taken directly from Connes [Co3, page 126].

<u>Proposition</u> 7.11 [Connes].

a) $\wp^m \circ \wp^n \subset \wp^{m+n}$ for all m,n.

b)   If $P \in \wp^0(E, E')$, then the family $\{P_x : x \in X\}$ extends to a bounded intertwining operator $L^2(G, \lambda, s^*(E)) \longrightarrow L^2(G, \lambda, s^*(E'))$.

c)   If $P \in \wp^m(\mathbb{C}, \mathbb{C})$, m<0, then $P \in C^*_r(G(X))$.

d)   If $P \in \wp^m(E, E')$ with $m < -p/2$, then its associated distribution kernel K is measurable on G(X) with

$$\text{Sup}_y \int \|K(\gamma^{-1})\|^2_{HS} d\nu^y(\gamma) < \infty.$$

Proof (Connes). a) Suppose first that $n = -\infty$. One may assume that $P' \in \wp^m$ corresponds to a continuous family $P \in \wp^m_c(\Omega, E, E')$ with $\Omega \cong L \times N$. A partition of unity argument shows that we may study functions f (with associated multiplication operators M_f) supported on

$$W' \cong L \times L' \times N,$$

where $\Omega' \cong L' \times N$ compatibly with $\Omega \cong L \times N$. The kernel associated to PM_f is of the form

$$K_1(t, t'', n) = \int K(t, t', n) f(t', t'', n) a \,|\, dt' \,|$$

and is tangentially smooth. Thus PM_f is smoothing, and this implies that $\wp^m \wp^{-\infty} \subset \wp^{-\infty}$. For the general case assume that $P' \in \wp^m$, $Q' \in \wp^n$ arise from $\wp_c(\Omega, E, E')$, where $\Omega \cong L \times N$, and then invoke the classical argument. In particular, this shows that $\wp^0(G(X), E, E)$ is an algebra.

b)   Assume that the operator is of the form P' for $P \in \wp^0_c(\Omega, E, E')$. The assertion follows from the inequality

$$\|P'_x\| \leqslant \text{Sup } \|P_n\|.$$

c)   This follows from the natural inclusion

$$C^*_r(\Omega) \longrightarrow C^*_r(G)$$

and the continuity of the map given by $n \longrightarrow P_n$.

d)   It suffices to prove the assertion for P', with $P \in \wp^m_c(\Omega, E, E')$. One has

$$K(t,t',n) = \int e^{i<t-t',\xi>} a(t,\xi,n) d\xi$$

where

$$\|a_{t,n}\|_2^2 = \int |a(t,\xi,n)|^2 |d\xi|$$

is uniformly bounded; i.e.,

$$|a(t,\xi,n)| \leq c(1 + |\xi|)^m.$$

Then the Parseval equality shows that

$$\int |K(t,t',n)|^2 dt' = \|a_{t,n}\|_2^2$$

is uniformly bounded. □

Let $P \in \mathcal{P}^m(G(X),E,E')$ be a tangential pseudodifferential operator from E to E'. We define its *principal symbol* $\sigma_m(P)$ to be that of the operator s(P) (which acts on bundles over X, rather than on bundles over G.) If P is a smoothing operator with associated kernel K, then s(P) is the operator associated with the kernel function

$$K'(y,x) = \Sigma K(\gamma) \in E_x^* \otimes E_y \text{ (sum over all } \gamma: x \longrightarrow y)$$

This is indeed a smoothing operator and its principal symbol $\sigma_m(P)$ is zero for all m. It follows that σ_m induces a homomorphism

$$\sigma_m: \mathcal{P}^m(G(X),E,E') \longrightarrow \Gamma_{\tau}(S^* F, \text{Hom}(E,E')).$$

One defines ellipticity of P by the invertibility of $\sigma_m(P)$ which is the same as the ellipticity of s(P).

Proposition 7.12 **[Co3, page 128]**. Suppose that $P \in \mathcal{P}^m(E,E')$ is a tangentially elliptic pseudodifferential operator. Then there exists a tangentially elliptic pseudodifferential operator $Q \in \mathcal{P}^{-m}(E',E)$ such that $PQ-\text{id}_{E'}$ and $QP-\text{id}_E$ are compactly smoothing.

Proof (Connes). Let $\{\Omega_i\}$ be a finite open cover of X by coordinate charts of the form $\Omega_i \cong L_i \times N_i$. Let $\{\phi_i\}$ be a tangentially smooth partition of unity subordinate to this cover. Let C be a compact neighborhood of $G^0 \subset G(X)$ such that for each i,

$$\{\gamma \in C : s(\gamma) \subset \text{Supp } \phi_i'\} \subset W_i = L_i \times L_i \times N_i$$

where $\phi_i' \in C^\infty_{TC}(\Omega_i)$ has value 1 on the support of ϕ_i and s is the source map. We may suppose that Supp $P \subset C$.

For each i, define M_i to be the tangential operator from E to E' given by multiplication by $\phi_i' \circ s$. The distribution K_i associated to PM_i is supported in W_i, so there exists $P_i \in \mathcal{P}^m_c(\Omega_i, E, E')$ such that $P_i' = PM_i$. The usual multiplicative property of principal symbols implies that

$$\sigma_m(P_i') = \sigma_m(P)\phi_i'$$

so that P_i is tangentially elliptic on the support of ϕ_i. We must show that there exists $Q_i \in \mathcal{P}^{-m}_c(\Omega_i, E', E)$ such that $P_iQ_i - \phi_i$ is compactly smoothing.

Since P_i is elliptic on the support of ϕ_i with total symbol p and principal symbol $p_m \in S^m$ there exists some $q \in S^{-m}$ with $p_m q - \phi_i$ smoothing. Define q_k inductively by $q_0 = q$ and

$$q_k = -q \cdot \Sigma \, d_\xi^\alpha p \cdot D_\xi^\alpha q_j / \alpha! \in S^{-m-k}$$

where the sum is taken over all α, j, k with $j < k$ and $|\alpha| + j = k$. Let $\tilde{Q}_i \in \mathcal{P}^{-m}$ with total symbol $q_0\phi_i' + q_1\phi_i' + \dots$. This defines $\tilde{Q}_i \in \mathcal{P}^{-m}$ so that $\sigma(P_i\tilde{Q}_i - \phi_i) \sim 0$ on Supp(ϕ_i). Similarly we could solve $\sigma(\hat{Q}_iP_i - \phi_i) \sim 0$ for $\hat{Q}_i \in \mathcal{P}^{-m}$. We compute:

$$\sigma(\phi_i\tilde{Q}_i - \phi_i\hat{Q}_i) = \sigma(\phi_i\tilde{Q}_i - \hat{Q}_iP_i\tilde{Q}_i) + \sigma(\hat{Q}_iP_i\tilde{Q}_i - \phi_i\tilde{Q}_i)$$

$$= \sigma((\phi_i - \hat{Q}_iP_i)\tilde{Q}_i) + \sigma((\hat{Q}_iP_i - \phi_i)\tilde{Q}_i)$$

so $\phi_i(\tilde{Q}_i - \hat{Q}_i) \sim 0$ modulo smoothing operators on the support of ϕ_i, which implies that $\tilde{Q}_i$ and $\hat{Q}_i$ agree modulo smoothing operators. Their distributional kernels are compactly supported since ϕ_i' is compactly supported. Set $Q_i = \tilde{Q}_i \phi_i$; then $P_i Q_i - \phi_i$ is compactly smoothing. Set $Q = \Sigma \ M_i Q_i'$. Then $PQ - I_{E'}$ is compactly smoothing, which implies the result. $\square$

Corollary 7.13 [Co3, page 128]. Suppose that P_1, $P_2 \in \mathcal{P}^m(G(X),E,E')$ with P_2 elliptic. Then there is a constant $c < \infty$ such that

$$\|P_{1,x}\xi\| \leqslant c(\|P_{2,x}\xi\| + \|\xi\|)$$

for all $x \in X$ and for all $\xi \in C_c^\infty(G^x)$.

Proof (Connes). Let $Q_2 \in \mathcal{P}^{-m}(G(X),E',E)$ with $Q_2 P_2 - id_E$ smoothing. As $P_1 Q_2 \in \mathcal{P}^0$ (by 7.11d), there is a constant $c_1 < \infty$ with

$$\|P_1 Q_2(P_2 \xi)\| \leqslant c_1 \|P_2 \xi\|$$

for each $\xi \in C_c^\infty(G^x)$. As $P_1(Q_2 P_2 - id_E)$ is smoothing, one has

$$\|P_1 Q_2 P_2 \xi - P_1 \xi\| \leqslant c_2 \|\xi\|$$

for each $\xi \in C_c^\infty(G^x)$, which implies the result. $\square$

Remark 7.14 (Connes). We note two special cases of this corollary. First, suppose that P_2 is the identity. Then

$$\|P_{1,x}\xi\| \leqslant c(2\|\xi\|)$$

so that P_1 is a bounded operator. Second, suppose that $P_2 = (1 + \Delta)^m$, some power of the identity plus the tangential Laplacian. Then the corollary implies that

$$\| P_{1,x} \xi \| \leqslant \| (1 + \Delta)^m \xi \| + \| \xi \|.$$

In particular,

$$\| (1 + \Delta)^k \xi \|_s \leqslant \| \xi \|_{s+2} + \| \xi \|_s$$

for any k.

      Corollary 7.13 implies that if $P \in \mathscr{P}^s(E,E)$ is tangentially elliptic, then P defines a bounded invertible G(X)-operator

$$P: W^s_\tau \longrightarrow Dom(P)$$

where $Dom(P)$ has norm $\| \xi \| + \| P\xi \|$. This implies that each $Q \in \mathscr{P}^m(E,E')$ extends for each s to a bounded G(X)-invariant operator

$$Q: W^{s+m}_\tau(E) \longrightarrow W^s_\tau(E').$$

Proposition 7.15 [Co3].

a)   Let $U = L \times N$ be a distinguished coordinate patch, let $P \in \mathscr{P}^m_c(U,E,E')$, and let

$$P': W^{s+m}_\tau(E) \longrightarrow W^s_\tau(E')$$

be the canonical extension. Then there is a constant $b > 0$ (independent of P) such that

$$\| P' \|_{W^{s+m}_\tau, W^s_\tau} \leqslant b \sup_n \| P_n \|_{W^{s+m}, W^s}.$$

b)   Let ν be an invariant transverse measure with associated trace ϕ_ν on $W^*(G(X),\tilde{\mu})$. Then each $T \in W^*(G(X),\tilde{\mu})$ which has a continuous extension to

$$W_\tau^{-s}(E) \longrightarrow W_\tau^s(E)$$

for some $s > p$ (the dimension of the leaves) is in the domain of ϕ_ν and there is a constant c, independent of T, such that

$$|\phi_\nu(T)| \leq c\|T\|_{W^{-s},W^s}.$$

Proof. a) If $m = 0$ then this estimate follows as in the proof of Proposition 7.11b. In general, fix some s' and consider the tangential operator

$$Q = (1 + \Delta)^{-s'/2m}P(1 + \Delta)^{-s/2m}$$

where Δ is the tangential Hodge-Laplace operator $\Delta = \{\Delta_n\}$, Δ_n defined over $L \times \{n\}$, formed from the underlying tangential Riemannian connection. Then

$$\|Q\| = c\|P\|_{W^s,W^{s'}}.$$

If Q were in $\wp^0$ then the argument would be complete, but this is not so in general. However, we may uniformly approximate the distributional kernel of Q by kernels K_j supported on compact neighborhoods of the diagonal $\{(x,x)\} \times N$. Let T_j be the associated operator to K_j. Then $T_j \in \wp^0$, so that

$$\|T_j\| \leq \sup \|T_{j,n}\|$$

by the earlier estimate and the T_j uniformly approximate Q, which completes the argument.

b) There is some $S \in W^*(G(X),\widetilde{\mu})$ such that

$$T = (1 + \Delta)^{-s/2m}S(1 + \Delta)^{-s/2m}$$

with $\|S\| = \|T\|_{-s,s}.$ Proposition 7.11 implies that S has finite trace.

227

So it suffices to show that $(1 + \Delta)^{-s/m}$ is in the domain of ϕ_ν. Corollary 7.12 implies that there is a tangential pseudodifferential operator P of order $-s$ with

$$(1 + \Delta)^{-s/m} \leqslant P^*P.$$

So it suffices to show that $\phi_\nu(P^*P)$ is finite. Let K_P denote the distributional kernel of P. Restrict to a leaf ℓ. Proposition 1.12 implies that $(P^*P)_\ell$ is a locally traceable operator with local trace given by

$$\mu_{(P^*P)_\ell} = \int_y |K_P(y,x,n)|^2 d\lambda(y)\lambda(x).$$

Thus

$$\phi_\nu(P^*P) = \int_X \mu_{(P^*P)_\ell} d\nu = \int_X \int_y |K_P(y,x,n)|^2 d\lambda(y)\lambda(x)d\nu$$

which is finite by (7.11d) and the fact that

$$\int \|K_P\|_{HS}^2 d\nu < \infty. \qquad \square$$

Recall from (7.11b) that each pseudodifferential operator

$$P \in \wp^0(E,E)$$

extends to a bounded operator on the Hilbert field $L_\tau^2(G(X)) = \{L^2(G^x)\}$ with norm given by

$$\|P\| = \sup_{x \in X} \|P^x\|,$$

where P^x acts on $L^2(G^x)$. These form the *-algebra $\wp^0(G(X),E,E)$ which contains $C_{\tau c}^\infty(G(X))$ as a two-sided ideal. Taking closures we obtain a C^*-algebra $\bar{\wp}$ called the (closed) *pseudodifferential operator algebra* with closed two-sided ideal $C_r^*(G(X))$. (In fact $C_r^*(G(X))$ depends on E, but we suppress this for simplicity.) Recall that S^*F denotes the cosphere bundle of the foliated space.

Proposition 7.16 (Connes [Co3, page 138]). The tangential principal symbol map

$$\sigma: \wp^0(G(X),E,E) \longrightarrow \Gamma(S^*F, \, \text{End}_\tau(E))$$

is a surjective $*$-homomorphism. It extends to a surjection of C^*-algebras and induces a canonical short exact sequence of C^*-algebras

$$0 \longrightarrow C_r^*(G(X)) \longrightarrow \bar{\wp}^0 \overset{\sigma}{\longrightarrow} \Gamma(S^*F,\text{End}(E)) \longrightarrow 0$$

Proof. That σ is surjective is proved in the classical setting in [Pa, cf. p. 269, 246] by the construction of a continuous linear section. The general idea is to use partition of unity arguments to reduce down to the case of trivial vector bundles over open balls in Euclidean space, and then to explicitly write down the section. All this generalizes in an obvious way to our setting. It suffices, then, to compute $\ker(\sigma)$. It is clear that $\text{Ker}(\sigma)$ contains $C_r^*(G(X))$, so it suffices to prove the opposite inclusion. Note that since σ has a continuous linear section, any $T \in \wp^0$ with $\|\sigma(T)\|$ small has small spectral radius in $\bar{\wp}^0/C_r^*(G(X))$. The proposition then follows from the following Lemma (with $A = C_r^*(G(X))$, $B = \text{Ker}(\sigma)$).

Lemma 7.17. Let $\wp$ be a dense $*$-subalgebra of a C^*-algebra $\bar{\wp}$ and let $A \subset B \subset \bar{\wp}$ be ideals. Suppose that the following condition holds:

(*)   If $x \in \wp$ with $|x|$ small in $\bar{\wp}/B$ then the spectral radius $\rho(x)$ is small in $\bar{\wp}/A$.

Then $A = B$.

Proof. Let $\wp_A \subset \bar{\wp}/A$ be the (dense) image of $\wp$ and similarly for $\wp_B \subset \bar{\wp}/B$. Let

$$\psi: \wp_A \longrightarrow \wp_B$$

be the obvious surjection. If $x \in \wp_A$ with $\psi(x) = 0$ then $|\psi x| = 0$ in $\bar{\wp}/B$ by (*). Then $\rho(x) = 0$ in $\bar{\wp}/A$ and $x = 0$; thus ψ is injective and so an isomorphism. Let

$$\phi = \psi^{-1}: \wp_B \longrightarrow \wp_A.$$

Then ϕ is a bounded map, by (*), and it extends to

$$\bar{\phi}: \bar{\wp}/B \longrightarrow \bar{\wp}/A.$$

It is easy to see that $\bar{\phi}$ is the inverse to the natural projection $\bar{\psi}: \bar{\wp}/A \longrightarrow \bar{\wp}/B$, so $\bar{\psi}$ is an isomorphism, and $A = B$. □.

Note that if P is a smoothing operator of order 0 which is not compactly smoothing then it might not be in $\bar{\wp}$ and in particular not in $C_r^*(G(X))$. Such operators are, however, in the Breuer ideal of compact operators (cf. proof of 6.35) in $W^*(G(X), \tilde{\mu})$ as we shall see (cf. 7.37). Similarly, if P is (say) compactly smoothing with distribution kernel which is measurable but not continuous then the same conclusion holds.

The previous proposition enables us to extend the definition of tangential ellipticity to any P in the closure of $\wp^0$ by declaring P to be *tangentially elliptic* if $\sigma(P)$ is invertible.

The short exact sequence (7.16) induces a long exact sequence in K-theory and, in particular, there is a natural connecting homomorphism

$$\partial: K_1(\Gamma(S^*F, \text{End}(E))) \longrightarrow K_0(C_r^*(G(X)).$$

If P is a tangential, tangentially elliptic pseudodifferential operator of order zero, then its tangential principal symbol $\sigma_0(P)$ is invertible and hence defines a class in $K_1(\Gamma(S^*F, \text{End}(E)))$. Apply the connecting homomorphism ∂ and one obtains the index class

$$\mathrm{ind}(P) \in K_0(C_r^*(G(X))).$$

We remind the reader that the content of the Connes–Skandalis index theorem is to identify this class, while the content of the Connes index theorem is to identify the class

$$c(\mathrm{ind}(P)) \in \bar{H}_\tau^P(X),$$

where c is the partial Chern character.

The next step in the argument is to relate the index class $\mathrm{ind}(P)$ to the families of kernels $\mathrm{Ker}(P_L)$ and $\mathrm{Ker}(P_L^*)$, to the associated families of local traces, and to the associated von Neumann algebra projections.

Fix an invariant transverse measure ν and form the associated von Neumann algebra $W^*(G(X),\tilde{\mu})$ with trace ϕ_ν. It is clear from the construction that there is a natural map

$$\beta: \bar{\wp} \longrightarrow W^*(G(X),\tilde{\mu})$$

whose image is weakly dense. Let $\pi: C_r^*(G(X)) \longrightarrow W^*(G(X),\tilde{\mu})$ be the canonical map and let

$$\pi_*: K_0(C_r^*(G(X))) \longrightarrow K_0^f(W^*(G(X),\tilde{\mu}))$$

be the induced homomorphism. Recall that $\iota_P = \{\iota_P^x\}$ is the index measure of P and that $\mathrm{ind}(P) \in K_0(C_r^*(G(X)))$ is the image of the tangential principal symbol of P.

Proposition 7.18. Let $P \in \bar{\wp}^0$ be a tangentially elliptic operator. Then

a) $\mathrm{Ker}(\beta P)$ and $\mathrm{Ker}(\beta P^*)$ are <u>finite</u> projections in $W^*(G(X),\tilde{\mu})$, so that

$$[\mathrm{Ker}(\beta P)] - [\mathrm{Ker}(\beta P^*)] \in K_0^f(W^*(G(X),\tilde{\mu})).$$

b) $\kappa_*(\text{ind}(P)) = [\text{Ker}(\textit{B}P)] - [\text{Ker}(\textit{B}P^*)]$.

c) $c(\text{ind}(P)) = [\ell_P] \in \bar{H}_\tau^P(X)$.

d) $\text{ind}_\nu(P) \equiv \int \ell_P d\nu = \phi_\nu([\text{Ker}(\textit{B}P)] - [\text{Ker}(\textit{B}P^*)])$.

Proof. This is immediate from 6.35.   □

Corollary 7.19. Let $P \in \bar{\wp}^0$ be a tangentially elliptic operator. Then

$$\text{ind}_\nu(P) = \phi_\nu([\text{Ker}(\textit{B}P)] - [\text{Ker}(\textit{B}P^*)])$$

depends only upon the homotopy class of the principal symbol of P in $K_0(C_r^*(G(X)))$.

Proof. This is immediate from (7.18) and the fact that $\text{ind}(P) \in K_0(C_r^*(G(X)))$ depends only upon the homotopy class of the tangential principal symbol $\sigma_0(P)$ of P.   □

This completes our introduction to abstract tangential pseudodifferential operators.

B.   DIFFERENTIAL OPERATORS AND FINITE PROPAGATION

The most natural operators on foliated spaces are parametrized versions of the classical differential operators. These operators are unbounded, and it is necessary to exercise some care in promoting them to bounded operators in defining an index. There are at least two possible technical approaches. Connes prefers to use methods from geometric asymptotics. We have chosen to use finite propagation techniques, in part because of their lovely simplicity, and in part because we have been impressed by their efficacy as demonstrated, e.g., by Taylor [Tay2], Cheeger, Gromov, and Taylor [CGT], and more recently by John Roe [Ro3].

Definition 7.20. Let D be a first order differential operator over a

noncompact complete manifold with self adjoint principal symbol $\sigma_1(D)$. The propagation speed of D is defined by

$$c(x) = \text{Sup}\{\,\|\sigma_1(x,\xi)\| : \|\xi\| = 1\}.$$

If $c(x) \leqslant c$ then D is said to have *finite propagation speed* (cf. [CGT], [Tay2], [Ch], [Ro3]). In that case, solutions to the hyperbolic system

$$\left(\frac{\partial}{\partial t} + iD\right)u = 0$$

exist ([Fr]) and propagate at speeds bounded by c.

Recall that if D is a densely defined operator then the *formal adjoint* D^t of D is defined by $(D^t u, v) = (u, Dv)$. If $D = D^t$ then D is *formally self-adjoint*. In general, the closure $\bar{D}$ of D satisfies $\bar{D} \subset D^t$. A symmetric operator T is *essentially self adjoint* provided that $\bar{T}$ is self-adjoint, or equivalently, T^t is symmetric, in which case $\bar{T} = T^t$.

Theorem 7.21 **(Chernoff [Ch] Lemma 2.1).** Suppose that $D: \Gamma(E) \longrightarrow \Gamma(E)$ is a first order (not necessarily elliptic) differential operator over a noncompact complete manifold and suppose that D is formally self-adjoint and has finite propagation speed with a uniform bound $c(x) \leqslant c < +\infty$. Then D is essentially self-adjoint and, more generally, D^k is essentially self-adjoint for all k. Thus for any bounded Borel function on $\mathbb{R}$, f(D) is defined as a bounded operator on $L^2(E)$.

Proof. We repeat Chernoff's proof. Fix a positive integer k and let $A = D^k$. It suffices to show that there is no non-trivial solution to the eigenvalue equation $A^t u = \pm iu$; that is, there is no non-zero choice for u such that

$$\langle u, v \rangle \pm \langle u, Av \rangle = 0$$

for all $v \in Dom(A)$.

Suppose that $A^t u = iu$. We want to show that $u = 0$. Let $v \in C_c^\infty(M)$. Then $U_t = e^{itA}$ extends to a unitary operator on L^2. Define $F(t)$ by

$$F(t) = \langle U_t v, u \rangle.$$

The function F is bounded on $\mathbb{R}$ since U_t is unitary. The k'th derivative $F^{(k)}$ of $F(t)$ is given by

$$F^{(k)}(t) = \langle i^k D^k U_t v, u \rangle = \langle i^k A U_t v, u \rangle =$$

$$= \langle i^k U_t, A^t u \rangle = -i^{k+1} F(t).$$

Hence $F(t)$ is a linear combination of exponential functions $e^{\alpha t}$ where α runs through the solutions of the equation $\alpha^k = -i^{k+1}$. So none of the α's is pure imaginary. As F is bounded, this implies that F is identically zero, so that $\langle U_t v, u \rangle = 0$. Finite propagation implies that U_t restricts to an isomorphism $C_c^\infty(M) \longrightarrow C_c^\infty(M)$. Thus $\langle C_c^\infty(M), u \rangle = 0$ and so $u = 0$ as required. A similar argument applies to the solutions of $A^t u = -iu$. This establishes the theorem. $\square$

Pick some point $x \in X$. The map

$$\mathbb{R}^p \cong F_x \xrightarrow{\ \exp_x\ } X$$

maps some open p–ball B about the origin to a chart of ℓ_x, the leaf which contains x. Choose an orthonormal base for F_x and extend the map to

$$F_x \times N \xrightarrow{\ \exp_x\ } X$$

to obtain a "tangential normal coordinate system" at x. It is determined uniquely up to an element of $C(N, O(p))$, where $O(p)$ is the orthogonal group. Choose an orthonormal basis for S_x, the fibre of the bundle S at x. For $y \in \exp(B \times N) \cap \ell$ there is a well–defined isomorphism $S_x \longrightarrow S_y$ given by parallel transport. Thus a basis is determined for the sections of S over the patch. So fix a choice of basis at S_x; the resulting system is called a *canonical coordinate system*. Letting $\{e_i\}$ denote the basis vector-fields on exp(B×N) which correspond to the canonical coordinate system on S, then the tangential Levi-Civita connection acts by

$$\nabla_i(e_j) = \Sigma \; \Gamma^k_{ij} e_k \quad \text{and} \quad \nabla_i(s_\alpha) = \Sigma \; \Gamma^\beta_{i\alpha} s_\beta.$$

Definition 7.22. (X,S) has *bounded geometry* if

1) X has positive tangential injectivity radius; that is, there is a nonempty open ball $B \subset \mathbb{R}^p$ which is injected by the exponential map at every point of X,

2) For each leaf, the Christoffel symbols of the tangential connection on X lie in a bounded set of the Frechet space $C^\infty(B)$, and

3) For each leaf, the Christoffel symbols of the tangential connection of the bundle S lie in a bounded set of $C^\infty(B)$.

Proposition 7.23. Let M be a smooth Riemannian manifold with a C^∞ bounded geometry covered by open sets $\{U_j\}$ with exponential coordinate charts on each U_j of fixed radius c. Let D be an elliptic differential operator of positive order whose coefficients are bounded in C^∞ with a uniform ellipticity estimate. Then D and its formal adjoint D^t act as unbounded operators on $L^2(M)$ with domain $C_0^\infty(M)$, and the closure of D^t is the Hilbert space adjoint D^* of D.

Proof. (This proof was kindly supplied to us by M. Taylor.) We define D and D^t as unbounded operators on $L^2(M)$ with domain $Dom(D) = Dom(D^*) = C_0^\infty(M)$. We aim to prove that the closure of D^t is D^*. Suppose first that the order of D is even. Recall from (7.21) that all powers of the Laplace operator Δ are essentially self-adjoint, since

M is complete. Since by definition,

$$u \in Dom(D^*) \text{ iff}$$

$$|(u, Dv)| \leqslant C_u \|v\|_{L^2} \text{ for all } v \in C_o^\infty(M),$$

using local elliptic regularity of D^t, we can state

$$u \in Dom \ (D^*) \text{ iff}$$

$$u \in L^2(M), \ u \in W_{loc}^{2m}(M), \text{ and } D^t u \in L^2(M),$$

where D^t is a priori applied to u in the distributional sense. Since the weak and strong extensions of Δ^m coincide, we can say both that

$$u \in Dom(\Delta^m) \text{ iff}$$

$$u \in L^2(M), \ u \in W_{loc}^{2m}(M), \text{ and } \Delta^m u \in L^2(M),$$

and that

(*) $u \in Dom(\Delta^m)$ iff $u \in L^2(M)$ and for a sequence

$$v_j \in C_o^\infty(M), \ v_j \longrightarrow u \text{ in } L^2(M)$$

$$\text{then } \Delta^m(v_j - v_k) \longrightarrow 0 \text{ in } L^2(M) \text{ as } j,k \longrightarrow \infty.$$

Now elliptic estimates bound L^2 norms of $D^t u$ over a ball $V_j \subset M$ ($V_j \subset U_j$, say of radius $c_o/2$), in terms of L^2 norms of $\Delta^m u$ and of u over U_j (with bounds independent of j) and conversely, one has a bound on L^2 norms of $\Delta^m u$ over V_j in terms of L^2 norms of $D^t u$ and u over U_j. One can suppose the V_j cover M and that the U_j do not have too many overlaps, so we deduce

$$Dom(D^*) = Dom(\Delta^m).$$

From here it is easy to complete the proof. Indeed, given $u \in Dom(D^*) = Dom(\Delta^m)$, we know by (*) that there exist $v_j \in C_o^\infty(M)$ such that $v_j \longrightarrow 0$ in $L^2(M)$ and $\Delta^m(v_j - v_k) \longrightarrow 0$ in $L^2(M)$, as $j,k \longrightarrow \infty$. The boundedness hypotheses on the coefficients of D^t, together with elliptic estimates, imply

$$\|D^t w\|_{L^2(M)} \leq C \|\Delta^m w\|_{L^2(M)} + C \|w\|_{L^2(M)}, \quad w \in C_o^\infty(M).$$

Thus,

$$\|D^*(v_j - v_k)\|_{L^2} \leq C \|\Delta^m(v_j - v_k)\|_{L^2} + C \|v_j - v_k\|_{L^2} \longrightarrow 0$$

$$\text{as } j,k \longrightarrow \infty$$

and the theorem is established for D of even order.

It remains to consider the case when D is of odd order. Let P_0 be the closure of D^t, the minimal extension of D^t and let $P_1 = D^*$, the Hilbert space adjoint, which is the maximal extension of D^t. Clearly $P_0 \subset P_1$. Let $A = P_0^* P_0$ and $B = P_1^* P_1$. By von Neumann's theorem, A and B are self-adjoint and

$$Dom(A^{1/2}) = Dom(P_0), \qquad Dom(B^{1/2}) = Dom(P_1).$$

However A and B are extensions of the even order elliptic operator $D^t D$. The previous case implies that $A = B$. Thus $Dom(P_0) = Dom(P_1)$ and we are through. $\square$

If D is a tangential differential operator then $Ker(D) = \{Ker(D_\ell)\}$ forms a measurable field of Hilbert spaces. If (a suitable closure of) D is locally traceable along the leaves then there is also associated a tangential measure (cf. 4.11)

$$\mu_{Ker(D)} = \{\mu_{Ker(D_\ell)}\},$$

where $\mu_{\text{Ker}(D_\ell)}$ is the local dimension (defined in Chapter I, after 1.8) of the orthogonal projection onto the subspace $\text{Ker}(D_\ell)$. Similarly there is a natural tangential measure $\mu_{\text{Ker}(D^*)}$. These measures would appear to depend upon the choice of closure of D. This problem is disposed of by the following Corollary.

Corollary 7.24. Let X be a compact foliated space with some fixed tangential Riemannian metric and let D be a tangentially elliptic differential operator. Then the (leafwise) closure of the (leafwise) formal adjoint of D is the (leafwise) Hilbert space adjoint of D. Thus D has a unique closure. Hence ker D and ker D* are uniquely defined Hilbert fields, and $\mu_{\text{Ker}(D)}$ and $\mu_{\text{Ker}(D^*)}$ are uniquely defined tangential measures.

Proof. This follows immediately from the preceeding proposition and the observation that if ℓ is a leaf in X then ℓ is a Riemannian manifold with bounded geometry as required. □

It still remains to define the index of a tangential differential operator of positive order. The most natural definition at this point is to form an index measure

$$\iota_D = \mu_{\text{Ker}(D)} - \mu_{\text{Ker}(D^*)}$$

which is unique, by 7.24, and let the index be the total mass of this measure:

$$\text{ind}_\nu(D) = \int \iota_D d\nu.$$

As this stands it is not at all clear how this corresponds to the index of order zero operators and the canonical pseudodifferential operator extension, nor is it clear how to compute. We turn to these matters next.

Let D be a tangential, tangentially elliptic differential operator of positive order m from sections of E to sections of E'. Then D extends to a densely defined unbounded operator $D = \{D_\ell\}$ of

$$L^2_\tau(\tilde{E}) \longrightarrow L^2_\tau(\tilde{E}')$$

where $\tilde{\ell}$ is the holonomy covering of the leaf ℓ, $\tilde{E}$, $\tilde{E}'$ are the pullbacks of the bundles to the holonomy groupoid $G = G(X)$ and L^2 denotes the corresponding Hilbert fields obtained by pulling the bundles back to $G(X)$, lifting the action of D, and then restricting. The operator D has a unique leafwise closure, by Corollary 7.24, which for convenience we also denote by D. By standard functional analysis, $(1 + D^*D)$ is a positive operator which is bounded below and hence has an inverse $(1 + D^*D)^{-1}$ which is a <u>bounded</u> operator $L^2_\tau(\tilde{E}'_\ell) \longrightarrow L^2_\tau(\tilde{E}_\ell)$. Recall that

$$Dom(D^*D) = \{\phi: \phi \in Dom(D), D\phi \in Dom(D^*)\}.$$

Then $(1 + D^*D)$ has a square root $(1 + D^*D)^{1/2}$ by standard functional analysis. The spectral theorem implies that

$$Dom((1 + D^*D)^{1/2}) = Dom((D^*D)^{1/2}).$$

As $A = (D^*D)^{1/2}$ is the positive part of the polar decomposition $D = UA$ (U partial isometry), $Dom(A) = Dom(D)$ and so

$$Dom((1 + D^*D)^{1/2}) = Dom(D).$$

This implies that the operator $(1 + D^*D)^{-1/2}$ has range equal to $Dom(D)$. Since the operator $(1 + D^*D)^\alpha$ is onto for each $\alpha > 0$, the operator $(1 + D^*D)^{-1/2}$ is defined on all of $L^2(\tilde{E})$. Thus

$$L = D(1 + D^*D)^{-1/2}$$

makes sense and is bounded by direct composition. In polar form, $L = UB$. That is L has the same polar part U as $D = UA$; D has been replaced by $D(1 + D^*D)^{-1/2}$, a bounded version of D, and

$$B = A(1 + A)^{-1} = [D^*D(1 + D^*D)^{-1}]^{1/2}.$$

Note that $L = UB$ and $D = UA$ have the same kernel. The closure of the ranges is likewise the same. Hence

$$Ker(L) = Ker(D) \quad \text{and} \quad Ker(L^*) = Ker(D^*)$$

in the von Neumann algebra $W^*(G(X),\tilde{\mu})$. If we knew that L were in $\wp^0$ or even in $\bar{\wp}^0$ then we would know that these projections were ν-finite and that the ν-index of D was just the ν-index of L. One can establish this in greater generality using methods of Connes, but we specialize to first order operators.

Theorem 7.25 Taylor [Tay, Ch. XII] and Roe [Ro3]. Let $D = \{D_\ell\}$ be a tangential, tangentially elliptic and tangentially formally self-adjoint operator. Lift each D_ℓ to its holonomy covering $D_{\tilde{\ell}}$. Let f be a bounded Borel function, so that $f(D_{\tilde{\ell}})$ is defined by the spectral theorem. Let $f(D) \equiv \{f(D_{\tilde{\ell}})\}$ act on the canonical Hilbert field $L^2_{\mathcal{T}}(G(X))$ of $C^*_r(G(X))$. Then:

1)   If f is a Schwartz function with Fourier transform $\hat{f}$, then

$$f(D) = (1/2\pi)\int \hat{f}(t)e^{itD}dt$$

where the integral is understood to be in the weak sense along the leaves.

2)   If D is first order with finite propagation speed on each leaf and $f \in C_0(\mathbb{R})$ with Fourier transform $\hat{f} \in C^\infty_c(\mathbb{R})$ then $f(D) \in C^\infty_{\mathcal{T}c}(G(X))$.

3)   If D is first order with finite propagation speed on each leaf and $f \in C_0(\mathbb{R})$ then $f(D) \in C^*_r(G(X))$.

Proof. Part 1) is proved by [Tay, Ch. XII]. For parts 2) and 3) see [Ro3, Theorem 2.1 and Corollary 2.2.].   □

Corollary 7.26 (Roe). Let D be a first order tangential tangentially elliptic and tangentially formally self-adjoint differential operator from sections of E to sections of E' with uniformly bounded propagation speed on all leaves. Define

$$L = D(1 + D^2)^{-1/2}.$$

Then L is a bounded operator on L^2, and $L \in \bar{\wp}^0$.

Proof. It suffices to prove that $L \in \bar{\wp}^0$. Let

$$f(x) = x(1 + x^2)^{-1/2}.$$

Then

$$f'(x) = (1 + x^2)^{-3/2}$$

and

$$f''(x) = O(|x|)^{-3} \text{ at } \infty.$$

Regard f as a tempered distribution (i.e., as a functional on the Schwartz space) and let $g(\xi)$ be the Fourier transform of f. Then g is itself a tempered distribution, and

$$(f')^{\wedge} = i\xi g(\xi) \qquad (f'')^{\wedge} = -\xi^2 g(\xi).$$

Thus $\xi g(\xi)$ is a function and f'' is in L^2, which implies that $(f'')^{\wedge} = -\xi^2 g(\xi)$ is bounded, so that $g(\xi) = O(\xi^{-2})$ at ∞. Write $g = g_1 + g_2$, where g_1 has support very near 0 and $g_2 \in L^1(\mathbb{R})$. Then

$$f(D) = \int g_1(\xi)e^{i\xi D}d\xi + \int g_2(\xi)e^{i\xi D}d\xi.$$

The inverse Fourier transform of g_2 belongs to $C_0(\mathbb{R})$ by the Riemann-Lebesgue lemma, so the second term is in $C_r^*(G(X))$. The first term is properly supported, by the finite propagation speed

241

condition, and it is a pseudodifferential operator by the argument of [Tay, Theorem 1.3, p. 296]; thus it belongs to $\wp^0$, and hence $f(D) \in \bar{\wp}^0$. □

The preceeding Corollary shows us how to fit classical first order tangentially elliptic operators into the general framework of tangential pseudodifferential operators presented in Section A. For arbitrary higher order differential operators we adopt an alternate strategy: we work directly at the von Neumann algebra level.

Proposition 7.27. Let T be a tangential, tangentially elliptic pseudodifferential operator of order $m > 0$, and let Δ be the tangential Hodge–Laplace operator associated to the bundle of T. Define $P = (1+\Delta)^{-m/2}T$. Then $P \in \bar{\wp}^0$, $\sigma_0(P)$ is homotopic to $\sigma_m(T)$, and

$$\phi_\nu(\text{Ker } P) = \phi_\nu(\text{Ker } T)$$

and

$$\phi_\nu(\text{Ker } P^*) = \phi_\nu(\text{Ker } T^*)$$

in $W^*(G(X)), \tilde{\mu})$.

Proof. Since $P = (\text{invertible})T$, $\sigma_0(P)$ is homotopic to $\sigma_m(T)$. It suffices to prove that $P \in \bar{\wp}^0$ and $\phi_\nu(\text{Ker } P^*) = \phi_\nu(\text{Ker } T^*)$. For the first, note that Δ is the 0-form component of $D = d + d^*$ extended to the bundle via the connection. Use the leafwise finite propagation speed property of the operator D to write

$$(1 + \Delta)^{-m/2}T = \int \hat{g}_1(\xi)e^{i\xi D}Td\xi + \int \hat{g}_2(\xi)e^{i\xi D}Td\xi$$

$$= g_1(D)T + g_2(D)T.$$

Then $g_1(D)T \in \wp^0$ as in (7.26). The operator $g_2(D)$ is smoothing and tends to 0 if we make g_1 have large support. Thus $\|g_2(D)T\|_{L^2}$

$\longrightarrow 0$ and thus $P \in \bar{\varphi}^0$.

It remains to prove that $\phi_\nu(\text{Ker } P^*) = \phi_\nu(\text{Ker } T^*)$ in W^*. If π is the orthogonal projection onto Ker T^* then the orthogonal projection onto Ker P^* is given by

$$\pi' = (1 + \Delta)^{-m/2}\pi(1 + \Delta)^{m/2}.$$

We should like to say that π and π' have the same trace. **This is** not immediate, since $(1 + \Delta)^{m/2}$ is an unbounded operator. The ellipticity estimate in general takes the form

$$|(1 + \Delta)^{m/2}\phi| \leqslant |T^*\phi| + c|\phi|.$$

If $\phi \in \text{Ker}(T^*)$ then

$$|(1 + \Delta)^{m/2}\phi| \leqslant c|\phi|.$$

and thus $(1 + \Delta)^{m/2}$ is bounded on $\text{Ker}(T^*)$, and $\pi(1 + \Delta)^{m/2}$ is bounded. Thus

$$\phi_\nu((1 + \Delta)^{-m/2}\pi(1 + \Delta)^{m/2})$$

$$= \phi_\nu((1 + \Delta)^{-m/2}\pi\pi(1 + \Delta)^{m/2})$$

$$= \phi_\nu(\pi(1 + \Delta)^{m/2}(1 + \Delta)^{-m/2}\pi) = \phi_\nu(\pi). \quad \square$$

This completes our general study of tangential differential operators with finite propagation speed. These results will be used in Section C, which deals with a special class of tangential differential operators which are closely tied to the geometry of foliated spaces.

C.   DIRAC OPERATORS AND THE McKEAN–SINGER FORMULA

We turn now to the study of generalized Dirac operators and asymptotics. The goal of this section is the McKean–Singer formula

(7.39) which is the bridge to the asymptotic development of the index $\text{ind}_\nu(D)$.

Assume for the rest of the chapter that X is a compact foliated space with oriented foliation bundle F which is equipped with the Levi–Civita tangential connection (5.18) and associated tangential Riemannian metric. Each leaf of X is a complete Riemannian manifold with bounded geometry. Suppose that V is a real inner product space. We write Cliff(V) for its associated *Clifford algebra*. The Clifford algebra is universal with respect to linear maps $j\colon V \longrightarrow A$, where A is a real unital algebra and

$$(v,v)1 + (jv)^2 = 0,$$

and this characterizes the algebra. More concretely, Cliff(V) may be regarded as the free associative unital algebra on the basis vectors $\{e_k\}$ of V modulo the relations

$$e_i e_j + e_j e_i = 0 \ (i \neq j) \text{ and } e_i^2 = -1 \quad \text{all } i,j.$$

Let $\text{Cliff}^C(V) = \text{Cliff}(V) \otimes_{\mathbb{R}} \mathbb{C}$ be the complexified algebra. If $E \longrightarrow X$ is any real Riemannian vector bundle, then $\text{Cliff}^C(E)$ is the associated bundle of Clifford algebras. In particular, if X is a foliated space with a tangential Riemannian metric then we may form $\text{Cliff}^C(X) \equiv \text{Cliff}^C(F)$, where F is the tangent bundle of the foliated space.

Definition 7.28. If S is a bundle of left modules over $\text{Cliff}^C(X)$, then S is a *tangential Clifford bundle* if it is equipped with a tangential Hermitian metric and compatible tangential connection such that

a)   if $e \in F_x$ then $e\colon S_x \longrightarrow S_x$ is an isometry.

b)   if $\phi \in \Gamma_\tau(\text{Cliff}^C(X))$, $s \in S$, then

$$\nabla(\phi s) = \phi\nabla(s) + (\nabla\phi)s.$$

244

If S has an involution which anticommutes with the Clifford action of tangent vectors, then it is a __graded__ Clifford bundle. Associated to a tangential Clifford bundle S is a natural first order differential operator $D = D_S$ called the *(generalized) Dirac operator*. It is defined to be the composition

$$\Gamma_\tau(S) \longrightarrow \Gamma_\tau(F^*\!\otimes S) \longrightarrow \Gamma_\tau(F\otimes S) \longrightarrow \Gamma_\tau(S)$$

where the first map is given by the tangential connection, the second by the tangential Riemannian metric, and the third by the Clifford module structure on S. In an orthonormal basis $\{e_1,...,e_p\}$ for F_x one may write

$$(Ds)_x = \Sigma\, e_k(\nabla_k s)_x.$$

If S is graded then D is similarly graded; it interchanges sections of the positive and negative eigenbundles of the involution. In Chapter VIII we shall show that this definition encompasses the operators of primary interest in the proof (and in many applications) of the index theorem.

Lemma 7.29. The Dirac operator is formally self–adjoint on each leaf.

Proof. Fix tangentially smooth sections r, s of S, one of which is compactly supported. Let α be the tangential 1-form

$$\alpha(v) = -(r, vs)$$

where vs is the module action of s on v. Let $e_1,...,e_p$ be a normal basis of vector fields near x. Then

$$(Dr,s)_x = \Sigma(e_i\nabla_i r, s)_x$$

$$= -\Sigma(\nabla_i r, e_i s)_x$$

$$= -\Sigma[\nabla_i(r, e_i s) - (r, e_i\nabla_i s)]$$

$$= (d^* \alpha)_X + (r, Ds)_X$$

and hence $\langle Dr, s \rangle = \langle r, Ds \rangle$ by integration, as required.   □

Form the Hilbert field $L^2_\tau(S)$ by completing the tangentially smooth sections of S along each leaf in the norm determined by the tangential Riemannian metric.

Proposition 7.30.   The Dirac operator is essentially self adjoint regarded as an operator on the Hilbert field $L^2_\tau(S)$.   Thus (by 7.21) if f is a bounded Borel function on $\mathbb{R}$ then f(D) is defined as a bounded operator on $L^2_\tau(S)$.

Proof.   This follows immediately from 7.29 and 7.21.   □

Let R: $\Lambda^2 F^* \longrightarrow$ End(S) be the tangential curvature operator associated to the tangential connection on S.   Define R' ∈ End(S) by

$$R'(s) = (1/2)\Sigma \ e_i e_j R(e_i \wedge e_j)s$$

with respect to the orthonormal basis.

Proposition 7.31 (**Weitzenbock formula**).   For $s \in \Gamma_\tau(S)$, $D^2 s = \nabla^* \nabla s + R's$.

Proof.   Work in normal coordinates at a point.   Then the result is formal:

$$D^2 s = \Sigma \ e_i \nabla_i (e_j \nabla_j s)$$

$$= -\Sigma \ \nabla_j \nabla_i s + \sum_{i < j} e_i e_j (\nabla_i \nabla_j - \nabla_j \nabla_i) s$$

$$= \nabla^* \nabla s + R's.   \square$$

This formula allows us to analyze the coefficients of the Dirac operator. Suppose that (X,S) has bounded geometry. Using tangential normal coordinates near a point of X, one may regard the operator D^2 as a partial differential operator along the leaves of B×N acting on matrix-valued functions. By the Weitzenbock formula, D^2 may be written as

(7.32) $$-\sum_{i,j} g_{ij}(x)(\partial/\partial x^i)(\partial/\partial x^j) + \Sigma_j(a_j\partial/\partial x^j) + b$$

where the a_j and b are matrix-valued tangentially smooth functions on B×N which are constant in n and which, by virtue of the bounded geometry, may be estimated independently of the particular point of X chosen. In particular one sees that D^2 is a tangentially elliptic operator with principal symbol $-\xi^2$. As the origin of the tangential normal coordinate system varies, the operators D^2 form a bounded family of tangentially elliptic operators with the same tangential principal symbol.

Suppose that f is a function on $\mathbb{R}^p$ supported within B/2. Then f may be regarded as defined on B×N and D^2 may be regarded as acting on f. The elliptic estimate 7.14 applied to D^2 gives

(7.33) $$\|f\|_{k+2} \leqslant C(\|f\|_k + \|D^2f\|_k)$$

for some constant C and the usual Sobolev norms. Moreover, since det(g) is bounded away from zero locally and (by the compactness of X) globally with a global lower bound, the tangential principal symbol of D^2 is bounded away from zero with a global lower bound, so C may be chosen uniformly on X. This makes it possible to prove a Sobolev embedding theorem for X.

Definition 7.34. Suppose $r \geqslant 0$. The *uniform* C^r *space* $UC^r_\tau(S)$ is the bundle obtained by taking over each leaf the Banach space of all C^r_τ sections s of S (restricted to the leaf) such that the norm

$$\|\,\|s\|\,\|_r = \sup\{\, |\nabla_{v_1}...\nabla_{v_q} s(x)|\,\}$$

is finite, where sup is over all $x \in X$ and over all choices $v_1,...,v_q$

$(0 \leqslant q \leqslant r)$ of unit tangent vectors at x.

Theorem 7.35 [Ro0 5.20]. Let k be an even integer with $k > r + p/2$. Then $W_\tau^k(S)$ is included continuously in $UC_\tau^r(S)$.

Proof. We may immediately restrict attention to some leaf ℓ. [The constant involved in the elliptic estimate is continuous from leaf to leaf.] Choose some $s \in W^k(\ell)$. Then

$$\| \|s\| \| = \sup_{(B,\phi)} \|\phi s\|_{UC^r(\ell)}$$

where $B \subset \mathbb{R}^p$, $\phi:B \longrightarrow [0,1]$ is a smooth function supported on $(B/2)$ with $\phi \equiv 1$ on $(B/4)$ and ϕs is regarded as a function on $(B/2)$. Then

$$\|\phi s\|^2_{UC^r(\ell)} = \|\phi s\|^2_{C^r(\mathbb{R}^p)}$$

$$\leqslant \|R\| \|\phi s\|_{W^k(\mathbb{R}^p)}$$

by the classical Sobolev embedding

$$\leqslant c_1 \|s\|_{W^k((B/2))}$$

$$\leqslant c_2 \|s\|_{W^k(\ell)}$$

and so $\| \|s\| \| \leqslant (\text{const}) \|s\|_{W^k(\ell)}$ as required. $\square$

Theorem 7.36 [Ro0 5.21]. Suppose that X is a compact foliated space and that P is a tangential, tangentially elliptic differential operator on the module S. Let $f \in \mathcal{S}(\mathbb{R})$, the Schwartz space. Then the operator f(P), which is defined (leafwise) by the spectral theorem is a tangential smoothing operator and its distribution kernel

$$K_f \in \Gamma_\tau(S^* \otimes \Lambda F^* \otimes S)$$

is uniformly bounded.

Proof. Since f is a Schwartz function, the functions $x \longrightarrow |x|^k f(x)$ are bounded on $\mathbb{R}$ for any k, which implies by the spectral theorem that the operators $P^k f(P)$ are bounded on $L^2(\ell, S)$ for each leaf ℓ. Thus f(P) maps $W^k(S)$ into $W^{k+n}(S)$ for any k and n. If follows in the usual way that f(P) is a smoothing operator when restricted to each leaf. Thus f(P) is tangentially smoothing. As for the uniform bound, let $x \in X$ and $v \in S_x^*$; let $\epsilon_{x,v}$ be the distributional section defined by

$$\epsilon_{x,v}(s) = \langle s_x, v \rangle$$

for $s \in \Gamma_\tau(S)$. Then $\epsilon_{x,v} \in UC^r(S)^* \subset W^{-k}(S)$ (by 7.35) and so

$$K_f(v,-) = f(P)\epsilon_{x,v} \in W^k(S) \subset UC^r(S)$$

by (7.35) again. This implies that K_f is uniformly bounded. □

Proposition 7.37. Let P be a self-adjoint tangentially elliptic differential operator of any positive order on a module S with bounded geometry, and let $f \in C_0(\mathbb{R})$. Then f(P) is in the Breuer ideal [Bre] of compact operators in $W^*(G(X), \tilde{\mu})$. Similarly, if $P \geq 0$ then e^{-tP} is in the Breuer ideal.

Proof. By a continuity and density argument we may assume that f is compactly supported and smooth. Then 7.36 gives a distribution kernel for f(P) which is bounded on G, and similarly for $f(P)^2$. This implies that

$$\text{Tr}(f(P)^2) = \int |K_{f(P)}(x,x)|^2 d\nu < \infty$$

so that $f(P)^2$ is trace class. Thus f(P) is Hilbert-Schmidt, so in particular, f(P) is compact. The same argument applies to e^{-tP} for P positive since the function e^{-tz} may be replaced by a function which dies quickly off $\mathbb{R}^+$ and which agrees with e^{-tz} on $\mathbb{R}^+$. □

It is now possible to prove a generalization of the

McKean–Singer formula, which provides a bridge to the asymptotic expansion.

Proposition 7.38. Fix an invariant transverse measure ν on X and let

$$D: \Gamma_{\tau}(E) \longrightarrow \Gamma_{\tau}(E')$$

be a tangentially elliptic differential operator of order $m > 0$ on modules with bounded geometry. Then for each $t > 0$, the operators

$$e^{-tD^*D} \text{ and } e^{-tDD^*}$$

are tangential smoothing operators whose distribution kernels are functions which are uniformly bounded. The corresponding elements in $W^*(G(X),\tilde{\mu})$ have finite trace, and

(7.39) $$\text{ind}_{\nu}(D) = \phi_{\nu}([e^{-tD^*D}] - [e^{-tDD^*}]).$$

Proof. The first part of the theorem is immediate from (7.37). To establish formula (7.39), we argue as follows. Write $D = UA$ in polar form, where $A = (D^*D)^{1/2}$ and U is a partial isometry. Then U^*U is the projection onto $(\text{Ker } D)^{\perp}$, UU^* is the projection onto $(\text{Ker } D^*)^{\perp}$, and

$$U^*(DD^*)U = U^*UAAU^*U = ID^*DI = D^*D$$

so U is an equivalence between corresponding spectral projections of D^*D and DD^* for any Borel subset of $(0,+\infty)$. Thus

$$\phi_{\nu}([e^{-tD^*D}] - [e^{-tDD^*}]) =$$

$$= \phi_{\nu}([\beta \text{ Ker } D] - [\beta \text{ Ker } D^*]) =$$

$$= \text{ind}_{\nu}(D). \quad \square$$

The next step in our development is the reformulation of the McKean–Singer formula (7.39) in terms of superoperators, and the resulting asymptotic expansion of the heat kernel.

D.   SUPEROPERATORS AND THE ASYMPTOTIC EXPANSION

In this section the McKean–Singer formula (7.39) is rephrased in terms of superoperators.   Then symbols which depend upon a complex parameter are introduced and the asymptotic expansion (Theorem 7.48) for $\text{ind}_\nu(D)$ is developed.   This leads to the formula (7.48), which expresses $\text{ind}_\nu(D)$ as the total mass of the tangentially smooth p-form $\omega_D(g,E)$.   More detailed study of this form for particular Dirac operators leads in Chapter VIII to the proof of the index theorem.

Definition 7.40. A *graded vector space* is a vector space of the form $V = V^+ \oplus V^-$ thought of as the eigenspace decomposition of an involution of V. A *superoperator* is an operator $T: V \longrightarrow V$. A superoperator has an obvious 2 by 2 matrix decomposition and is said to be *grade-preserving* if $T_{ij} = 0$ for $i \neq j$. A trace ϕ on $\mathscr{L}(V^+)$ and $\mathscr{L}(V^-)$ extends to a *supertrace* ϕ^S on V by

$$\phi^S(T) = \phi(T_{11}) - \phi(T_{22}).$$

If $D: V^+ \longrightarrow V^-$ then define its *associated superoperator* $\hat{D}$ by

$$\hat{D} = \begin{bmatrix} 0 & D^* \\ D & 0 \end{bmatrix}^2 = \begin{bmatrix} D^*D & 0 \\ 0 & DD^* \end{bmatrix} : V \longrightarrow V.$$

Then $\hat{D}$ is a positive grade-preserving superoperator of order twice the order of D.   The utility of this construction is illustrated by the following corollary.

Corollary 7.41. Fix an invariant transverse measure of X and let

$$D: \Gamma_\tau(E^+) \longrightarrow \Gamma_\tau(E^-)$$

be a tangentially elliptic differential operator on modules with bounded geometry with associated superoperator $\hat{D}$. Then

$$\mathrm{ind}_{\nu}(D) = \phi_{\nu}^{s}(e^{-t\hat{D}}).$$

Proof. This is immediate from 7.39 and the definition of the supertrace. $\square$.

Next we develop the machinery of complex symbols in order to produce the asymptotic expansion of $\phi_{\nu}^{s}(e^{-t\hat{D}})$. Let us assume given a tangentially elliptic differential operator D of order $m/2$ with positive definite principal symbol. Let $\hat{D}$ be the associated superoperator of order m; say $\hat{D}: \Gamma_{\tau}(E) \longrightarrow \Gamma_{\tau}(E)$. Let $\mathcal{C}$ be a fixed curve of distance ≥ 1 from the positive real axis as shown.

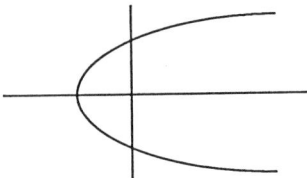

For $\mathfrak{s} \in \mathcal{C}$, the operator $(\hat{D}-\mathfrak{s})^{-1}$ is defined by the spectral theorem and has norm $|(\hat{D}-\mathfrak{s})^{-1}| \leq 1$. The map

$$\mathfrak{s} \longrightarrow |(\hat{D}-\mathfrak{s})^{-1}|$$

is continuous for $\mathfrak{s} \in \mathcal{C}$, and hence we may write

$$e^{-t\hat{D}} = (2\pi i)^{-1}\int_{\mathcal{C}} e^{-t\mathfrak{s}}(\hat{D}-\mathfrak{s})^{-1}d\mathfrak{s}.$$

Extend $(\hat{D}-\mathfrak{s})^{-1}$ to graded tangential Sobolev spaces. Note that $|\mathfrak{s}-x|^{-1} \leq 1$ for $\mathfrak{s} \in \mathcal{C}$, $x \in \mathbb{R}^{+}$, so that

$$|(\hat{D}-\mathfrak{s})^{-1}f|_{0} \leq c|f|_{0}.$$

Then

$$|(\hat{D}-\mathfrak{s})^{-1}f|_{km} \leqslant c\{|\hat{D}^k(\hat{D}-\mathfrak{s})^{-1}f|_0 + |(\hat{D}-\mathfrak{s})^{-1}f|_0\}$$

(by [Gi3] 1.3.5)

$$\leqslant c\{|\hat{D}^{k-1}f|_0 + |\mathfrak{s}\hat{D}^{k-1}(\hat{D}-\mathfrak{s})^{-1}f|_0 + |f|_0\}$$

$$\leqslant c\{|f|_{km-m} + |\mathfrak{s}||(\hat{D}-\mathfrak{s})^{-1}f|_{km-m}\}.$$

If $k = 1$ we obtain

$$|(\hat{D}-\mathfrak{s})^{-1}f|_m \leqslant c(1 + |\mathfrak{s}|)|f|_0$$

and then by induction

$$|(\hat{D}-\mathfrak{s})^{-1}f|_{km} \leqslant c(1 + |\mathfrak{s}|)^{k-1}|f|_{km-m}.$$

Interpolate to obtain

Proposition 7.42. Let $\hat{D}$ be as above. Then given s, there is a $k = k(s)$ and $c = c(s)$ so that

$$|(\hat{D}-\mathfrak{s})^{-1}f|_s \leqslant c(1 + |\mathfrak{s}|)^{k-1}|f|_s.$$

for all $\mathfrak{s} \in \mathcal{C}$.

Definition (7.43). $q(x,\xi,\mathfrak{s},n) \in S^k(\mathfrak{s})(U{\times}N)$ is a *symbol of order k depending on* $\mathfrak{s}$ if q is smooth in $(x,\xi,\mathfrak{s})$, continuous in n, has compact x-support in U, is holomorphic in $\mathfrak{s}$ and if there are estimates

(*) $\quad |D_x^\alpha D_\xi^\beta D_\mathfrak{s}^\gamma q| \leqslant C_{\alpha,\beta,\gamma}(1 + |\xi| + |\mathfrak{s}|^{1/m})^{k-|\beta|-m|\gamma|}.$

Say that q is *homogeneous of order k in* $(\xi,\mathfrak{s})$ if

$$q(x,t\xi,t^m\mathfrak{s}) = t^k(x,\xi,\mathfrak{s}) \text{ for } t \geqslant 1.$$

Homogeneity implies the decay condition (*). Grading in this manner corresponds to regarding $\mathfrak{s}$ as having order m. It follows that $(p_m - \mathfrak{s})^{-1} \in S^{-m}(\mathfrak{s})$. Further, the spaces $S^*(\mathfrak{s})$ are closed under multiplication and differentiation in the usual manner. In particular, given $n > 0$, there is some $k(n) > 0$ such that if Q has symbol in $S^k(\mathfrak{s})$ then it induces $Q(\mathfrak{s})$: $W^{-n} \longrightarrow W^n$ and

(7.44) $$|Q(\mathfrak{s})|_{-n,n} \leqslant C(1+|\mathfrak{s}|)^{-n}$$

Let $\wp_{\mathfrak{s}}^k(U{\times}N)$ denote those grade-preserving pseudodifferential superoperators

$$C_{\mathcal{T}c}^{\infty}(U{\times}N) \longmapsto C_{\mathcal{T}c}^{\infty}(U{\times}N)$$

with symbol in $S^k(\mathfrak{s})$ and x-support in U. For $\mathfrak{s}$ fixed, $Q(\mathfrak{s}) \in \wp^k(U{\times}N)$. If h: $U{\times}N \longrightarrow \tilde{U}{\times}\tilde{N}$ is a tangentially smooth homeomorphism then h induces a map $\wp_{\mathfrak{s}}^k(U{\times}N) \longrightarrow \wp_{\mathfrak{s}}^k(\tilde{U}{\times}\tilde{N})$ which respects principal symbols, so that it makes sense to speak of operators with complex symbol on $C_{\mathcal{T}c}^{\infty}(X)$ or even of operators on bundles over X. Finally, define $\wp_{\mathfrak{s}}^k(G(X))$ as before-, that is, as finite linear combinations of pseudodifferential operators from patches lifted from $\wp_{\mathfrak{s}}^k(U{\times}N))$ and compactly smoothing operators.

We wish to find an approximation for $(\hat{D}-\mathfrak{s})^{-1}$ which is a pseudodifferential operator. Fix some finite open cover $\{V_i\}$ of X by distinguished coordinate patches and let η_i be a subordinate tangentially smooth partition of unity. On a coordinate patch, let $\hat{D}$ have total symbol $p_0 + ... + p_m$. Let $p_j' = p_j$ for $j<m$ and $p_m' = p_m - \mathfrak{s}$. Then

$$\sigma(\hat{D}-\mathfrak{s}) = \sum_{j=0}^{m} p_j'$$

so $\hat{D}-\mathfrak{s}$ is tangentially elliptic. Use the equation

$$\sigma(\eta_i R(\mathfrak{s})(\hat{D}-\mathfrak{s})) - I \sim 0$$

to give a local solution for $R_i(\mathfrak{s})$, with symbol $n_i r_0 + \ldots + r_{n_0}$, n_0 large. Precisely, set

$$r_0 = n_i(p_m - \mathfrak{s})^{-1} \text{ and}$$

$$r_n = -r_0 \Sigma d_\xi^\alpha r_j D_x^\alpha p_k{}'/\alpha!$$

where we sum over $j < n$ and $|\alpha| + j + m - k = n$. Define $R(\mathfrak{s}) = \Sigma n_i R_i(\mathfrak{s})$. The principal tangential symbol of $R(\mathfrak{s})$ is $(p_m - \mathfrak{s})^{-1}$, so $R(\mathfrak{s})$ is a parametrix for $\hat{D}$. We have established the following proposition.

Proposition 7.45.   Let $\hat{D} \in \mathcal{P}_\mathfrak{s}^m(E, E)$ be a tangentially elliptic grade–preserving differential superoperator. Then n_0 may be selected sufficiently large so that $(\hat{D} - \mathfrak{s})^{-1}$ is approximated arbitrarily well by the parametrix $R(\mathfrak{s})$ in the operator norm as $\mathfrak{s} \longrightarrow \infty$. That is,

$$|\zeta(\hat{D} - \mathfrak{s})^{-1} - R(\mathfrak{s})\mathfrak{J}f|_k \le c_k(1 + |\mathfrak{s}|)^{-k}|f|_{-k}$$

$$\text{for } \mathfrak{s} \in \mathcal{C}, \, f \in \Gamma_\tau(E). \qquad \Box$$

Proposition 7.45 implies that

$$\|R(\mathfrak{s})(\hat{D} - \mathfrak{s}) - \mathrm{id}_E\|_{-k,k} \le c_0(1 + |\mathfrak{s}|)^{-k-1}$$

and hence, for $t < 1$,

$$\|R(\mathfrak{s}/t) - (\hat{D} - \mathfrak{s}/t)^{-1}\|_{-k,k} \le c_0(1 + |\mathfrak{s}/t|)^{-k-1}.$$

Define $E(t) \in \mathcal{P}$ by

$$E(t) = (2\pi i)^{-1}\int_{\mathcal{C}} R(\mathfrak{s})e^{-\mathfrak{s}t}d\mathfrak{s} =$$

$$= (2\pi i)^{-1}\int_{\mathcal{C}} R(\mathfrak{s}/t)e^{-\mathfrak{s}}d\mathfrak{s}/t.$$

Then

255

$$\|e^{-t\widehat{D}} - E(t)\|_{-k,k} \leqslant$$

$$\leqslant (2\pi)^{-1}\int_{\mathcal{C}} |e^{-\varsigma}|(1 + |\varsigma/t|)^{-k-1}|d\varsigma/t|$$

by the analyticity in ς. Thus

$$t^{-k}\|e(t) - e^{-t\widehat{D}}\|_{-k,k} \leqslant$$

$$\leqslant (2\pi)^{-1}\int_{\mathcal{C}} |e^{-\varsigma}|(t + |\varsigma|)^{-k-1}|d\varsigma|$$

is bounded at $t \longrightarrow 0$. If $k > p/2$ then $\|\cdot\|_{-k,k}$ bounds the uniform norm and hence

$$\phi_\nu(e^{-t\widehat{D}}) - \phi_\nu(E(t)) = O(t^k).$$

So it suffices to find an asymptotic expansion for $\phi_\nu(E(t))$. Define

$$E^i(t) = (2\pi i)^{-1}\int_{\mathcal{C}} e^{-\varsigma t}R_i(\varsigma)d\varsigma.$$

Then

$$\phi_\nu(E(t)) = \Sigma_i \phi_\nu(E^i(t)) \qquad \text{(finite sum)}$$

so it suffices to expand $E^i(t)$.

Recall that the total symbol of $R(\varsigma)$ is denoted $\{r_j\}$. Define

$$e_{j,t}(x,\xi) = (2\pi i)^{-1}\int_{\mathcal{C}} e^{-t\varsigma}r_j(x,\xi,\varsigma)d\varsigma.$$

Then $E^i(t)$ is a pseudodifferential operator with total symbol $e_{0,t} + \ldots + e_{n_0,t}$, and $e_{j,t} \in S^{-\infty}$ for all t. Let $E^i_j(t)$ be the operator associated to $e_{j,t}$. Then $E^i_j(t)$ is represented by a distribution kernel $K_{j,t}$ defined by

$$K_{j,t}(x,y,n) = \int e^{i(x-y)\xi} e_{j,t}(x,\xi,n)d\xi.$$

Thus

$$K_{j,t}(x,x,n) = t^{(j-p)/m} e_j(x,n)$$

where

$$e_j(x,n) = (2\pi i)^{-1}\iint_{\mathcal{C}} e^{-\zeta} r_j(x,\xi,\zeta,n)d\zeta d\xi.$$

Note that each $e_j(x,n)$ is tangentially smooth and is transversely bounded. If m is even then $e_j(x,n) = 0$ for j odd.

Define

$$\lambda_j(\widehat{D})_{(x,n)} = e_j(x,n)dvol(x)$$

and

$$\lambda_j(\widehat{D}) = \{\lambda_j(\widehat{D})_{(x,n)}\}.$$

Then each $\lambda_j(\widehat{D})$ is a signed tangential measure

$$\lambda_j(\widehat{D}) = \lambda_j(D) - \lambda_j(D^*)$$

which depends on the (X,F), D, the tangential Riemannian metric g, but not on t.

Proposition 7.46.   Each $\lambda_j(\widehat{D})$ is locally bounded.

Proof.   Note that $\lambda_j(\widehat{D})_\ell$ is given by

$$\lambda_j(\widehat{D})_\ell = e_j(x,n)dvol(x) \qquad (x,n) \in \mathbb{R}^p \times N.$$

It is clear that $\lambda_j(\widehat{D})^x$ has a signed Borel density on $r^{-1}(x)$ for each x. So it suffices to show that the Radon–Nikodym derivative of $\lambda_j(\widehat{D})$ with respect to the standard tangential measure is bounded on any set

257

$r^{-1}(C)$, C compact. This derivative is of course just the function $e_j(x,n)$. So the problem reduces to proving that $e_j(x,n)$ is locally bounded.

Recall that

$$e_j(x,n) = (2\pi i)^{-1} \iint e^{-\zeta} r_j(x,\xi,\zeta,n) d\zeta d\xi$$

where the r_j are the homogeneous parts of the total symbol of $\hat{D}_\zeta$ given explicitly above, and where $p_m + p_{m-1} + \ldots$ is the total symbol of $\hat{D}$. Note further that

$$e_{j,t}(x,\xi,n) = (2\pi i)^{-1} t^{-k} \int_C e^{-t\zeta} \frac{d^k}{d\zeta^k} r_j(x,\xi,\zeta,n) d\zeta.$$

Since $\frac{d^k}{d\zeta^k} r_j$ is homogeneous of degree $-m-j-km$ in (ξ,ζ), it follows that e_j is a smoothing operator on each leaf for any $t > 0$ (though not necessarily compactly supported) and hence $\phi_\nu(e_j)$ is finite, as required. $\square$

This establishes the following theorem.

Theorem 7.47 [Co3]. Fix (X,F,g), an invariant transverse measure ν and a tangential, tangentially elliptic differential operator D from E to E' where (X,E) and (X,E') have bounded geometry. Let $\hat{D}$ be the associated superoperator, and let ϕ_ν be the associated trace on $W^*(G(X),E,\tilde{\mu})$. Then for $j \geq -p$ there is a family of signed tangential measures $\lambda_j(\hat{D})$ on X (which depend on (X,g,F,D) but not on t) and an asymptotic expansion

(7.48) $$\phi_\nu^s(e^{-t\hat{D}}) \sim \sum_{j \geq -p} t^{j/2p} \int_X \lambda_j(\hat{D}) d\nu.$$

Corollary 7.49. In the notation of 7.47,

$$\text{ind}_\nu(D) = \int (\lambda_0(D) - \lambda_0(D^*)) d\nu.$$

Proof of 7.49. Fix some $t > 0$. Then

$$\text{ind}_\nu(D) = \emptyset^s_\nu(e^{-t\hat{D}}) = \sum_{j \geq -p} \int_X (\lambda_j(D) - \lambda_j(D^*)) d\nu.$$

As the left hand side is independent of t, each term in the summation which involves non-trivial powers of t must vanish. The remaining term (corresponding to j = 0) then gives the index. ◻

Corollary 7.50. The tangential measure λ_j is homogeneous of weight j/2p in the coefficients of D. That is,

$$\lambda_j(\mathfrak{s}D) = \mathfrak{s}^{j/2p}\lambda_j(D).$$

Further, in case m = 1, then $(\det(a))^{1/2}\lambda_j(D)$ is a polynomial in the coefficients of D relative to x, their derivatives relative to x, and $\det^{-1}(a)$, where det(a) is the leading term of the quadratic form. ◻

Write

(7.51) $$\omega_D(g,E) = (\lambda_0(D) - \lambda_0(D^*)) \, |\, d\lambda\,|$$

for the tangentially smooth p-form which corresponds to D. Note that this form is measurable but not necessarily continuous transversely. In Chapter VIII this form will be identified for certain classical operators and this will enable us to complete the proof of the Index Theorem.

CHAPTER VIII: THE INDEX THEOREM

In this chapter we compute the index of a tangentially elliptic pseudodifferential operator on a compact foliated space.

Let X be a compact foliated space with leaves of dimension p and foliation bundle F which we assume oriented and equipped with a tangentially smooth oriented tangential Riemannian structure g. Let D be a tangential, tangentially elliptic pseudodifferential operator on (bundles over) X. For each leaf ℓ the spaces $\operatorname{Ker} D_\ell$ and $\operatorname{Ker} D_\ell^*$ are well defined by 7.23 and are locally finite dimensional with local index measure

$$\iota_{D_\ell} = {}^\mu D_\ell - {}^\mu D_\ell^*.$$

We define the analytic index of D to be the tangential measure

$$\iota_D = (\iota_D^x).$$

where $\iota_D^x = \iota_{D_\ell}$ for $x \in \ell$. If D is a differential operator of positive order m then Proposition 7.24 implies that the index measure ι_D is still well-defined. Alternatively, the operator $(1 + \Delta)^{-m/2}D$ is in the closure of the order zero pseudodifferential operator algebra and has the same tangential principal symbol and same index class in the von Neumann algebra and thus may be used to replace D.

The topological index of D is defined as $\operatorname{ch}_\tau(D)\operatorname{Td}_\tau(X)$, where

$$\operatorname{ch}_\tau(D) = (-1)^{p(p+1)/2} \, \Phi_\tau^{-1} \operatorname{ch}_\tau(\sigma(D)).$$

(The peculiar introduction of signs is explained in Atiyah-Singer III [ASIII], p. 557.).

With these preliminaries in hand, we may state the Connes Index Theorem.

Theorem 8.1. (Connes [Co3]). Let D be a tangential, tangentially elliptic pseudodifferential operator on X and let ν be an invariant transverse measure. Then

$$\text{ind}_\nu(D) \equiv \int \iota_D d\nu = \int \text{ch}_\tau(D) \text{Td}_\tau(X) d\nu.$$

Our study of invariant transverse measures and tangential cohomology enables us to reformulate Theorem 8.1 so as to avoid mention of transverse measures.

Theorem 8.2. Let D be a tangential, tangentially elliptic operator on X. Then

$$[\iota_D] = [\text{ch}_\tau(D)\text{Td}_\tau(X)]$$

as classes in $\bar{H}^p_\tau(X)$.

Theorem 8.2 implies Theorem 8.1 since an invariant transverse measure ν corresponds uniquely to a homomorphism $\bar{H}^p_\tau(X) \longrightarrow \mathbb{C}$. If X has no invariant transverse measures then $\bar{H}^p_\tau(X) = 0$ and so the statement is simply $[\iota_D] = 0$.

The proof follows a well-known path; we establish Theorem 8.1 for twisted signature operators and then argue homotopy-theoretically that this suffices to prove (8.1). Theorem 8.1 and the Riesz representation theorem (4.27) immediately imply Theorem 8.2. As an introduction to the techniques we first consider the de Rham and signature operators.

Define complex vector bundles E^k_X and E^k_G by

$$E^k_X = \Lambda^k(F^*_{\mathbb{C}})$$

$$E^k_G = \Lambda^k(s^*F^*_{\mathbb{C}}) = s^*E^k_X.$$

$$E_X = \oplus E^k_X \qquad E_G = \oplus E^k_G.$$

The de Rham operator d is defined briefly in Chapter VII; we recall additional detail here. Let $f \wedge dx_I$ be a tangentially smooth k-form on G(X) in local coordinates; i.e., a tangentially smooth section of the bundle E_G^k. Then exterior differentiation

$$d(f \wedge dx_I) = \sum_{i=1}^{p} \frac{\partial f}{\partial x_i} \, dx_i \wedge dx_I$$

induces a natural map

$$d: \Gamma_\tau(E_G) \longrightarrow \Gamma_\tau(E_G)$$

which agrees with the usual exterior derivative on each leaf G^X of $G(X)$. Thus d yields an unbounded operator on the Hilbert field

$$d: L^2(G(X), E_G) \longrightarrow L^2(G(X), E_G).$$

If ξ is some vector then explicit computation yields

$$\xi \wedge dx_I = \sum \xi_j \wedge dx_I$$

so that the principal symbol $\sigma(d)$ of d is given locally by

(8.3) $\qquad\qquad \sigma(d)(x,\xi)(v) = \xi \wedge v \qquad v \in F_{\mathbb{C}}^*.$

The associated symbol sequence

$$\cdots \longrightarrow E_G^k \xrightarrow{\;(\xi \wedge \;)_*\;} E_G^{k+1} \longrightarrow \cdots$$

is an exact sequence for each $\xi \neq 0$ and hence d is a tangentially elliptic operator. To summarize:

Proposition 8.4. The de Rham operator

$$d: L^2(G(X), E_G) \longrightarrow L^2(G(X), E_G)$$

is a densely defined unbounded tangential, tangentially elliptic operator with tangential principal symbol given by

$$\sigma(d)(x, \xi)(v) = \xi \wedge v.$$

The orientation of F induces a natural bundle isomorphism

$$* : E_G^k \longrightarrow E_G^{p-k}$$

which is given locally in terms of an orthonormal basis $\{e_1, \ldots, e_s\}$ by $*(e_I) = \pm e_J$ where I and J are complementary multi-indices and the sign is determined by the parity of the permutation (I, J). This induces the Hodge inner product on sections $\Gamma_\tau(E_G)$:

$$\langle u, v \rangle = \int_{G(X)} u \wedge *\bar{v} d\mu.$$

With respect to this inner product the de Rham operator has a formal adjoint δ given on $(k+1)$-forms by

$$\delta_{k+1} = (-1)^{pk+p+1} * d_k *$$

where d_k denotes the restriction of d to k-forms. As $*$ induces an isometry on forms, its symbol is invertible, so that $\sigma(\delta) = \pm\sigma(*)\sigma(d)\sigma(*)$ is also invertible. Hence δ is a tangentially elliptic operator.

The *Hodge-Laplace operator* Δ is defined by

(8.4) $$\Delta = d\delta + \delta d : \Gamma_\tau(E_G) \longrightarrow \Gamma_\tau(E_G)$$

and its restriction to E_G^i is denoted Δ_i.

We note that

$$(d+\delta)^2 = d^2 + d\delta + \delta d + \delta^2$$

$$= \Delta \quad \text{since } d^2 = \delta^2 = 0$$

so that $(d + \delta)$ is a first order tangentially elliptic operator. Furthermore,

$$Ker(\Delta_i) = Ker((d+\delta)_i)$$

since

$$<\delta\omega,\omega> = <(d+\delta)^2\omega,\omega>$$

$$= <(d+\delta)\omega, (d+\delta)\omega>$$

$$= |(d+\delta)\omega|^2.$$

Proposition 8.5.   The Hodge-Laplace operator is an essentially self-adjoint tangential, tangentially elliptic operator.

Proof: This proposition follows from (7.21), but we prefer to give a more direct proof.   Let ℓ be a leaf of $G(X)$.   Restrict d to ℓ and consider the resulting commuting diagram

$$
\begin{array}{ccc}
\Gamma_\tau(E_G\big|_\ell) & \xrightarrow{\quad d \quad} & \Gamma_\tau(E_G\big|_\ell) \\
\| & & \| \\
\Gamma(\Lambda^*((T\ell)^*_{\mathbb{C}})) & \xrightarrow{\quad d \quad} & \Gamma(\Lambda^*((T\ell^*_{\mathbb{C}}))) \\
\downarrow & & \downarrow \\
L^2(\ell,(s^*\alpha)^*) & \cdots\xrightarrow{T_\ell}\cdots & L^2(\ell,(s^*\alpha)^*)
\end{array}
$$

where α is the density associated to (X,g) restricted to ℓ.   Let T_ℓ be the closure of d.   Then $Im(T_\ell) \subseteq Ker(T_\ell)$ and the operator

$$T_\ell^* T_\ell + T_\ell T_\ell^*$$

is a self-adjoint operator on $L^2(\ell,(s^*\alpha))$ which extends the operator $\Delta\big|_\ell$.   This implies that the restriction of Δ to each leaf is a self-adjoint elliptic operator.   □

The locally finite-dimensional space $Ker(\Delta_i)_\ell$ is just the

space of *square-integrable harmonic forms* of degree i on
ℓ. The local dimension of this space is well-defined up to
equivalence. since changing the metric g on F results in a change by
similarity. Define the *Betti measure* β_i to be the tangential
measure given by

$$\beta_i^x = \text{local dimension of } \text{Ker}(\Delta_i)_\ell \equiv {}^u\text{Ker}(\Delta_i)_\ell$$

where $x \in \ell$. The Betti numbers relative to some invariant transverse
measure ν are given by $\int \beta_i d\nu$.

Note that the Hodge *-operator induces a natural isomorphism
of Hilbert fields

$$\text{Ker}(\Delta_i) \cong \text{Ker}(\Delta_{p-i})$$

and hence the Betti measures β_i and β_{p-i} coincide. Define the
tangential Euler characteristic to be the signed tangential
measure $\sum_{i=0}^{p} (-1)^i \beta_i$. If p is odd then

$$\sum_{i=0}^{p} (-1^i \beta_i = (\beta_0 - \beta_p) + (\beta_1 - \beta_{p-1}) + ... + (\frac{\beta_{p-1}}{2} - \frac{\beta_{p+1}}{2})$$

$$= 0$$

since $\beta_i = \beta_{p-i}$. Let us assume, then. that p is even.

Let D be the restriction of $(d+\delta)$ to even forms, so that
$D: \Gamma_\tau(\oplus E_G^{2i}) \longrightarrow \Gamma_\tau(\oplus E_G^{2i+1})$. Then the local trace of D_ℓ is just
$(\Sigma(-1)^i \beta_i)^x$ for $x \in \ell$. Furthermore. D is a Dirac operator in the
sense of Chapter VII. so that we may apply the heat equation
argument, as follows.

Theorem 8.6. (Gauss–Bonnet)

a)   For any invariant transverse measure ν,

$$\int \sum_{i=0}^{p} (-1)^i \beta_i \, d\nu = \int \mathrm{Pf}(K/2\pi) d\nu.$$

b) $\left[\sum\limits_{i=0}^{p} (-1)^i \beta_i \right] = [\mathrm{Pf}(K/2\pi)] = e^{\tau}(X)$ (the tangential

Euler class) in $H_{\tau}^p(X)$.

Proof. The index of the operator D may be expressed in two ways. On the one hand, the local index $\iota_{D_{\ell}}$ is given by

$$\iota_{D_{\ell}} = \Sigma(-1)^i \beta_i^{\times}$$

more or less by definition. On the other hand, D is a first order tangential tangentially elliptic operator and Theorem 7.47 implies that

$$\mathrm{ind}_{\nu}(D) = \int \omega_D(g,E) d\nu$$

where ω_D is a tangentially smooth p–form which corresponds to D. Restrict $\omega_D(g,E)$ to a leaf ℓ. Then the local index theorem of Atiyah–Bott–Patodi [ABP] implies that

$$\omega_D(g,E)_{\ell} = \mathrm{Pf}(K/2\pi)_{\ell}.$$

Thus

$$\int d\Sigma(-1)^i \beta_i^{\times} d\nu = \int \mathrm{Pf}(K/2\pi) d\nu.$$

This holds for each invariant transverse measure ν, and hence

$$\Sigma(-1)^i \beta_i = \mathrm{Pf}(K/2\pi)$$

as classes in $\bar{H}^p(X)$. $\square$

Corollary 8.7. (Connes [Co3]) Let X be a foliated compact manifold with leaves of dimension 2. Let F be oriented and equipped with a tangentially smooth Riemannian metric. Fix some invariant transverse measure ν. Suppose that $\int K d\nu > 0$ (i.e., the ν-average curvature of X is strictly positive). Then X must have a closed leaf, and in fact, the set of closed leaves has positive ν measure.

Proof. The Gauss-Bonnet theorem and the curvature assumption imply that

$$\int (\beta_0 - \beta_1 + \beta_2) d\nu > 0.$$

Suppose that there is no generic closed leaf. Then $\text{Ker}(\Delta_0)|_\ell = 0$ for ν-almost every leaf, and hence $\int \beta_0 d\nu = 0$. By duality we have $\int \beta_2 d\nu = 0$. Thus $-\int \beta_1 d\nu > 0$ which is a contradiction since, $\int \beta_1 d\nu \geq 0$. $\square$

We move next to the signature operator. Assume that $p = 2q$ is even. Then there is a natural involution $t: E_X \longrightarrow E_X$ given by $t = (-1)^{q*}1$. This decomposes E_G to $E_G = E_G^+ \oplus E_G^-$, the $\pm$ eigenspaces. The elliptic operator $(d+\delta): \Gamma_\tau(E_G) \longrightarrow \Gamma_\tau(E_G)$ anticommutes with t and hence restricts to an operator

(8.8) $$A: \Gamma_\tau(E_G^+) \longrightarrow \Gamma_\tau(E_G^-)$$

given by $A = (d+\delta)\big|_{\Gamma_\tau(E_G^+)}$. This is the tangential *signature operator*. The tangential principal symbol of A_ℓ is the restriction of the symbol of the elliptic operator $(d+\delta)_\ell$ to $\Gamma_\tau(E_G^+|_\ell)$ which is invertible, so A is tangentially elliptic. Graded as in (8.8), A is a Dirac operator in the sense of Chapter VII.

The involution t restricts to an involution of $\text{Ker}(d+\delta) \cong \text{Ker}(\Delta)$ since t anticommutes with $(d+\delta)$. The $\pm$ eigenspace decomposition is simply the decomposition of Hilbert fields

$$\text{Ker}(\Delta) \cong \text{Ker}(A) \oplus \text{Ker}(A^*).$$

Decompose $\text{Ker}(\Delta)$ further as

$$Ker(\Delta) = \overset{p}{\underset{i=0}{\oplus}} Ker(\Delta_i).$$

The subspace $Ker(\Delta_k) \oplus Ker(\Delta_{p-k})$ is t-invariant for each k, $0 \leqslant k \leqslant q$, and there is a unitary equivalence

$$[Ker(\Delta_k) \oplus Ker(\Delta_{p-k})]^+_\ell \longrightarrow [Ker(\Delta_k) \oplus Ker(\Delta_{p-k})]^-_\ell$$

given by $x+t(x) \rightsquigarrow x-t(x)$. Thus only the middle dimension $Ker(\Delta_q)$ contributes to the index. If $u,v \in Ker(\Delta_q) \subset H^q_\tau(X)$ then $u \wedge v \in H^p_\tau(X)$ and for any invariant transverse measure ν, $\int u \wedge v d\nu \in \mathbb{R}$. This gives a natural bilinear form on $Ker(\Delta_q)$ and it is reasonable to think of ι_A as the signature measure of this bilinear form, since

$$(8.9) \quad \int \iota_A d\nu = \int (+1 \text{ eigenspace of t}) d\nu - \int (-1 \text{ eigenspace of t}) d\nu \equiv$$
$$\equiv Sign(X, \nu).$$

If $X = M$ foliated as one leaf then $Ker(\Delta_i) \cong H^i(M)$ and $Sign(X, \nu)$ is exactly the signature of M.

Suppose that $p = 2q = 4r+2$. Then $t = \pm i*$ and so $** = -id$. Thus $*$ is a real transformation on $Ker(\Delta_q)$. The $\pm$ eigenspaces of $*$ (and hence of t) are conjugate via the map

$$a \otimes 1 - (*a) \otimes i \rightarrow a \otimes 1 + (*a) \otimes i$$

and thus $Ker(\Delta_q)^+$ is unitarily equivalent to $Ker(\Delta_q)^-$ and ι_A is the zero measure. The topological index also vanishes. Thus the index theorem holds trivially. So we restrict attention to the case $p = 4r$.

Recall that Hirzebruch L-polynomials L_k are polynomials of degree 4k in the Pontrjagin classes which are given by the splitting principle as

$$\sum_k L_k = \Pi \left(\frac{x_j}{\tanh x_j} \right).$$

The first few polynomials (in the tangential cohomology setting) are as follows:

$$L_0 = 1$$

$$L_1 = \tfrac{1}{3}p_1^{\mathcal{T}}$$

$$L_2 = \frac{1}{45}(7p_2^{\mathcal{T}} - p_1^{\mathcal{T}^2})$$

$$L_3 = \frac{1}{945}(62p_3^{\mathcal{T}} - 13p_1^{\mathcal{T}}p_2^{\mathcal{T}} + 2p_1^{\mathcal{T}3})$$

Lemma 8.10. $L_r(p_1^{\mathcal{T}},...,p_r^{\mathcal{T}}) = ch_{\mathcal{T}}(A)Td_{\mathcal{T}}(F_{\mathbb{C}})\big|_p$, where $\omega\big|_p$ denotes the component of ω in dimension p.

We omit this calculation; a proof may be found in Shanahan [Sh] §3.

Theorem 8.11 (Hirzebruch Signature Theorem). Let X be a compact oriented foliated space with leaves of dimension 4r. Let $L_r(p_1^{\mathcal{T}},...,p_r^{\mathcal{T}})$ denote the Hirzebruch L-polynomial in $\Omega_{\mathcal{T}}^p(X)$. Then

a)   For any invariant transverse measure ν,

$$Sign(X,\nu) = \int L_r(p_1^{\mathcal{T}},...,p_r^{\mathcal{T}})d\nu.$$

b)   The index class ι_A of the signature operator is equal to $L_r(p_1^{\mathcal{T}},...,p_r^{\mathcal{T}})$ in $\bar{H}_{\mathcal{T}}^p(X)$.

Proof. The index of the operator A may be expressed in two ways. On the one hand, the local trace ι_{D_ℓ} is given by

$$\iota_{D_\ell} = Sign(X,\nu)_\ell$$

as explained above. On the other hand, D is a first order tangential tangentially elliptic operator and Theorem 7.47 implies that

$$ind_\nu(A) = \int \omega_A(g,E)d\nu$$

where ω_A is a tangentially smooth p-form which corresponds to A. Restrict $\omega_A(g,E)$ to a leaf ℓ. Then the local index theorem of Atiyah-Bott-Patodi [ABP] implies that

$$\omega_A(g,E)_\ell = L_r(p_1^\mathcal{T}, \ldots, p_r^\mathcal{T})_\ell.$$

Thus

$$\mathrm{Sign}(X, \nu) = \int L_r(p_1^\mathcal{T}, \ldots, p_r^\mathcal{T}) d\nu.$$

This holds for each invariant transverse measure ν, and hence

$$\iota_A = L_r(p_1^\mathcal{T}, \ldots, p_r^\mathcal{T})$$

as classes in $\bar{H}_\mathcal{T}^p(X)$. □

With this preparation in hand we consider the twisted signature operators. Let X be a compact foliated space with leaves of dimension $p = 2q$ and oriented foliation bundle F. Let V be a tangentially smooth complex vector bundle with a tangential Hermitian structure and tangential connection Δ_V. The bundle $E_X \otimes_\mathbb{C} V$ carries a twisted de Rham differential

$$d_V \colon \Gamma_\mathcal{T}(E_X \otimes_\mathbb{C} V) \longrightarrow \Gamma_\mathcal{T}(E_X \otimes_\mathbb{C} V)$$

given by

$$d_V(u \otimes v) = du \otimes v + (-1)^i u \wedge \Delta_V v$$

where $u \in \Gamma_\mathcal{T}(\Lambda^i(F_\mathbb{C}^*))$, $v \in \Gamma_\mathcal{T}(V)$, and $\wedge$ is the external pairing. The map $*$ acts as $*(u \otimes v) = (*u) \otimes v$. The involution $t \colon E_X \longrightarrow E_X$ extends to a natural involution of $E_X \otimes V$ which fixes $1 \otimes V$. The twisted differential and the involution lift to $E_G \otimes V$, as usual.

Let δ_V denote the formal adjoint of d_V. Then $\delta_V = -*d_V*$, and

$$(d_V + \delta_V)t = -t(d_V + \delta_V).$$

Decompose $E_G \otimes V$ into $\pm$ eigenspaces with respect to t. Then the operator restricts to an operator

(8.12) $$A_V: \Gamma_\tau((E_G \otimes V)^+) \longrightarrow \Gamma_\tau((E_G \otimes V)^-)$$

called the *twisted signature operator*. This is a Dirac operator in the sense of Chapter VII.

Define

$$\mathcal{L}(X) = \Sigma \; 2^{-2s} L_s(p_1^\tau, \ldots, p_r^\tau).$$

Lemma 8.13.

$$ch(V) \cdot 2^q \mathcal{L}(X)\Big|_p = ch(A_V)Td_\tau(X)\Big|_p$$

This is a fairly involved purely topological calculation whose proof we omit (cf. Shanahan [Sh]).

Theorem 8.14. Let X be a compact foliated space with oriented leaves of dimension $p = 2q = 4r$. Let V be a complex vector bundle over X. Then

a) (Twisted Signature Theorem). For every invariant transverse measure ν,

$$\int \iota_{A_V} d\nu = \int ch_\tau(V) \cdot 2^q \mathcal{L}(X) d\nu.$$

b) $[\iota_{A_V}] = [ch_\tau(V) \cdot 2^q \mathcal{L}(X)\big|_p] \in \bar{H}_\tau^p(X).$

d) The index theorem holds for twisted signature operators.

Proof: The argument is virtually identical to the argument in Theorem 8.11. □

The remaining task before us is to demonstrate how knowledge of the index of twisted signature operators (that is to say, of certain natural tangential differential operators) implies the index theorem for all tangential pseudodifferential operators. We begin with the following lemma.

Lemma 8.15. Suppose that the index theorem holds for pseudodifferential operators of order zero. Then it holds for pseudodifferential operators of all orders.

Proof. There is nothing to prove for operators of negative order, since such operators lie in the kernel of the tangential principal symbol map on $\bar{\wp}^0$. Suppose that T is a tangential, tangentially elliptic pseudodifferential operator of positive order m. Let $\hat{T}$ be the associated superoperator of order 2m. Then $\hat{T}$ is tangential, tangentially elliptic and formally self-adjoint of order 2m, with

$$\operatorname{ind}_\nu(T) = \phi_\nu^s(\hat{T}) \quad \text{and} \quad \sigma_m(T) \simeq \sigma_{2m}(\hat{T}).$$

Let $P = (1 + \Delta)^{-m}\hat{T}$. Then Proposition 7.27 implies that $P \in \bar{\wp}^0$ with

$$\phi_\nu^s(\hat{T}) = \operatorname{ind}_\nu(P) \quad \text{and} \quad \sigma_{2m}(\hat{T}) \simeq \sigma_0(P).$$

The index theorem for P then implies the index theorem for T. $\square$

We turn next to the case of a tangential, tangentially elliptic pseudodifferential operator of order zero on a compact foliated space X with leaves of dimension p and oriented foliation bundle F. We wish to reduce to the case where p is even. For this purpose we briefly consider the multiplicative properties of the topological and analytical indices.

Suppose that X_1 and X_2 are compact foliated spaces as above and V_i, W_i are tangentially smooth Hermitian bundles over X_i. Let $X = X_1 \times X_2$ and define bundles V and W over X by

$$V = (V_1 \otimes V_2) \oplus (W_1 \otimes W_2)$$

$$W = (W_1 \otimes V_2) \oplus (V_1 \otimes W_2).$$

Let D_i be tangential, tangentially elliptic pseudodifferential operators of positive order. (If D_i are non–positive we modify as in Atiyah–Singer I [ASI], p. 528-9.) Define $D_1 \# D_2$ by the matrix

$$\begin{pmatrix} D_1 \otimes I_{V_2} & -I_{W_1} \otimes D_2^* \\ I_{V_1} \otimes D_2 & D_1^* \otimes I_{W_2} \end{pmatrix}$$

Then $D_1 \# D_2$ is tangential and tangentially elliptic.

Let $\iota_D^{top} = ch_\tau(D)Td_\tau(X)$.

Proposition 8.16. Let ν_i be invariant transverse measures on X_i, and let $\nu_1 \times \nu_2$ be the product measure. Then

a) $$\int \iota_{D_1 \# D_2}^{top} \, d(\nu_1 \times \nu_2) = \left[\int \iota_{D_1}^{top} d\nu_1 \right] \left[\int \iota_{D_2}^{top} d\nu_2 \right].$$

b) $$\int \iota_{D_1 \# D_2} d(\nu_1 \times \nu_2) = \left[\int \iota_{D_1} d\nu_1 \right] \left[\int \iota_{D_1} d\nu_2 \right]$$

Proof: The multiplicative property of the classical index (cf. Seeley [Pa], p. 217-228) implies that

$$\iota_{D_1 \# D_2}^{top} \Big|_\ell = \iota_{D_1}^{top} \Big|_{\ell_1} \times \iota_{D_2}^{top} \Big|_{\ell_2}$$

where $\ell = \ell_1 \times \ell_2$ is a leaf of $X_1 \times X_2$. Thus

$$\iota_{D_1 \# D_2}^{top} = \iota_{D_1}^{top} \times \iota_{D_2}^{top}.$$

Then

$$\int \iota_{D_1 \# D_2}^{top} d(\nu_1 \times \nu_2) = \int (\iota_{D_1}^{top} \times \iota_{D_2}^{top}) d(\nu_1 \times \nu_2)$$

$$= \left[\int \iota_{D_1}^{top} d\nu_1 \right] \left[\int \iota_{D_2}^{top} d\nu_2 \right]$$

as required.   A similar argument holds for $\iota_{D_1 \# D_2}$.   □

Note that we do not claim that the multiplicative property holds at the level of $H_\tau^{p_1}(X_1) \times H_\tau^{p_2}(X_2) \longrightarrow H_\tau^{p_1 + p_2}(X_1 \times X_2)$ since not every invariant transverse measure on $X_1 \times X_2$ is determined by product measures.   However this is certainly true if $X_2 = M$ (one leaf).

<u>Corollary</u> 8.17.   If the index theorem 8.1 holds for all X with p even then it holds for p arbitrary.

<u>Proof</u>:   Suppose that the index theorem holds for all X with p even and suppose given a foliated space X with p odd.   Then $X \times S^1$ is foliated with leaves of even dimension (p+1).   Let T be the operator on the circle defined by

$$T e^{inx} = \begin{cases} e^{i(n+1)x} & n \geq 0 \\ e^{inx} & n < 0. \end{cases}$$

Then $T = e^{ix}P + (1-P)$ where P: $L^2(S^1) \longrightarrow H^2(S^1)$ is the orthogonal projection.   Thus P and (hence) T are pseudodifferential operators of order zero, and

$$\sigma(T)(x, \xi) = \begin{cases} e^{ix} & \xi > 0 \\ 1 & \xi < 0 \end{cases}$$

so T is elliptic.   A direct check shows Ker T = 0, Cok T is generated by constant functions, and so ind(T) = -1.   The map

$K^0(F_X) \longrightarrow K^0(F_{X \times S^1})$ given by multiplication by the symbol of T is an isomorphism by Bott periodicity. Thus every symbol class in $K^0(F_{X \times S^1})$ has the form $\sigma(P)\sigma(T)$ for some pseudodifferential operator on X. As the topological and analytical indices are multiplicative by (8.16), the theorem follows. $\square$

Let X be a compact foliated space with leaves of dimension $p = 2q$ and F oriented. Fix an invariant transverse measure ν. The functions

$$D \to \int \iota_D d\nu$$

and

$$D \to \int ch_\tau(D) Td_\tau(X) d\nu$$

depend only upon the class of the tangential principal symbol $\sigma(D)$ and extend to $\mathbb{R}$-linear functions $K^0(F) \otimes \mathbb{R} \longrightarrow \mathbb{R}$. This is clear for the topological index. For the analytic index we must show that

$$\iota_{D_1 \oplus D_2} = \iota_{D_1} + \iota_{D_2}.$$

This is immediate if one thinks of $D_1 \oplus D_2$ as $\begin{bmatrix} D_1 & 0 \\ 0 & D_2 \end{bmatrix}$. Further, the two functions agree on the classes of the symbols $[\sigma(A_V)]$ of the twisted signature operators.

We note that (8.15), (8.17), and the following (8.18) together imply the index theorem 8.1.

Proposition 8.18. Suppose that X is a compact foliated space with oriented foliation bundle F of dimension $p = 2q$. Then the classes $[\sigma(A_V)]$ span the vector space $K^0(F) \otimes \mathbb{R}$.

In fact we shall prove the following more general proposition which Atiyah [At2] refers to in the case of X foliated by a single leaf as the Global Bott theorem.

Proposition 8.19. Let X be a compact foliated space with oriented foliation bundle F of dimension $p = 2q$. For any open subset Y of X let F_Y be the restriction of F to Y and define

$$\theta = \theta_Y \colon K^0(Y) \otimes \mathbb{R} \longrightarrow K^0(F_Y) \otimes \mathbb{R}$$

by

$$\theta(V) = [\sigma(A_V)].$$

Then θ is an isomorphism.

We begin by clarifying the map θ.

Lemma 8.20. If V is a tangentially smooth complex vector bundle over X then $[\sigma(A_V)] = V \cdot [\sigma(A)]$ in $K^0(F)$. Thus θ is given by multiplication by the symbol of the signature operator:

$$\theta(V) = V \cdot [\sigma(A)]$$

and hence extends to a transformation

$$\theta \colon K^*(Y) \otimes \mathbb{R} \longrightarrow K^*(F_Y) \otimes \mathbb{R}$$

which is natural with respect to inclusions of subspaces and boundary maps.

Proof: The twisted signature complex factors as $\Gamma_\tau(V) \otimes \Lambda^*(F_{\mathbb{C}}^*)$.
□

Lemma 8.21. Suppose that θ_Y is an isomorphism whenever F_Y is a trivial bundle. Then θ_Y is an isomorphism for all Y and Proposition 8.19 holds.

Proof: Let Y be an open subspace of X. Cover X by a finite open cover X_1, .X_t such that each F_{X_i} is a trivial bundle. Then of course F trivializes when restricted to each $Y_i = Y \cap X_i$ and on $(Y_1 \cup ... \cup Y_i) \cap Y_{i+1}$ for each i. The maps θ_{Y_i} and $\theta_{(Y_1 \cup ... \cup Y_i) \cap Y_{i+1}}$ are isomorphisms by assumption. A Mayer-Vietoris argument and a finite induction completes the argument. $\square$

In order to complete the proof of the Index Theorem 8.1 then, we are reduced to considering the case where $X = \ell^{2q} \times N$ is a product foliation. A Mayer-Vietoris argument on ℓ shows that we may reduce further to $X = \mathbb{R}^{2q} \times N$. that is, we must show that the map

$$\theta: K^*(\mathbb{R}^{2q} \times N) \otimes \mathbb{R} \longrightarrow K^*(\mathbb{R}^{2q} \times (\mathbb{R}^{2q} \times N)) \otimes \mathbb{R}$$

is an isomorphism.

Next we consider the diagram

$$
\begin{array}{ccc}
K^*(\mathbb{R}^{2q} \times N) \otimes \mathbb{R} & \xrightarrow{\theta_{\mathbb{R}^{2q} \times N}} & K^*(\mathbb{R}^{2q} \times \mathbb{R}^{2q} \times N) \otimes \mathbb{R} \\
\Big\uparrow{\scriptstyle \alpha \otimes 1} & & \Big\uparrow{\scriptstyle \alpha \otimes 1} \\
K^*(\mathbb{R}^{2q}) \otimes K^*(N) \otimes \mathbb{R} & \xrightarrow{\theta_{\mathbb{R}^{2q}} \otimes 1} & K^*(\mathbb{R}^{2q} \times \mathbb{R}^{2q}) \otimes K^*(N) \otimes \mathbb{R}
\end{array}
$$

where α denotes the Künneth map (an isomorphism over $\mathbb{R}$). The diagram commutes by the naturality of θ. So it suffices to demonstrate that the map

$$\theta_{\mathbb{R}^{2q}}: K^*(\mathbb{R}^{2q}) \otimes \mathbb{R} \longrightarrow K^*(\mathbb{R}^{2q} \times \mathbb{R}^{2q}) \otimes \mathbb{R}$$

is an isomorphism. The groups are isomorphic by the Bott periodicity map β.

Lemma 8.22. $\theta_{\mathbb{R}^{2q}} = 2^q \beta$. and hence $\theta_{\mathbb{R}^{2q}}$ is an isomorphism.

The proof of this lemma involves careful consideration of the

Dirac operator on $\mathbb{R}^{2q}$ and some classical representation theory. We omit the proof and refer the reader to Atiyah [At2] for details.

This completes the proof of Proposition 8.19 and hence the proof of the Index Theorem 8.1. □

APPENDIX A: THE $\bar{\partial}$-OPERATOR

By S. Hurder

CONTENTS

The purpose of this Appendix is to discuss the conclusion of the foliation index theorem in the context of foliations whose leaves are two-dimensional. Such foliations provide a class of reasonably concrete examples, which while they are certainly not completely representative of the wide range of foliations to which the theorem applies, are sufficiently complicated to warrant special attention, and also have the benefit of possessing the smallest leaf dimension for which the leaves have interesting topology. There is another, more fundamental reason for studying these foliations, and that is the observation that given any leafwise C^{∞}-Riemannian metric on a two-dimensional foliation, $\mathcal{F}$, there is a corresponding complex-analytic structure on leaves making $\mathcal{F}$ into a leafwise complex analytic foliation. Thus, two dimensional foliations automatically possess a Teichmüller space, and for each point in this space of complex structures, there is an associated Dirac operator along the leaves. The foliation index theorem then assumes the role of a Riemann-Roch Theorem for these complex structures.

We begin in §A1 with a discussion of the average Euler characteristic of Phillips-Sullivan, which is the prototype for the topological index character of the foliation index theorems for surfaces. In §A2, the index theorem is reformulated for the $\bar{\partial}$-operator along the leaves of a leafwise-complex foliation. The Teichmüller spaces for two-dimensional foliations are discussed in

§A3, and a few remarks about their properties are given. In §A4, some homotopy questions concerning the K-theory of the symbols of leafwise elliptic operators are discussed, with regard to the determination of all possible topological indices for a fixed foliation. Finally, §A5 describes <u>some</u> of the "standard" foliations by surfaces, especially of three-manifolds, and the calculation of the foliation indices for them.

The reader will observe that this Appendix concentrates upon topological aspects of the foliation index theorem and serves as an elaboration upon Connes' example of a foliation by complex lines on the four-manifold $\mathbb{C}/\Lambda_1 \times \mathbb{C}/\Lambda_2$ described in §A3. A key point of this example is that the meaning of the analytic index along the leaves can also be explicitly described in terms of functions with prescribed zeros-and-poles and a growth condition. For the foliations we consider, such an explicit description of the analytic index is much harder to describe, and would take us too far afield, but must be considered an interesting open problem, especially with regards to the Riemann-Roch nature of the foliation index theorem.

§A1. <u>Average Euler Characteristic</u>

The index theorem for the de Rham complex of a compact even dimensional manifold, M, yields the Chern-Gauss-Bonnet formula for its Euler characteristic, which is equal to the alternating sum of the Betti numbers of M. In a likewise fashion, it was shown in Chapter VIII that the foliation index theorem for the tangential de Rham complex of a foliated space yields an alternating sum of "Betti measures". When the transverse measure ν has a special form, i.e., it is defined by an averaging sequence, the d-foliation index can also be interpreted as the ν-average Euler characteristic of the leaves. We examine this latter concept more closely, for it provides a prototype for the calculation of the topological index in the general foliation index theorem. First, recall the integrated form of Theorem 8.6c):

<u>Theorem</u> A1.1 (d-Foliation Index Theorem). Let ν be a transverse invariant measure for a foliation $\mathcal{F}$ of a foliated space X, with

$C_\nu \in H_p^T(X)$ the associated Ruelle–Sullivan homology class of ν. Let d be the the de Rham operator on the tangential de Rham complex of $\mathcal{F}$. Assume the tangent bundle FX is oriented, with associated Euler form $e^T(X)$. Then

(A1.2) $\qquad x(\mathcal{F},\nu) \equiv \displaystyle\int_X \iota_d \cdot d\nu = \int_X e^T(X)d\nu = \langle e^T(X),c_\nu\rangle.$

The left–hand–side of (1.2) is interpreted in Chapter VIII as the alternating sum of the ν-dimensions of the L^2-harmonic forms on the leaves of $\mathcal{F}$. To give a geometric interpretation of the right–hand–side of (A1.2), we require that ν be the limit of discrete regular measures:

Definition A1.3. An *averaging sequence* [GP] for $\mathcal{F}$ is a sequence of compact subsets $\{L_j \mid j=1,2,...\}$ where each L_j is a submanifold with boundary of some leaf of $\mathcal{F}$, and $\dfrac{\text{vol}(\partial L_j)}{\text{vol}(L_j)} \longrightarrow 0$.

(The sets $\{L_j\}$ may belong to differing leaves as j varies, and we are assuming a Riemannian metric on FX has been chosen and fixed.)

The sequence $\{L_j\}$ is *regular* if the submanifolds ∂L_j of X have bounded geometry (i.e., there is a uniform bound on the sectional curvatures, the injectivity radii and the second fundamental forms of the ∂L_j).

For X compact, the measure ν_L associated to an averaging sequence is defined, on a tangential measure λ, by the rule

$$\int_X \lambda d\nu_L = \lim_{j\to\infty} \frac{1}{\text{vol}(L_j)} \cdot \int_{L_j} \lambda,$$

where, if necessary, we pass to a subsequence of the $\{L_j\}$ for which the integrals converge in a weak-* topology. The closed current associated to ν_L determines an asymptotic homology class denoted by $C_L \in H_p(X;\mathbb{R})$.

We say a transverse invariant measure ν is *regular* if ν

$= \nu_L$ for some regular averaging sequence $\{L_j | j=1,2,\dots\}$.

Not all invariant transverse measures arise from averaging sequences, but there are many examples where they do, the primary case being foliations with growth restrictions on the leaves. Choose a Riemannian metric on FX. Its restriction to a leaf $L \subset X$ of $\mathcal{F}$ defines a distance function and volume form on L. Pick a base point $x \in L$ and let $g(r,x)$ be the volume of the ball of radius r centered at x. We say L has:

polynomial growth of degree $\leq n$ if $\lim\sup\limits_{r \longrightarrow \infty} \dfrac{g(r,x)}{r^n} < \infty$

subexponential growth if $\lim\sup\limits_{r \longrightarrow \infty} \dfrac{1}{r} \log g(r,x) = 0$

non-exponential growth if $\lim\inf\limits_{r \to \infty} \dfrac{1}{r} \log g(r,x) = 0$

exponential growth if $\lim\inf\limits_{r \longrightarrow \infty} \dfrac{1}{r} \log g(r,x) > 0.$

For X compact, the growth type of L is independent of the choice of metric on FX and the basepoint x.

For a leaf L with non-exponential growth, there is a sequence of radii $r_j \longrightarrow \infty$ for which the balls L_j of radius r_j centered at X form an averaging sequence [Pl1]. In this case, all of the sets L_j are contained in the same leaf L. For X compact and the foliation of class C^2, these sets L_j can be modified to make them regular as well.

For a foliation $\mathcal{F}$ with even-dimensional leaves and a regular measure ν, the d-Index Theorem becomes

$$\chi(\mathcal{F},\nu) = \lim_{j \to \infty} \frac{1}{\text{vol}(L_j)} \int_{L_j} e^{\tau}(X).$$

By the Gauss-Bonnet theorem,

$$\int_{L_j} e^{\tau}(X) = e(L_j) + \int_{\partial L_j} \epsilon_j$$

282

where $e(L_j)$ is the Euler characteristic of L_j and ϵ_j is a correction term depending on the Riemannian geometry of ∂L_j. The assumption that the submanifolds $\{\partial L_j\}$ have uniformly bounded geometry implies there is a uniform estimate

$$\left| \int_{\partial L_j} \epsilon_j \right| \leq K \cdot \text{vol}(\partial L_j).$$

Therefore, in the limit we have

(A1.4) $$\chi(\mathcal{F}, \nu) = \lim_{j \to \infty} \frac{e(L_j)}{\text{vol}(L_j)}$$

and the right side of (A1.4) is called the *average Euler characteristic* of the averaging sequence $\{L_j\}$. Phillips-Sullivan [PS] and Cantwell-Conlon [CC1] use this invariant of a non-compact Riemannian manifold to give examples of quasi-isometry types of manifolds which cannot be realized as leaves of foliations of a manifold X with $H_p(X, \mathbb{R}) = 0$.

Consider three examples of open 2-manifolds (cf. [PS]), whose metric is defined by the given embedding in E^3. Each of the following, with their quasi-isometry class of metrics, can be realized as leaves of some foliation of some 3-manifold, but the first two cannot be realized (with the given quasi-isometry class of metrics) as leaves in S^3.

(A1.5) <u>Jacob's Ladder</u>

The growth type of L is linear, and the average Euler characteristic of L is $\dfrac{1}{vol(H)}$.

(A1.6) <u>Infinite Jail Cell Window</u>

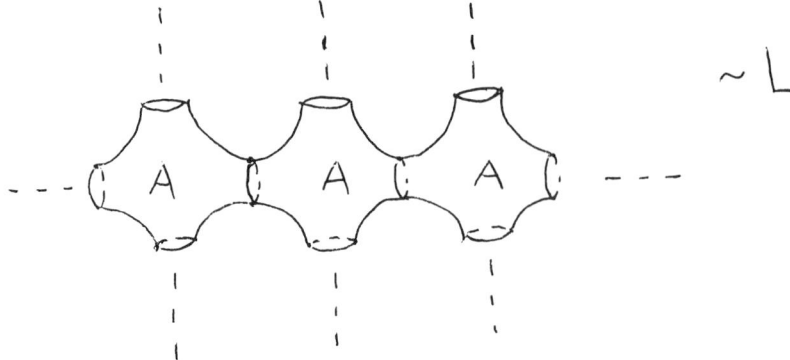

The growth of L is quadratic, and average Euler characteristic of L is $\dfrac{2}{vol\ A}$.

(A1.7) <u>Infinite Loch Ness Monster</u>

The growth of L is quadratic, but the average Euler characteristic is zero.

The construction of the average Euler characteristic for

surfaces suggests that a similar geometric interpretation can be given for the topological index of other differential operators. For the $\bar{\partial}$-operator of complex line foliations, this is indeed true, as discussed further in §3.

§A2.   The $\bar{\partial}$-Index Theorem and Riemann-Roch

We next examine in detail the meaning of the foliation index theorem for the tangential $\bar{\partial}$-operator. Let $\mathcal{F}$ be a foliation of a foliated space X and assume the leaves of $\mathcal{F}$ are complex manifolds whose complex structure varies continuously in X. That is, in Definition 2.1 of Chapter II, we assume that foliation charts $\{\varphi_x\}$ can be chosen for which the composition $t_y \circ \varphi_x^{-1}(\cdot, n)$ is holomorphic for all n, and n $\longmapsto$ $t_y \circ \varphi_x^{-1}(\cdot, n)$ is continuous in the space of holomorphic maps.

Let k be such that the dimension of the leaves of $\mathcal{F}$ is p = 2k.

A continuous vector bundle E $\longrightarrow$ X is *holomorphic* if for each leaf L $\subset$ X with given complex structure, the restriction E$|$L $\longrightarrow$ L is a holomorphic bundle. As before, FX is the tangent bundle to the leaves of $\mathcal{F}$, and this is holomorphic in the above sense. Let $A^{r,s} \longrightarrow$ X be the bundle of smooth tensors of type (r,s):

$$A^{r,s} = \wedge^{r,s}(F_{\mathbb{C}}X^*).$$

Given a holomorphic bundle E, let $A^{r,s} \otimes$ E denote the (r,s)-forms with coefficients in E. Assume that E has an Hermitian inner product, and then set

$$L^2(\mathcal{F},E) = \underset{x \in X}{\oplus} L^2(L_x, E|L_x),$$

where L_x is the leaf of $\mathcal{F}$ through x, $E|L_x$ is the restriction of the Hermitian bundle E to L_x, and we then take the L^2-sections of E over L_x with respect to a Lebesgue measure on L_x inherited from a Riemannian metric on FX. Note that $L^2(\mathcal{F},E)$ is in general neither a subspace nor a quotient of $L^2(X,E)$, the global L^2-sections of E over X.

For E a leafwise-holomorphic bundle, the leafwise $\bar{\partial}$-operator for $\mathcal{F}$ has a densely defined extension to

$$\bar{\partial} \otimes E: L^2(\mathcal{F},A^{r,s} \otimes E) \longrightarrow L^2(\mathcal{F},A^{r,s+1} \otimes E)$$

which is tangentially elliptic. Let $\ker_s(\bar{\partial} \otimes E)$ denote the kernel of

$$\bar{\partial} \otimes E: L^2(\mathcal{F},A^{0,s} \otimes E) \longrightarrow L^2(\mathcal{F},A^{0,s+1} \otimes E).$$

An element $\omega \in \ker_s(\bar{\partial} \otimes E)$ is a form whose restriction to each leaf L is a smooth form of type $(0,s)$ satisfying $\bar{\partial}(\omega|L) = 0$. Furthermore, for each $s \geqslant 0$, $\ker_s(\bar{\partial} \otimes E)$ is a locally finite dimensional space over X (cf. Chapter I). For an invariant transverse measure ν, the total ν-density of the $(0,s)$-solutions ω to the equation $\bar{\partial} \otimes E(\omega) = 0$ is $\dim_\nu \ker_s(\bar{\partial} \otimes E)$, and we set

$$\dim_\nu \ker(\bar{\partial} \otimes E) = \sum_{s=0}^{k} \dim_\nu \ker_s(\bar{\partial} \otimes E).$$

Similar arguments apply to the adjoint $\bar{\partial}^*$, and with the notation of Chapter IV we have

$$\int_X \iota_{\bar{\partial} \otimes E} \, d\nu = \dim_\nu \ker(\bar{\partial} \otimes E) - \dim_\nu \ker(\bar{\partial}^* \otimes E)$$

Theorem A2.1 ($\bar{\partial}$-Index Theorem). Let ν be an invariant transverse measure for a complex foliation $\mathcal{F}$ of X, $C_\nu \in H^\tau_{2k}(X;\mathbb{R})$ the associated Ruelle-Sullivan homology class, and $\mathrm{Td}_\tau(X) = \mathrm{Td}(FX \otimes \mathbb{C})$ the tangential Todd class for $\mathcal{F}$. Then

(A2.2) $$\int_X \iota_{\bar{\partial} \otimes E} \, d\nu = \langle \mathrm{ch}(\bar{\partial} \otimes E) \, \mathrm{Td}_\tau(X), \, C_\nu \rangle$$

The left-hand-side of (A2.2) is identified with the *arithmetic genus* of $\mathcal{F}$ with coefficients in E.

$$\chi(\bar{\partial} \otimes E, \nu) = \sum_{i=0}^{k} (-1)^i \dim_\nu H^i(\mathcal{F};E)$$

where $H^i(\mathcal{F};E) \equiv \ker_i(\bar{\partial} \otimes E)/\ker_i(\bar{\partial}^* \otimes E)$ is a locally-finite dimensional space over X. The number $\dim_\nu H^i(\mathcal{F};E)$ measures the density of this cohomology group in the support of ν, and generalizes the ν-Betti numbers of the operator d.

On the right-hand-side of (A2.2), the term $ch(\bar{\partial} \otimes E)$ is the Chern character of the K-theory class determined by the complex $A^{0,*} \otimes E$. There is a standard simplification of the cup product $ch(\bar{\partial} \otimes E)Td(FX \otimes \mathbb{C})$, which yields:

Corollary A2.3. $\chi(\bar{\partial} \otimes E, \nu) = \langle ch(E) \; Td_\tau(FX), \; C_\nu \rangle$.

Proof. Use the splitting principle and the multiplicativity of the Chern and Todd characters. For details, see [Sh]. □

Corollary A2.3 is exactly the classical Riemann-Roch Theorem in the context of foliations. The arithmetical genus $\chi(\bar{\partial} \otimes E, \nu)$ is the ν-density of the alternating sum of the dimensions of the $\bar{\partial}$-closed L^2-forms on the leaves of $\mathcal{F}$. The right-hand side is a topological invariant of E, FX and C_ν. For a given measure ν, one can hope to choose the bundle E so that $\chi(\bar{\partial} \otimes E, \nu) \neq 0$, guaranteeing the existence for ν-a.e. leaf L of $\mathcal{F}$ of $\bar{\partial}$-closed L^2-forms on L with coefficients in E.

§A3. Foliations by Surfaces (Complex Lines or k=1)

Let X be a compact foliated space with foliation $\mathcal{F}$ having leaves of dimension p = 2. For example, we may take X = M to be a smooth manifold and assume TM admits a 2-plane subbundle F. Then by Thurston [Th2], F is homotopic to a bundle FM which is tangent to a smooth foliation of M by surfaces.

Lemma A3.1. Let $\mathcal{F}$ be a two dimensional foliation of X with FX orientable. Then every Riemannian metric g on FX canonically determines a continuous complex structure on the leaves of $\mathcal{F}$. That is, the pair $(\mathcal{F},g)$ determines a complex foliation of X.

Proof. Define a J-operator J_g on FX to be rotation by $+\pi/2$ with respect to the given metric g and the orientation. For each leaf L, the structure $J_g\big|L$ is integrable as the leaf is two-dimensional hence uniquely defines a complex structure on L. Furthermore, by the parametrized Riemann mapping theorem [Ah], there exist foliation charts for $\mathcal{F}$ with each $t'\circ\varphi_x(\cdot,n)$ holomorphic and continuous in the variable n. □

We remark that if $\mathcal{F}$ has a given complex structure, J, then a metric g can be defined on FX for which $J_g = J$. Thus, the construction of Lemma 2.1 yields all possible complex structures on $\mathcal{F}$. This suggests the definition of the Teichmüller space of a 2-dimensional foliation $\mathcal{F}$ of a space X. We say two metrics g and g' on FX are *holomorphically equivalent* if there is a homeomorphism $\phi: X \longrightarrow X$ mapping the leaves of $\mathcal{F}$ smoothly onto themselves, and $\phi^* g'$ is conformally equivalent to g. We say that g and g' are *measurably holomorphically equivalent* if there is a measurable automorphism ϕ of X mapping leaves of $\mathcal{F}$ smoothly onto leaves of $\mathcal{F}$, and $\phi^*(g')$ is conformally equivalent to g by a measurable conformal factor on X.

Definition A3.2. The *(measurable) Teichmüller space* $T(X,\mathcal{F})$ (respectively, $T_m(X,\mathcal{F})$) is the set of (measurable) holomorphic equivalence classes [g] of metrics on FX.

When $\mathcal{F}$ consists of one leaf, this reduces to the usual Teichmüller space of a surface. When $\mathcal{F}$ is defined by a fibration $X \longrightarrow Y$ with fibre a surface Σ, let $T(\Sigma)$ be the Teichmüller space of Σ, then $T(X,\mathcal{F}) = C^0(Y,T(\Sigma))$ is infinite dimensional. The more interesting question is to study $T(X,\mathcal{F})$ for an ergodic foliation $\mathcal{F}$. There are constructions of foliated manifolds, due to E. Ghys, which show that $T(X,\mathcal{F})$ can be infinite-dimensional, even for $\mathcal{F}$ ergodic.

As an analogue of the Phillips-Sullivan Theorem in §1, one can ask if given a surface Σ with complex structure J_Σ, and given a compact manifold X, does there exist a foliation $\mathcal{F}$ of X and [g] $\in$ $T(X,\mathcal{F})$ with Σ a leaf of $\mathcal{F}$ so that the complex structure induced on Σ by [g] coincides with J_Σ? The average Euler characteristic of Σ still provides an obstruction to solving this problem, when Σ has non-exponential growth type, but the additional requirement that Σ have a prescribed complex structure should force other obstructions to arise. This would be especially interesting to understand for Σ of exponential growth-type, where no obstructions are presently known. Related to this is a problem posed by J. Cantwell and L. Conlon, which is when does there exist a metric on FX for which every leaf has constant negative curvature? Surprisingly, a complete solution is given for codimension-one, proper foliations in their preprint "Leafwise Hyperbolicity of Proper Foliations" (1986).

We now turn to consideration of the $\bar{\partial}$-Index Theorem for a foliation by complex lines, and derive an analogue of the average Euler characteristic.

Lemma A3.3. Let $\mathcal{F}$ be a complex line foliation of X. Then

(A3.4) $\qquad \chi(\bar{\partial} \otimes E, \nu) = \langle c_1(E), C_\nu \rangle + \frac{1}{2} \chi(\mathcal{F}, \nu)$

Proof. The degree two component of $\mathrm{ch}_\tau(E)\mathrm{Td}_\tau(X)$ is

$c_1(E) + \frac{1}{2} c_1(FX)$. □

Our goal is to give a geometric interpretation of the term $\langle c_1(E), C_\nu \rangle$ in (A3.4) similar to the average Euler characteristic.

Let $\kappa \longrightarrow \mathbb{C}P^N$ be the canonical bundle over the complex projective N-space. For large N, there exists a tangentially smooth map $f_E: X \longrightarrow \mathbb{C}P^N$ with $f_E^* \kappa = E$. (We say that f_E classifies E.) Let $H \subset \mathbb{C}P^N$ be a hypersurface dual to the first Chern class $c_1 \in H^2(\mathbb{C}P^N)$ of κ. For convenience, we now assume X is a C^1-manifold and $\mathcal{F}$ is also C^1. The complex structure on $\mathcal{F}$ orients its leaves, and the complex structure on $\mathbb{C}P^N$ orients the normal bundle to H. A connection on $\kappa \longrightarrow \mathbb{C}P^N$ pulls back under f_E to a connection on $E \longrightarrow X$, so $f_E^*(c_1) = c_1(E)$ holds both for cohomology classes and on the level of forms. Furthermore, a C^1-perturbation of f_E results in a C^0-perturbation of the form $c_1(E)$.

Given a regular averaging sequence $\{L_j\}$, for each $i \geqslant 1$ choose a C^1-perturbation f_j of f_E so that $f_j(L_j)$ is transverse to H, and $f_j^*(c_1)$ converges uniformly to $c_1(E)$. We say a point $x \in L_j \cap f_j^{-1}(H)$ is a *zero* of E if $f_j(L_j)$ is positively oriented at $f_j(x)$, and a *pole* if the orientation is reversed. Let $Z(L_j)$ and $P(L_j)$ denote the corresponding set of zeros and poles in L_j. Then elementary geometry shows

$$\int_{L_j} c_1(E) = \#Z(L_j) - \#P(L_j) + \epsilon_j,$$

where the error term ϵ_j is proportional to $\mathrm{vol}(\partial L_j)$. This uses that $\{\partial L_j\}$ has uniformly bounded geometry. Combined with Lemma 3.3. we obtain:

Proposition A3.5. Let X be a C^1-manifold and assume $\mathcal{F}$ is a C^1-holomorphic foliation by surfaces. For $\nu = \nu_L$ given by a regular averaging sequence $\{L_j\}$,

$$\chi(\bar{\partial} \otimes E, \nu) = \lim_{j \to \infty} \frac{\#Z(L_j)}{\mathrm{vol}(L_j)} - \lim_{j \to \infty} \frac{\#P(L_j)}{\mathrm{vol}(L_j)} + \frac{1}{2} \chi(\mathcal{F}, \nu)$$

$$= \begin{bmatrix} \text{average density of} \\ \text{zeros of E} \end{bmatrix} - \begin{bmatrix} \text{average density of} \\ \text{poles of E} \end{bmatrix}$$

$$+ \frac{1}{2} \begin{bmatrix} \text{average Euler} \\ \text{char} \end{bmatrix}.$$

Consider the case of a foliation of a 3-manifold X by surfaces. Let $\{\gamma_1,...,\gamma_d\}$ be a collection of d embedded closed curves in X which are transverse to $\mathcal{F}$. and $\{n_1,...,n_d\}$ a collection of non-zero integers. This data defines a complex line bundle $E \longrightarrow X$. and for a leaf L the restriction $E|L$ is associated to the divisor $\sum_{i=1}^{d} n_i \cdot (\gamma_i \cap L)$. Let ν be an invariant measure. Then Proposition A3.5 takes on the more precise form:

Proposition A3.6.

$$\dim_\nu H^0(\mathcal{F};E) - \dim_\nu H^1(\mathcal{F};E)$$

$$= \sum_{i=1}^{d} n_i \cdot \nu(\gamma_i) + \frac{1}{2} \chi(\mathcal{F}, \nu).$$

Proof. $c_1(E)$ is dual to the 1-cycle $\displaystyle\sum_{i=1}^{d} n_i \cdot \gamma_i$. □

If $\nu = \nu_L$ is defined by an averaging sequence $\{L_j\}$, then $\nu(\gamma_i)$ is precisely the limit density of $(\gamma_i \cap L_j)$ in L_j, so Proposition A3.6 relates the ν-dimension of L^2-harmonic forms on the leaves of $\mathcal{F}$ with the average density of the zeros and poles of E. This is exactly what a Riemann-Roch Theorem should do. The latitude in choosing E for a given $\mathcal{F}$ means one can often ensure that either $H^0(\mathcal{F};E)$, the L^2-meromorphic functions on the leaves of $\mathcal{F}$ with order $\geq \sum n_i \cdot \gamma_i$, or the corresponding space of meromorphic 1-forms $H^1(\mathcal{F};E)$ has positive ν-density. This type of result is of greatest interest when the complex structures of the leaves of $\mathcal{F}$ can be prescribed in advance, as in Example A3.7 below.

For a complex line foliation $\mathcal{F}$ of an n-manifold X, given closed oriented submanifolds $\{V_1,...,V_d\}$ of codimension 2 in X transverse to $\mathcal{F}$, and integers $\{n_1,...,n_d\}$, there is a holomorphic line bundle E $\longrightarrow$ X corresponding to the divisor $\displaystyle\sum_{i=1}^{d} n_i \cdot V_i$. The existence of such closed transversals V_i to $\mathcal{F}$, and more generally of holomorphic vector bundles E $\longrightarrow$ X, is usually hard to ascertain. However, there is one geometric context in which such V_i always exists in multitude, the foliations given as in (2.2) of Chapter II. We briefly recall their construction.

Let Y be a compact oriented manifold of dimension n-2, Σ_g a surface of genus g, and $\rho: \Gamma_g \longrightarrow$ Diff(Y) a representation of the fundamental group $\Gamma_g = \pi_1(\Sigma_g)$. The quotient manifold $M = (\tilde{\Sigma} \times Y)/\Gamma_g$ has a natural 2-dimensional foliation transverse to the fibres of $\pi: M \longrightarrow \Sigma_g$. The leaves of $\mathcal{F}$ are coverings of Σ_g associated to the isotropy groups of ρ, and inherit complex structures from Σ_g. The d-index theorem for $\mathcal{F}$ can be deduced from Atiyah's L^2-index theorem for coverings [At3]. For the $\bar{\partial}$-index theorem, this is no longer the case. Also, note that the Teichmüller spaces of this class of foliations always has dimension at least that of Σ_g, as every metric on $T\Sigma_g$ lifts to a metric on FM. However, they need not have the same dimension, and $T(M,\mathcal{F})$ or $T_m(M,\mathcal{F})$ provide a very interesting geometric "invariant" of the representation ρ of Γ_g on Y.

For each point $x \in \Sigma_g$, the fibre $\pi^{-1}(x) \subset M$ is a closed orientable transversal to $\mathcal{F}$. To obtain further transversals, we assume the fibration $M \longrightarrow \Sigma_g$ is trivial, so there is a commutative diagram

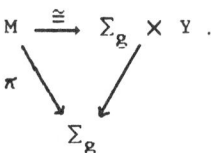

Note that the foliation $\tilde{\mathcal{F}}$ on $\Sigma_g \times Y$ induced from its identification with M will not, in general, be the product foliation. A transversal to $\mathcal{F}$ corresponds to a transversal to $\tilde{\mathcal{F}}$, and the latter can often be found explicitly.

Example A3.7. Consider the example described by Connes in [Co7]. Here, $\Sigma_1 = \mathbb{C}/\Gamma_1$ is a complex torus, as is $Y = \mathbb{C}/\Gamma_2$, for lattices Γ_1 and Γ_2 in $\mathbb{C}$. Let Γ_1 act by translations on $\mathbb{C}/\Gamma_2$, and form

$$M = (\mathbb{C} \times \mathbb{C}/\Gamma_2)/\Gamma_1 \cong (\mathbb{C}/\Gamma_1) \times (\mathbb{C}/\Gamma_2)$$

$$\pi \downarrow$$

$$\mathbb{C}/\Gamma_1$$

Connes takes $V_1 = 0 \times \mathbb{C}/\Gamma_2$ and $V_2 = \mathbb{C}/\Gamma_1 \times 0$ as the transversals in $\Sigma_1 \times Y$. Neither V_1 nor V_2 is homotopic to a fibre $\pi^{-1}(x)$ so the $\bar{\partial}$-index theorem for E associated to the divisor V_1-V_2 is not derivable from the L^2-index theorem for coverings. For ν the Euclidean volume on $\mathbb{C}/\Gamma_2$, Connes remarks that

$$\chi(\bar{\partial} \otimes E, \nu) = \text{density } \Gamma_2 - \text{density } \Gamma_1.$$

292

so the dimension of the space of L^2-harmonic functions on almost every leaf $\mathbb{C} \subset M$ with divisor $\mathbb{C} \cap (V_1 - V_2)$ is governed by the density of the lattices Γ_1 and Γ_2. Again, this is exactly the role of a Riemann-Roch Theorem, where for foliations the degree of a divisor is replaced with the average density of the divisor.

§A4. Geometric K-Theories

The examples described at the end of §A3 for the $\bar{\partial}$-operator suggest that to obtain analytical results from the foliation index theorem, it is useful to understand the possible topological indices of leafwise elliptic operators. In the examples above, the ν-topological indices are varied by making choices of "divisors" which pair non-trivially with the foliation cycle C_ν. As a consequence, various spaces of meromorphic forms are shown to be non-trivial. To obtain similar results for a general foliation, $\mathcal{F}$, it is useful to determine the range of topological indices of leafwise elliptic operators for $\mathcal{F}$. In this section, we briefly describe the formal "calculation" of these indices in terms of K-groups of foliation groupoids. In some cases, these abstract results can be explicitly calculated, giving very useful information. The reader is referred to the literature for more detailed discussions. One other point is that the foliation index theorem equates the analytic index with the evaluation of a foliation cycle on a K-theory class; these evaluations can be much easier to make, than to fully determine the topological K-theory of the foliation. In this section, and in §A5, we will examine more carefully the values of the topological index paired with a foliation current for some classes of foliations.

Recall from Definition 2.20 of Chapter II the *holonomy groupoid*, or *graph*, G(X) associated to the foliated space X. A point in G(X) is an equivalence class $[\gamma_{xy}]$ of paths γ: $[0,1] \longrightarrow$ X with $\gamma(0) = x$, $\gamma(1) = y$ and γ remains on the same leaf for all t. Two paths are identified if they have the same holonomy. G(X) is a topological groupoid with the multiplication defined by concatenation of paths.

Also associated to $\mathcal{F}$ and X is a groupoid $\Gamma(X)$, constructed by Haefliger in [Hae4]. The groupoid $\Gamma(X)$ coincides with one of the restricted groupoids $G_N^N(X)$ of Chapter II. Let $\{U_\alpha\}$ be a locally finite open cover of X by foliation charts such that $U_\alpha \cap U_\beta$ is contractible if non-empty. For each α, there is given a diffeomorphism $\phi_\alpha: U_\alpha \longrightarrow \mathbb{R}^p \times \mathbb{R}^q$ sending the leaves of $\mathcal{F}|U_\alpha$ to $\mathbb{R}^p \times$ pt. Define a transversal $T_\alpha = \phi_\alpha^{-1}(\{0\} \times \mathbb{R}^q) \subset U_\alpha$ for each α. By a judicious choice of the $\{U_\alpha\}$, we can assume the $\{T_\alpha\}$ are pairwise disjoint (cf. [HS]). Then set $N = \cup\, T_\alpha$, an embedded open q-submanifold of X. It is an easy exercise to show $G_N^N(X)$ coincides with the Haefliger groupoid $\Gamma(X)$ constructed from the foliation charts $\{\phi_\alpha: U_\alpha \longrightarrow \mathbb{R}^{p+q}\}$.

The inclusion $N \times N \subset X \times X$ induces an inclusion of topological groupoids $\Gamma(X) \subset G(X)$. The cofibre of the inclusion is modeled on the trivial groupoid $\mathbb{R}^p \times \mathbb{R}^p$, where all pairs (x,y) are morphisms. One thus expects the above inclusion to be an equivalence, and Haefliger shows in [Hae4] that this is indeed so:

Theorem A4.1 (Haefliger). The inclusion $\Gamma(X) \subset G(X)$ is a Morita equivalence of categories.

For any topological groupoid, $\mathcal{B}$, there is a classifying space $B\mathcal{B}$ of $\mathcal{B}$ structures, which is constructed using a modification of the Milnor join construction [Hae2,Mi]. Applying this to $G(X)$ yields the space $BG(X)$ which is fundamental for foliation K-theory (cf. Chapter 9 [Co7]). Applying the B-construction to $\Gamma(X)$, we obtain a space $B\Gamma(X)$ which is fundamental for the characteristic class theory of $\mathcal{F}$.

Corollary A4.2. Let $\mathcal{F},X$ be a foliated space. The inclusion $\Gamma(X) \subset G(X)$ induces a homotopy equivalence $B\Gamma(X) \simeq BG(X)$.

Thus, the topological invariants of $B\Gamma(X)$ and $BG(X)$ agree. Note the open contractible covering $\{U_\alpha\}$ of X defines a natural continuous map $X \longrightarrow B\Gamma(X)$. If all leaves of $\mathcal{F}$ are contractible, then this inclusion is a homotopy equivalence, so that the topological type of $BG(X)$ is the same as X. By placing weaker restrictions on the topological types of the leaves of $\mathcal{F}$, one can more generally deduce that the inclusion is an N-equivalence on homotopy groups,

[Hae4]. For the generic foliation, however, one expects that the space $BG(X)$ will have a distinct topological type from X, probably more complicated.

The space $B\Gamma(X)$ can be studied from a "universal viewpoint" by introducing the Haefliger classifying spaces. For the class of transversally C^r-differentiable foliations of codimension q, Haefliger defines a space $B\Gamma_q^{(r)}$, and there is a universal map

$$i_X: B\Gamma(X) \longrightarrow B\Gamma_q^{(r)}.$$

The cohomology groups of $B\Gamma_q^{(r)}$ then define universal classes which pull back to $B\Gamma(X)$ via $(i_X)^*$. The non-triviality of $(i_X)^*$ is then a statement about both the topology of $B\Gamma(X)$ and the inclusion i_X. A short digression will describe the situation for C^∞ foliations.

Let $B\Gamma_q$ denote the universal classifying space of codimension q C^∞-foliations. (It is important to specify the transverse differentiability of $\mathfrak{F}$, as the topology of $B\Gamma_q$ depends strongly on how much differentiability is required.) The composition

$$f_{\mathfrak{F}}: X \longrightarrow B\Gamma(X) \longrightarrow B\Gamma_q.$$

or more precisely its homotopy class, was introduced by Haefliger in order to "classify" the C^∞-foliations on a given X. The classification is modulo an equivalence relation which turns out to be *concordance* for X compact, and *integrable homotopy* for X open (cf. [Hae2]).

For $B\Gamma(X)$, the principal invariants are the characteristic classes: for a codimension q, C^∞-foliation there are universal classes (cf. [L])

$$\tilde{\Delta}_*: H^*(WO_q) \longrightarrow H^*(B\Gamma_q),$$

and for given $\mathfrak{F}$ on X we obtain its secondary classes via

$$\Delta_* = f_{\mathfrak{F}}^* \circ \tilde{\Delta}_*: H^*(WO_q) \longrightarrow H^*(X).$$

We next describe how the topology of BG(X) is related to the topological indices of leafwise elliptic operators of $\mathcal{F}$. For $\mathcal{F}$ a C^1-foliation of a manifold X, the groupoid G(X) has a natural map to GL(q,ℝ) obtained by taking the Jacobian matrix of the holonomy along a path $[\gamma_{xy}]$. This induces a map

$$BG(X) \longrightarrow BGL(q,\mathbb{R}),$$

which defines a rank q vector bundle $\xi \longrightarrow BG(X)$ whose pullback to X under $X \longrightarrow B\Gamma(X) \longrightarrow BG(X)$ is the normal bundle to $\mathcal{F}$. The ξ-twisted K-theory of BG(X) is defined as

$$K_*^{\xi}(BG(X)) \equiv K_*(B(\xi),S(\xi))$$

where $B(\xi)$ is the unit disc subbundle of $\xi \longrightarrow BG(X)$, and $S(\xi)$ is the unit sphere bundle.

Connes and Skandalis construct in [CS2] a map

$$\text{Ind}_t \colon K_*^{\xi}(BG(X)) \longrightarrow K_*(C_r^*(X)),$$

which they call the topological index map, via an essentially topological procedure which converts a vector bundle or unitary over BG(X) into an idempotent or invertible element over $C_r^*(X)$. Let F_1^*X denote the unit cotangent bundle to $\mathcal{F}$ over X. Then there is a natural map of K-theories, b: $K^1(F_1^*X) \longrightarrow K_0^{\xi}(BG(X))$, obtained from the exact sequence for the pair $(B(\xi),S(\xi))$. If $\mathcal{F}$ admits a transverse invariant measure ν, then there is a linear functional ϕ_ν on $K_0(C_r^*(X))$ (cf. 6.23), and the composition $\phi_\nu \circ \text{Ind}_t \circ b = \text{Ind}_\nu^t$, the topological measured index. That is, for D a leafwise operator with symbol class u = $[\sigma_D] \in K^1(F_1^*X)$,

$$\phi_\nu \circ \text{Ind}_t \circ b(u) = \langle \text{ch}(D)\text{Td}_\tau(X), C_\nu \rangle.$$

Connes and Skandalis also construct a direct map,

$$\text{Ind}_a: K^1(F_1^*X) \longrightarrow K_0(C_r^*(X)),$$

which they call the analytic index homomorphism, by associating to an invertible u the index projection operator over $C_r^*(X)$ of a zero-order leafwise elliptic operator whose symbol class. is u. Also, $\text{Ind}_\nu^a \equiv \phi_\nu \circ \text{Ind}_a(u)$ is the analytic index of this operator, calculated using the dimension function associated to ν. They then proved:

Theorem A4.3 (General Foliation Index Theorem). For any foliaion $\mathcal{F}$, there is an equality of maps

$$\text{Ind}_a = \text{Ind}_t \circ b: K^1(F_1^*X) \longrightarrow K_0(C_r^*(X)).$$

Note that Theorem A4.3 makes sense even when $\mathcal{F}$ possesses no invariant measures. If there is an invariant measure, ν, then by the above remarks, the theorem implies the ν-measured foliation index theorem proved in Chapters 7 and 8. Note also that this formulation of the index theorem shows that the possible range of the analytic traces of leafwise operators, with respect to a given invariant measure ν, are contained in the image of the map $\phi_\nu \circ \text{Ind}_t$: $K_0^\xi(BG(X)) \longrightarrow \mathbb{R}$. This is the meaning of the earlier statement that the topology of BG(X) dictates the possible analytic indices of leafwise operators, and motivates the study of BG(X). In fact, Connes has conjectured that this space has K-theory isomorphic to that of $C_r^*(X)$.

Conjecture A4.4. Suppose that all holonomy groups of $\mathcal{F}$ are torsion-free. Then Ind_t is an isomorphism.

It is known that Conjecture A4.4 is true if $\mathcal{F}$ is defined by a free action of a simply connected solvable Lie group on X, [Co7]. Also, for flows on the 2-torus and for certain "Reeb foliations" of 3-manifolds, the work of Torpe [To] and Penington [Pen] shows that conjecture (A4.4) holds.

Given a foliated manifold X with both FX and TX orientable, a natural problem, related to Conjecture A4.4, is to determine to what extent the composition

$$H_*(X) \cong K_*(X) \overset{\text{Thom}}{\cong} K_*^\xi(X) \longrightarrow K_*^\xi(BG(X)) \overset{\text{Ind}_t}{\longrightarrow} K_*(C_r^*(X))$$

is an isomorphism. We describe three quite general results on this, and then show that the $\bar{\partial}$-Index Theorem also sheds some light on this problem in particular cases.

Let G be a connected Lie group. A locally free action of G on X is *almost free* if given $g \in G$ with fixed point $x \in X$, either $g = \text{id}$ or the germ of the action of g near x is non-trivial. If $\mathcal{F}$ is defined by an almost free action of G on X, then $G(X) \cong X \times G$. If G is also contractible, then $X \longrightarrow BG(X)$ is a homotopy equivalence.

Theorem A5.5 (Connes [Co7]). Let $\mathcal{F}$ be defined by an almost free action of a simply connected solvable Lie group G on X. Then there is a natural isomorphism $K_*(X) \cong K_*(C_r^*(X))$.

For $B^p = \Gamma \backslash G/K$ a locally symmetric space of rank one with negative sectional curvatures, there is a natural action of the lattice Γ on the sphere at infinity ($\cong S^{p-1}$) of the universal cover G/K. The manifold $M = (G/K \times S^{p-1})/\Gamma$ can be identified with the unit tangent bundle $T^1 B$. The codimension $q = (p-1)$ foliation of M defined in Chapter II corresponds here with the Anosov (= weak stable) foliation of $T^1 B$.

Theorem A5.6 (Takai [Ta]). The index yields an isomorphism

$$K_*(M) \cong K_*(C_r^*(M))$$

For B a surface of genus ≥ 2, this result is due to Connes (Chapter 12, [Co7]).

The third result deals with the characteristic classes of C^∞-foliations. Recall from above that each class $[z] \in H^*(WO_q)$ defines a linear functional $\Delta_*[z]$ on $H_*(X)$. Connes has shown [Co10] that $[z]$ also defines a linear functional on $K_*(C_r^*(X))$, and these functionals are natural with respect to the map $H_*(X) \longrightarrow K_*(C_r^*(X))$. From this one concludes:

Theorem A4.7 (Connes [Co10]). Suppose there exists $[z] \in H^*(WO_n)$ and $[u] \in H_*(X)$ such that $\Delta_*[z]([u]) \neq 0$. Then $[u]$ is mapped to a non-trivial class in $K_*(C_r^*(X))$.

Theorem A4.7 shows that the characteristic classes of $\mathcal{F}$ can be used to prove certain classes in $H_*(X)$ inject into $K_*(C_r^*(X))$.

After these generalities, we consider foliations of 3-manifolds with an invariant measure ν given, and study the ν-topological index, $\text{Ind}_\nu^t(u)$, for $u \in K_1(X)$, which calculates the composition

$$K_1(X) \longrightarrow K_0^\xi(X) \cong K^1(F_1^*X) \xrightarrow{\text{Ind}_\nu^a} \mathbb{R}.$$

First, here is a general statement for such foliations. Recall that a simple closed curve γ in X transverse to $\mathcal{F}$ determines a complex line bundle E_γ over $\mathcal{F}$ with divisor $[\gamma]$. Take $\bar{\partial}$ along leaves and form $\bar{\partial} \otimes E_\gamma$, then this gives a map

$$H_1(X;Z) \longrightarrow K^1(F_1^*X)$$

$$[\gamma] \longrightarrow [\bar{\partial} \otimes E_\gamma]$$

and composing with Ind_a yields a map

$$\text{Ind}: H_1(X;Z) \longrightarrow K_0(C_r^*(X)).$$

Proposition A4.8. Let $\mathcal{F}$ be a codimension-one, C^1-foliation of a compact 3-manifold X. Assume both TX and FX are orientable.

a) Suppose ν is an invariant transverse measure with $C_\nu \neq 0$ in $H_2(X;\mathbb{R})$, and the support of ν does not consist of isolated toral leaves. (A toral leaf L is isolated if no closed transverse curve to $\mathcal{F}$ intersects L.) Then there exists a holomorphic line bundle $E \longrightarrow X$ such that $\text{Ind}_\nu(\bar{\partial} \otimes E) \neq 0$, and thus $\text{Ind}(\bar{\partial} \otimes E) \in K_0(C_r^*(X))$ is non-zero.

b)   Let $\{\nu_1,\ldots,\nu_d\}$ be a collection of invariant transverse measures such that the associated currents $\{C_1,\ldots,C_d\} \subset H_2(X;\mathbb{R})$ are linearly independent when evaluated on closed transversals to $\mathcal{F}$. Then there exist holomorphic line bundles $E_1,\ldots,E_d$ over X such that the elements $\{Ind(\bar{\partial} \otimes E_i) \mid i=1,\ldots,d\} \subset K_0(C_r^*(X))$ are linearly independent.

For example, it is not hard to show that if $\mathcal{F}$ has a dense leaf, and the currents $\{C_1,\ldots,C_d\} \subset H_2(X;\mathbb{R})$ of part b) are independent, then they are independent on closed transversals. Define $H(\Lambda) \subset H_2(X;\mathbb{R})$ to be the subspace spanned by the currents associated to the invariant measures for $\mathcal{F}$.

Corollary A4.9.   If $\mathcal{F}$ has a dense leaf, then there is an inclusion $H(\Lambda) \subset K_0(C_r^*(X)) \otimes \mathbb{R}$.

Proof of A4.8.   First assume there is a closed transverse curve γ to $\mathcal{F}$ which intersects the support of ν. Then $\nu(\gamma) \neq 0$, and we define $E = E_{n \cdot \gamma}$ and use (A3.4) to calculate $Ind_\nu(\bar{\partial} \otimes E_{n \cdot \gamma}) \neq 0$ for all but at most one value of n. If no such curve γ exists, then the support of ν must consist of compact leaves. One can show these leaves must be tori which are isolated and this contradicts the hypothesis that there is a non-isolated toral leaf in the support of ν. This proves a). The proof of b) is similar.   □

§A5.   Examples of Complex Foliations of 3-Manifolds

The geometry of foliations on 3-manifolds has been intensively studied. In this section, we select four classes of these foliations for study, and consider the $\bar{\partial}$-index theorem for each. Let M be a compact oriented Riemannian 3-manifold. Then M admits a non-vanishing vector field, and this vector field is homotopic to the normal field of some codimension one foliation of M. Moreover, M even has uncountably many codimension one foliations which are distinct up to diffeomorphism and concordance, [Th1]. This abundance of foliations on 3-manifolds makes their study especially appealing.

There are exactly two simply connected solvable Lie groups of dimension two. the abelian group R^2 and the solvable affine group on the line.

$$A^2 = \left\{ \begin{bmatrix} x & y \\ 0 & x^{-1} \end{bmatrix} \,\middle|\, x>0 \right\} \subset SL(2,\mathbb{R}).$$

A locally free action of R^2 or A^2 on a 3-manifold M defines a codimension one foliation with very special properties. The foliations defined by an action of R^2 have been completely classified: see (A5.1) and (A5.2) below. For $\pi_1 M$ solvable. the locally free actions of A^2 on M have been classified by Ghys-Sergiescu [GS] and Plante [Pl1]; see (A5.4) and (A5.5) below. For $\pi_1 M$ not solvable, some restrictions on the possible A^2-actions are known.

Note that Connes' Theorem A4.5 applies only when $\mathcal{F}$ is defined by an almost free action of R^2 or A^2. This assumption does not always hold in the following examples, so we must use the geometry of $\mathcal{F}$ to help calculate the image of the index map.

Throughout. M will denote a closed. oriented Riemannian 3-manifold and $\mathcal{F}$ an oriented 2-dimensional foliation of M.

(A5.1) Locally-Free R^2-Actions

Let $a \in SL(2,\mathbb{Z})$. which defines a diffeomorphism $\phi_a: T^2 \longrightarrow T^2$. and a torus bundle over S^1 by setting

$$M_a = T^2 \times \mathbb{R}/(x,t) \sim (\phi_a(x),t+1).$$

Theorem A5.1 [RRW]. Suppose M admits a locally free action of R^2. Then M is diffeomorphic to M_a for some $a \in SL(2,\mathbb{Z})$.

For $\mathcal{F}$ defined by an R^2-action. $\pi_1 M$ is solvable by Theorem A5.1. and $\mathcal{F}$ has no Reeb components. The foliated 3-manifolds with $\pi_1 M$ solvable and no Reeb components have been completely classified by Plante (Theorem 4.1 of [Pl2]: note that only his cases II. III or V are possible for an R^2-action).

For $\pi_1 M$ solvable, there is also a classification of the invariant measures for any $\mathcal{F}$ on M:

Theorem A5.2 (Plante-Thurston). If $\pi_1 M$ is solvable and $\mathcal{F}$ is transversally oriented, then the space of foliation cycles, $H(\Lambda) \subset H_2(M)$, has real dimension 1.

For $\mathcal{F}$ defined by an R^2-action, this implies there is a unique non-trivial projective class of cycles in $H_2(M)$ which arise from invariant transverse measures. Fix such an invariant measure ν.

For the d-index theorem, evaluation on C_ν yields the average Euler characteristic of the leaves in the support of ν. These leaves are covered by R^2, hence have average Euler characteristic zero, and T_ν anihilates the class Ind(d).

For the operator $\bar{\partial}$, we use formula (A3.4) to construct holomorphic bundles over M for which $T_\nu \circ \mathrm{Ind}(\bar{\partial} \otimes E) \neq 0$. The number of such bundles is controlled by the *period mapping* of ν. This is a homomorphism $P_\nu: H_1(M;\mathbb{Z}) \longrightarrow \mathbb{R}$ defined as $P_\nu(\alpha) = \nu(\gamma)$ where γ is a simple closed curve representing the homology class α. The rank of its image is called the *rank* of $(\mathcal{F}, \nu)$, denoted by $r(\mathcal{F})$. Note that $1 \leqslant r(\mathcal{F}) \leqslant 3$.

Proposition A5.3. The elements $\mathrm{Ind}(\bar{\partial} \otimes E) \in K_0(C_r^*(X))$, for $E \longrightarrow M$ a holomorphic line bundle, generate a subgroup with rank at least $r(\mathcal{F})$.

Proof. For each $\alpha \in \pi_1 M$ with $P_\nu(\alpha) \neq 0$, choose a simple closed curve γ in M representing α and transverse to $\mathcal{F}$. This is possible by Theorem A5.1 and the known structure of R^2-actions. Then take $E = E_\gamma$ as in §3 to obtain $T_\nu \circ \mathrm{Ind}(\bar{\partial} \otimes E) = \langle \mathrm{ch}(E).c_\nu \rangle = \nu(\gamma) = P_\nu(\gamma)$. This shows the map T_ν is onto the image of P_ν. $\square$

It is easy to see that $r(\mathcal{F}) = 3$ if and only if $\mathcal{F}$ is a foliation by planes. This coincides with the R^2-action being free, and then one knows by Theorem A4.5 that

$$\alpha \longmapsto \text{Ind}(\bar{\partial} \otimes E_\gamma)$$

is an isomorphism from $H_2(M;Z)$ onto the summand of $K_0(C_r^*(M))$

corresponding to the image of $H_2(M;Z) \subset K_0(M) \xrightarrow{\cong} K_0(C_r^*(M))$.

(A5.2) An R^2-Action on a Nilmanifold

Let N_3 be the nilpotent group of strictly triangular matrices in $GL(3,R)$:

$$N_3 = \left\{ \begin{bmatrix} 1 & a & b \\ 0 & 1 & c \\ 0 & 0 & 1 \end{bmatrix} \text{ such that } a,b,c \in R \right\}.$$

For each integer $n>0$, define a lattice subgroup

$$\Gamma_n = \left\{ \begin{bmatrix} 1 & p & r/n \\ 0 & 1 & q \\ 0 & 0 & 1 \end{bmatrix} \text{ such that } p,q,r \in Z \right\}.$$

Then $M = N_3/\Gamma_n$ is a compact oriented 3-manifold, and the subgroup

$$R^2 = \left\{ \begin{bmatrix} 1 & a & b \\ 0 & 1 & 0 \\ 0 & 0 & 1 \end{bmatrix} \right\} \text{ acts almost freely on } M \text{ via left translations. Also}$$

note M is a circle bundle over T^2, and $H_2(M;R) \cong R^2$. By Theorem A4.5, the index map is an isomorphism, so $K_0(C_r^*(M)) \cong Z^3$.

The curve representing the homology class of $\alpha = \begin{bmatrix} 1 & 0 & 0 \\ 0 & 1 & 1 \\ 0 & 0 & 1 \end{bmatrix} \in \pi_1 M$

is transverse to $\mathcal{F}$ and $P_\nu(\alpha) \neq 0$ for a transverse measure ν with $C_\nu \neq 0$. However, $\text{Ind}(\bar{\partial} \otimes E_\gamma)$ cannot detect the contribution to $K_0(C_r^*(M))$ from the curve defined by a fibre of $M \longrightarrow T^2$.

(A5.3) Foliations Without Holonomy

If for every leaf L of a foliation, $\mathcal{F}$, the holonomy along each closed loop in L is trivial, then we say $\mathcal{F}$ is without holonomy. In codimension-one, such foliations can be effectively classified up to topological equivalence. We discuss this for the case of C^2-foliations. By Sacksteder's Theorem (cf. [L]), a codimension-one, C^2-foliation without holonomy of a compact manifold admits a transverse invariant measure ν whose support is all of M. Moreover, there is foliation-preserving homeomorphism between M and a model foliated space, $X = (\tilde{B} \times S^1)/\Gamma$, where Γ is the fundamental group of a compact manifold B, $\tilde{B}$ is its universal cover with Γ acting via deck translations, and Γ acts on S^1 via a representation $\exp(2\pi i \rho)\colon \Gamma \longrightarrow SO(2)$, for $\rho\colon \Gamma \longrightarrow \mathbb{R}$. The foliation of X by sheets $\tilde{B} \times \{\theta\}$ has a canonical invariant measure, $d\theta$, and ν corresponds to $d\theta$ under the homeomorphism. Since the index invariants are topological, in this case we can assume that M is one of these models. For a 3-manifold this implies $B = \Sigma_g$ for Σ_g a surface of genus $g \geqslant 1$. The case $g = 1$ is a special case of examples (A5.1) above.

Let Λ denote the abelian subgroup of $\mathbb{R}$ which is the image of ρ. Denote by $r(\mathcal{F})$ the rank of Λ. It is an easy geometrical exercise to see that the group Λ agrees with the image of the evaluation map $[d\theta]\colon H_1(M;\mathbb{Z}) \longrightarrow \mathbb{R}$. Moreover, there exists simple closed curves $\{\gamma_1,...,\gamma_r\}$ in M transverse to $\mathcal{F}$ for which $\{P_\nu(\gamma_i)\}$ yields a $\mathbb{Z}$-basis for Γ. Form the holomorphic bundles $\{E_i\}$ corresponding to the $\{\gamma_i\}$, then the set $\{\mathrm{Ind}(\bar{\partial} \otimes E_i)\}$ generates a free subgroup of rank r in $K_0(C_r^*(M))$. Since $H_2(BG(M);\mathbb{R})$ has rank r, this implies

Proposition A5.4. The index map

$$K_0^\mathcal{T}(BG(M)) \longrightarrow K_0(C_r^*(M))$$

is a monomorphism.

These foliations have been analyzed in further detail by Natsume [N] where he shows that this map is also a surjection.

(A5.4) Solvable Group Actions

The locally free actions of A^2 on 3-manifolds has been studied by Ghys-Sergiescu [GS] and Ghys [Gh]:

Theorem A5.5 [Gh]. Let $\pi_1 M$ be solvable and suppose A^2 acts on M. Then M is diffeomorphic to a torus bundle M_a over S^1, and the monodromy map $a \in SL(2,\mathbb{R})$ has two distinct real eigenvalues.

Theorem A5.6 [Gh]. Suppose A^2 acts locally freely on M and preserves a smooth volume form. Then M is diffeomorphic to $SL(2,\mathbb{R})/\Gamma$ for some cocompact lattice in the universal covering group $\widetilde{SL(2,\mathbb{R})}$, and the action of A^2 on M is via left translations.

Proposition A5.7 [Gh]. Suppose $H_1(M) = 0$ and A^2 acts locally freely on M. Then the action preserves a smooth volume form on M.

Let us describe the foliation on $M_0 = T^2 \times \mathbb{R}/\phi_a$. Let $\bar{v} \in \mathbb{R}^2$ be an eigenvector with eigenvalue $\lambda > 0$. The foliation of $\mathbb{R}^3$ by planes parallel to the span of $\{(\bar{v} \times 0), (\bar{0} \times 1)\}$ is invariant under the covering transformations of $\mathbb{R}^3 \longrightarrow M_a$, so descends to a foliation $\mathcal{F}_\lambda$ on M_a. When $\lambda = 1$, the R^2-action on $\mathbb{R}^3$ defining the foliation there descends to an R^2 action on M_a, defining $\mathcal{F}_\lambda$. When $\lambda \neq 1$, the leaves of $\mathcal{F}_\lambda$ are defined by an action of A^2 on M_a.

For A^2-actions on M with $\pi_1 M$ not solvable, it seems reasonable to conjecture they must have the form given in Theorem A5.6.

If the action of A^2 preserves a volume form on M^3, then $\mathcal{F}$ is transversally affine [GS], so there can be no invariant measures for $\mathcal{F}$. In this case Theorem 8.6 of Chapter VIII reveals no information about $K_0(C_r^*(M))$. However, one has Connes' Theorem A4.5 since the A^2-action is almost free. To give an illustration, let $\Gamma \subset SL(2,\mathbb{R})$ be a cocompact lattice, and set $M = SL(2,\mathbb{R})/\Gamma$. The group A^2 acts via left translations and preserves a smooth volume form on M. Then $G(M) \cong M \times A^2$, $K_0^\tau(BG(M)) \cong K^0(M)$ and Ind: $K_0^\tau(BG(M)) \longrightarrow K_0(C_r^*(M))$ is an isomorphism. Note the foliation on M admits 2g closed transversals $\{\gamma_1,...,\gamma_{2g}\}$ which span $H_1(M)$. Form the corresponding bundles $E_i \longrightarrow M$, and consider the classes

$\{\mathrm{Ind}(\bar{\partial} \otimes E_i)\} \subset K_0(C_r^*(M))$. It is natural to ask whether these classes are linearly independent, and for a geometric proof if so.

(A5.5) Foliations With All Leaves Proper

A leaf $L \subset M$ is *proper* if it is locally closed in M. $\mathcal{F}$ is proper if every leaf is proper. The geometric theory of codimension-one proper foliations is highly developed (cf. [CC2], [HHi]). We recall a few general facts relevant to our discussion.

Theorem A5.8. Let $\mathcal{F}$ be a proper foliation of arbitrary codimension. Then the quotient measure space $M/\mathcal{F}$, endowed with the Lebesgue measure from M, is a standard Borel space.

Corollary A5.9. Let $\mathcal{F}$ be a proper foliation of arbitrary codimension. Then any ergodic invariant transverse measure for $\mathcal{F}$ with finite total mass is supported on a compact leaf.

Theorem A5.10. For a codimension one proper foliation $\mathcal{F}$, all leaves of $\mathcal{F}$ have polynomial growth, and the closure of each leaf of $\mathcal{F}$ contains a compact leaf.

Let $\mathcal{F}$ be a proper codimension-one foliation of M^3. Given a transverse invariant measure ν, we can assume without loss of generality that the support of ν is a compact leaf L. If L has genus ≥ 2, then there exists a closed transversal γ which intersects L, so $T_\nu \circ \mathrm{Ind}(\bar{\partial} \otimes E_{n \cdot \gamma}) \neq 0$ for all but at most one value of n. Thus, the class $[L] \in H_2(M;Z)$ corresponds to a non-trivial class $\mathrm{Ind}(\bar{\partial} \otimes E_{n \cdot \gamma}) \in K_0(C_r^*(M))$. If L is a 2-torus, then it is difficult to tell whether the homology class of L is non-zero, and if so, whether it generates a non-zero class in $K_0(C_r^*(M))$. There is a geometric criterion which yields an answer.

Theorem A5.11 (Rumler-Sullivan). Suppose M admits a metric for which each leaf of $\mathcal{F}$ is a minimal surface. Then every compact leaf of $\mathcal{F}$ has a closed transversal which intersects it.

Corollary A5.12. Suppose $\mathcal{F}$ is a proper and minimal foliation. For each ergodic invariant transverse measure ν, there is a holomorphic bundle $E_\nu \longrightarrow M$ such that $\mathrm{Ind}(\bar{\partial} \otimes E_\nu) \in K_0(C_r^*(M))$ is non-zero, and $\mathrm{Ind}_\nu(\bar{\partial} \otimes E_\nu) \neq 0$.

We cannot conclude from Corollary A5.12 that the elements $\{\mathrm{Ind}(\bar{\partial} \otimes E_\nu) \mid \nu \text{ ergodic}\}$ are independent. (Consider the product foliation $\Sigma_g \times S^1$.) However, if M has a metric for which every leaf is geodesic submanifold, then there are as many independent classes in $K_0(C_r^*(M))$ as there are independent currents $C_\nu \in H_2(M;\mathbb{R})$.

The Reeb foliation of S^3 is another relevant example of a proper foliation. It is not minimal, and $K_0(C_r^*(M)) \cong \mathbb{Z}$ so the toral leaf does not contribute (cf. [Pen] and [To]).

(A5.6) Foliations With Non-Zero Godbillon-Vey Class

There is exactly one characteristic class for codimension-one foliations (of differentiability at least C^2), the Godbillon-Vey class GV $\in H^3(M;\mathbb{R})$. Recall from §A4 that GV defines linear functionals on both $K_*(M)$ and $K_*(C_r^*(M))$, and these functionals agree under the map $K_*(M) \longrightarrow K_*(C_r^*(M))$. (We remark that the map GV: $K_*(C_r^*(M)) \longrightarrow \mathbb{R}$ is *not* natural -- it depends upon the choice of a smooth dense subalgebra of $C_r^*(M)$.) If GV $\neq 0$ in $H^3(M)$, then there is a class [u] $\in K_*(C_r^*(M))$ on which GV is non-trivial. From this we conclude that the composition

$$H_3(M;Z) \longrightarrow K_1^\xi(BG(M)) \longrightarrow K_0(C_r^*(M))$$

is injective.

The information on $K_1(C_r^*(M))$ obtained from GV is about all one knows for these foliations $\mathcal{F}_\alpha$ on M, which underlines the need for better understanding of how the geometry of a foliation is related to the analytic invariants in $K_0(C_r^*(M))$.

APPENDIX B: L^2 HARMONIC FORMS ON NON-COMPACT MANIFOLDS

By Calvin C. Moore, Claude Schochet,
and Robert J. Zimmer

If M is a compact oriented manifold then the Hodge theorem supplies a unique harmonic form associated to each de Rham cohomology class of M. If the compactness assumption is dropped then the situation becomes considerably more sensitive. In this appendix we demonstrate how to use the index theorem for foliated spaces to produce L^2 harmonic forms on the leaves of certain foliated spaces.

We begin by recalling the Hirzebruch signature theorem. If M is a compact oriented manifold of dimension 4r then its signature is defined to be the signature of the bilinear form on $H^{2r}(M)$ given by

$$(x,y) = \int_M x \vee y.$$

Recall that there is a signature operator A (cf. Chapter VIII), and the signature of the manifold, Sign(M), is the Fredholm index of this operator. If M has positive signature then $H^{2r}(M)$ must be non-trivial and must contain classes represented by harmonic forms. (An easy special case: take $M^{4r} = \mathbb{C}P^{2r}$. Then Sign(M) = 1, $H^{2r}(M) = \mathbb{R}$ and so $\mathbb{C}P^{2r}$ has harmonic 2r-forms.)

Let X denote a compact metrizable foliated space with leaves of dimension 4r and oriented foliation bundle F. Then there is a signature operator A = $\{A_\ell\}$ with local trace denoted here by a = $\{a_\ell\}$ and associated partial Chern character [a] $\in H_\tau^{4r}(X)$. For each invariant transverse measure ν we define the signature of X by

$$\text{Sign}(X,\nu) = <[a].[C_\nu]>.$$

where C_ν is the Ruelle–Sullivan current associated to ν. This is independent of the metric chosen but of course does depend upon ν.

The foliated space version of the signature theorem states that $[a] = [L_r]$. where L_r is the Hirzebruch L-polynomial in the tangential Pontrjagin classes of F. If $Sign(X,\nu) > 0$ then there are ν-non-trivial L^2 harmonic 2r-forms on X (that is to say, there are non-zero L^2 harmonic 2r-forms on some of the leaves of X, and the support of ν is positive on the union of these leaves.).

Here is our first result.

Theorem B1. Suppose that X is a compact oriented foliated space with leaves of dimension 4. Assume that X has a tangential Riemannian structure so that each leaf is isometric to the complex 2-disk B^2 (with its Poincaré metric). Let ν be an invariant transverse measure on X. Then $Sign(X,\nu) > 0$.

Corollary B2. The space X cannot be written as a product of foliated spaces.

Corollary B2 also follows from the (significantly more general) assertions of [Z3].

Corollary (of the proof) B3. The space B^2 has non-trivial L^2 harmonic 2-forms.

Remark. It may be enough to assume that each leaf of X is quasi-isometric to B^2.

Our proof of B1 is somewhat round-about. First we prove B1 in a very special case in the setting of automorphic forms. Then we prove the corollaries. Finally we deduce the general case of B1 from Corollary B3. The foliated spaces index theorem is used twice, in different directions.

Proof. Consider the following special case. Let G be the group of holomorphic automorphisms of B^2, let Γ be a cocompact torsionfree lattice in G. and let K be a maximal compact subgroup of G. Then B^2 is isometric to the homogeneous space G/K. The quotient space B^2/Γ is a compact complex manifold of real dimension 4. We assume that it is oriented. The lattice Γ may be chosen so that $\text{Sign}(B^2/\Gamma)$ is strictly positive; we assume this to be the case.

Let S be a smooth manifold upon which Γ acts without fixed points and suppose that S has a finite Γ-invariant measure μ. (For instance, take $S = G/\Gamma'$ for some suitable lattice Γ'.) Define an action of Γ on $B^2 \times S$ by

$$(b,s)\gamma = (b\gamma, s\gamma)$$

and let

$$X = (B^2 \times S)/\Gamma$$

denote the resulting orbit space. Then X is a compact space foliated by the images of the various maps

$$B^2 \times \{s\} \longrightarrow B^2 \times S \longrightarrow X$$

so each leaf is isometric to B^2. There is a natural projection

$$\pi: X \longrightarrow B^2/\Gamma$$

given by sending (b,s) to the image of b under the map $B^2 \longrightarrow B^2/\Gamma$. The restriction of π to each leaf ℓ is a covering map $\ell \longrightarrow B^2/\Gamma$. The tangent bundle F to the foliated space is simply the pullback of the tangent bundle of (the manifold) B^2/Γ by π.

Let us compare the signature theorems on X and on B^2/Γ. The Hirzebruch signature theorem (in this low-dimensional situation) reads

$$\text{Sign}(B^2/\Gamma) = \int \frac{1}{3} p_1(T(B^2/\Gamma)^*) \, dvol$$

where $dvol$ is the volume form on B^2/Γ and p_1 is the first Pontrjagin class. The Connes signature theorem applied to the signature operator A with respect to the invariant transverse measure ν corresponding to the invariant measure μ on S reads

$$\text{Sign}(X, \nu) = \int \frac{1}{3} p_1^\tau (F^*) d\nu.$$

As $F = \pi^*(T(B^2/\Gamma))$ and p_1 projects to p_1^τ under the map from de Rham to tangential cohomology, we have

$$\text{Sign}(B^2/\Gamma) = \int \frac{1}{3} p_1 (T(B^2/\Gamma)^*) dvol$$

$$= \int \frac{1}{3} p_1^\tau (F^*) d\nu$$

$$= \text{Sign}(X, \nu)$$

so that

$$\text{Sign}(X, \nu) = \text{Sign}(B^2/\Gamma) > 0.$$

Thus $\text{Sign}(X, \nu)$ is strictly positive, and in fact is a positive integer (if we properly normalize μ originally). This proves the theorem for this particular class of foliated spaces.

Next we establish Corollary B3. By definition of $\text{Sign}(X, \nu)$ we see that

$$\int a_\ell d\nu > 0$$

in our example above. Now each leaf ℓ is isometric to B^2 and the measure a_ℓ is the local trace of the signature operator on B^2, so that in this example the measure a_ℓ really does not depend upon ℓ. As

$$a_\ell = \text{Ker}(A_\ell) - \text{Ker}(A_\ell^*)$$

we see that $\text{Ker}(A_\ell)$ must be non-trivial for some leaves ℓ; thus the space B^2 must have non-trivial L^2 harmonic 2-forms. This proves Corollary B3.

We turn next to the general case of Theorem B1. Let X be a compact foliated space as in the statement of the theorem. Then the local trace $a = \{a_\ell\}$ of the signature operator is independent of the

311

leaf ℓ. Thus

$$\text{Sign}(X, \nu) = \int a_\ell \, d\nu > 0$$

by our earlier argument. Apply the foliation index theorem again (in the opposite direction) and we see that

$$\int \frac{1}{3} p_1^\tau(F^*) d\nu > 0$$

which implies that the class $[p_1^\tau(F^*)] \neq 0$ in $H_\tau^4(X)$.

Finally, X cannot split as a product of foliated spaces since that would imply that $\text{Sign}(X, \nu) = 0$. This completes the proof of B1, B2, and B3. $\square$

In order to generalize, one need only look at those properties of B^2 which were actually used in the proof. The key fact was that there was a lattice group Γ such that $B^2 \longrightarrow B^2/\Gamma$ was well-behaved, and such that B^2/Γ was a compact manifold with positive signature.

Definition B4. A <u>Clifford-Klein form</u> of a connected and simply connected Riemannian manifold B is a Riemannian manifold B' whose universal Riemannian covering is isomorphic to B.

A. Borel [Bor] has shown that a simply connected Riemannian symmetric space B always has a compact Clifford-Klein form. Let $\mathfrak{B}$ be the collection of spaces which are finite products of irreducible symmetric domains whose compact counterparts are

$$U(p+2r)/(U(p){\times}U(2r)),$$

$$SO(4k+2)/(SO(4k){\times}SO(2)),$$

$$E_6/(\text{Spin}(10){\times}T^1).$$

The space B^2 is in $\mathfrak{B}$ since B^2 is associated to the space

$$U(3)/(U(1)\times U(2)) \cong \mathbb{C}P^2.$$

If B is a simply connected Riemannian symmetric space then Sign(B') = 0 unless B ∈ $\mathfrak{B}$. If B ∈ $\mathfrak{B}$ then Sign(B') ⩾ 1, by Borel [Bor,§3]. This is all that we need.

Theorem B5. Let X be a compact oriented foliated space with leaves of dimension p. Suppose that X has a tangential Riemannian structure such that each leaf is isometric to some fixed B ∈ $\mathfrak{B}$. Then Sign(X,ν) > 0 for each invariant transverse measure ν.

Corollary B6. If B ∈ $\mathfrak{B}$ is a manifold of dimension 4r, then B has non-trivial L^2 harmonic 2r-forms.

We omit the proof, which is essentially the same as the special case B = B^2.

We turn next to the use of the Gauss-Bonnet theorem. Recall that if X is a compact oriented foliated space with leaves of dimension 2q then the index theorem applied to the de Rham operator yields

$$[\chi] = [K_\tau]/2\pi$$

in $H_\tau^*(X)$, where χ is the alternating sum of the Betti measures (8.6). Given an invariant transverse measure ν, the theorem reads

$$\chi(X,\nu) = \int K_\tau/2\pi \, d\nu,$$

where

$$\chi(X,\nu) = \int \chi d\nu$$

is the tangential Euler characteristic of (X,ν).

Suppose that G is a semisimple Lie group with maximal compact subgroup K and torsionfree lattice Γ. Let S be some compact smooth manifold upon which Γ acts without fixed points and let μ be a finite Γ-invariant measure on S. Let B = G/K and define

X = (B×S)/Γ (where Γ acts diagonally). Then X is a foliated manifold with leaves corresponding to the image of B×{s}. The space B/Γ is a compact smooth manifold and each leaf ℓ is a covering space for B/Γ. The space X has an invariant transverse measure ν corresponding to the measure μ on the global transversal S. For instance, if G = PSL(2,$\mathbb{R}$) then B is the upper half plane H and X is foliated by copies of H. (Note that H has constant negative curvature– it is homeomorphic but not isometric to $\mathbb{C}$.)

The Euler characteristic of B/Γ is given by the classical Gauss–Bonnet theorem:

$$\chi(B/\Gamma) = \int K/2\pi \ d\text{vol}$$

where K is the curvature form on B/Γ and dvol is the volume form on B/Γ.

Specialize to the case where each leaf has dimension 2. The Betti measure β_0 is always zero since there are no L^2 harmonic functions on non-compact manifolds. Duality implies that $\beta_2 = 0$. Thus the foliation Gauss–Bonnet theorem reduces to

$$\int -\beta_1 d\nu = \int K_T/2\pi \ d\nu$$

where K_T is the Gauss curvature along the leaves. Arguing just as in the proof of the special case of Theorem B1, we see that

$$\chi(B/\Gamma) = \int K/2\pi \ d\text{vol} = \int K_T/2\pi \ d\nu = \int -\beta_1 d\nu.$$

Assume that the surface B/Γ has genus greater than 2. Then $\chi(B/\Gamma)$ is negative and hence the Betti measure β_1 is strictly positive. Since leaves have dimension 2, we see that

$$\int \text{Ker}(d \oplus d^*)_{(1-\text{forms})} > 0.$$

In our example we are again integrating a constant function. Thus on the generic leaf ℓ = G/K there are non-trivial L^2 harmonic 1-forms.

If we continue as in the study of the signature operator, we can obtain the following theorem.

Theorem B7. Let X be a compact oriented foliated space with tangential Riemannian structure such that each leaf is isometric to the upper half plane. Let ν be an invariant transverse measure. Then the tangential Euler characteristic $\chi(X,\nu)$ is strictly positive and X has non-trivial L^2 harmonic 1-forms.

Remark. If the leaves have dimension greater than 2 then β_1 does not correspond so neatly to the Euler characteristic. For example, if the leaves have dimension 4 and $\chi(B/\Gamma) < 0$ then

$$\int (-\beta_1 + \beta_2 - \beta_3)d\nu < 0$$

so that

$$\int \beta_2 d\nu < \int (\beta_1 + \beta_3)d\nu.$$

As the left hand side must be non-negative, this implies that the integral of either β_1 or β_3 (and hence both of them, by duality) must be strictly positive. Thus there are L^2 harmonic 1 and 3-forms.

APPENDIX C: POSITIVE SCALAR CURVATURE
ALONG THE LEAVES

By Robert J. Zimmer

Mikhael Gromov and Blaine Lawson, in their classic paper [GL], use Dirac operators with coefficients in appropriate bundles and associated topological invariants to investigate whether or not a given compact non-simply connected manifold can support a metric of positive scalar curvature. In this appendix we consider the analogous problem for foliated spaces. We use appropriate tangential Dirac operators to investigate the existence of a tangential Riemannian metric with positive scalar curvature along the leaves of a compact foliated space. Gromov-Lawson use the $\hat{A}$-genus and the Atiyah-Singer index theorem; we shall use the tangential $\hat{A}$-genus and the Connes index theorem.

Let M be a compact oriented manifold of dimension p = 2d with associated Hirzebruch $\hat{A}$-class, $\hat{A}(M) \in H^{even}(M,\mathbb{R})$, a polynomial in the Pontrjagin classes. If M is a spin manifold, then there are the associated bundles of half-spinors $S^{\pm}(M)$, and an associated Dirac operator

$$D^+: \Gamma^\infty(S^+) \longrightarrow \Gamma^\infty(S^-).$$

The Atiyah-Singer theorem implies that index(D^+) vanishes unless p is divisible by 4, and in that case

$$index(D^+) = \hat{A}[M]$$

where $\hat{A}[M]$ is the $\hat{A}$-genus of M, i.e., $\hat{A}[M] = \langle\hat{A}(M),[M]\rangle$. If E is any Hermitian bundle over M (with a unitary connection) then, following Gromov-Lawson [GL], $S(M)\otimes E$ is called a *twisted spin bundle* over M. Associated to this bundle there is also an elliptic operator called the *twisted Dirac operator* D^+. The

Atiyah–Singer theorem now implies that

$$\text{index}(D^+) = (-1)^d \langle \text{ch}(E)\hat{A}(M), [M]\rangle.$$

Suppose now that X is a compact foliated space with oriented foliation bundle F and invariant transverse measure ν. Suppose further that F is a spin foliation, i.e., F has a spin structure. Then there is an associated bundle of spinors and for each leaf ℓ an associated Dirac operator D_ℓ^+ on the leaf and hence a tangentially elliptic operator $D^+ = \{D_\ell^+\}$. Then by Connes' theorem,

$$\text{Index}_\nu(D^+) = (-1)^d \langle \hat{A}_T(F), [C_\nu]\rangle$$

where $[C_\nu]$ is the homology class of the Ruelle–Sullivan current associated to ν. Define

$$\hat{A}_\nu[X] = \langle \hat{A}_T(F), [C_\nu]\rangle$$

the *tangential* $\hat{A}$-*genus of* X with respect to the invariant transverse measure ν. Note that if ker $D_\ell = 0$ as an unbounded operator on $L^2(\ell)$ for ν–a.e. ℓ, then $\hat{A}_\nu[X] = 0$.

Choose some tangential Riemannian metric on X and let κ denote the scalar curvature along the leaves. We say that the metric has *positive scalar curvature on the leaf* ℓ if $\kappa \geq 0$ on ℓ and if $\kappa > 0$ at some point of ℓ. If this is so, then by Lichnerowicz's computations, ker $D_\ell^2 = 0$, and since we are in L^2 and D_ℓ is formally self-adjoint, ker $D_\ell = 0$. Thus:

Proposition C1. Let X be a compact foliated space with foliation bundle F with a given spin structure, and let ν be an invariant transverse measure. If there exists a metric on X which has positive scalar curvature along ν–a.e. leaf, then $\hat{A}_\nu[X] = 0$.

Now suppose that E is an Hermitian bundle on X with a unitary tangential connection. For any leaf ℓ, let E_ℓ be the restriction of E to ℓ. Then there is a twisted spin bundle $S(F)\otimes E$ on X, and a twisted Dirac operator $D^+ = \{D_\ell^+\}$.   Once again, Connes' theorem implies:

Proposition C2. $\mathrm{Ind}_\nu(D^+) = <\mathrm{ch}_\tau(E)\hat{A}_\tau(F), [C_\nu]>$.

Thus if $\ker(D^+) = 0$, then $<\mathrm{ch}_\tau(E)\hat{A}_\tau(F), [C_\nu]> = 0$.
For each leaf ℓ, the equation

$$D_\ell^2 = \nabla_\ell^*\nabla_\ell + \kappa/4 + (R_0)_\ell$$

holds, where ∇ is a certain first order tangential operator, and R_0 is described as in [GL, Theorem 1.3], in terms of the Clifford multiplication and the tangential curvature tensor of E_ℓ.   An argument as in [GL, Theorem 1.3] yields the following proposition.

Proposition C3.   If $\kappa \geq 4(R_0)_\ell$ and $\kappa > r(R_0)_\ell$ at some point, then $\ker D_\ell = 0$.

In   particular,   this   would   yield   vanishing   of $<\mathrm{ch}_\tau(E)\hat{A}_\tau(F), [C_\nu]>$.

Definition C4.   Call a manifold M *expandable* if for each r, there is a smooth embedding of the Euclidean ball

$$e_r: B_r \longrightarrow \tilde{M}$$

(where $\tilde{M}$ is the universal cover of M) such that

$$ae_r(v) \geq v \qquad \text{for all } v \in TB_r.$$

Example C5.   The torus T^n is expandable.

Proposition C6 (Gromov–Lawson).

1)   A compact solvmanifold is expandable.

2)   A manifold of nonpositive curvature is expandable.

Definition C7 (slight modification of [GL]). A compact manifold M of dimension n is *enlargeable* if for each $c > 0$ there is a finite covering $M' \longrightarrow M$ and a c-contracting map $M' \longrightarrow S^n$ of non-zero degree.

Proposition C8. Let M be a compact expandable manifold and suppose that $\pi_1(M)$ is residually finite. Then M is enlargeable.

Theorem C9 (Gromov-Lawson).   Suppose that M is an enlargeable manifold of even dimension and suppose that some finite cover of M is a spin manifold. Then M has no metric of positive scalar curvature.

Corollary C10.   No compact solvmanifold and no manifold of non-positive curvature with a finite spin covering supports a metric of positive scalar curvature.

    We move to the context of foliated spaces.

Theorem C11. Let M be a compact enlargeable manifold of even dimension with a finite spin covering M'. Let $\pi_1(M)$ act on a space Y with an invariant measure ν (not necessarily smooth). Form the associated foliated bundle over M

$$Y \longrightarrow X = \tilde{M} \times_{\pi_1(M)} Y \longrightarrow M$$

so that each leaf is of the form $\tilde{M}/$(subgroup of $\pi_1(M)$). Then there is no tangential Riemannian metric on the foliated space X such that every leaf has everywhere positive scalar curvature.

Corollary C12. For a foliated bundle over any compact solvmanifold or over any manifold of non-positive curvature with a finite spin cover, the result holds.

Proof. For the solvmanifold case in odd dimension, cross with S^1 with

319

the foliation (leaf)$\times S^1$ over solvmanifolds.   □

Proof of Theorem C11.   Suppose that there were such a metric.   Let $0 < \kappa_0 \leq \min \kappa$ on almost all leaves.   Passing to finite covers yields the diagram

$$
\begin{array}{ccccc}
Y & \longrightarrow & X' & \longrightarrow & X \\
& & \downarrow{\scriptstyle \rho} & & \downarrow \\
& & M' & \longrightarrow & M
\end{array}
$$

with a c-contracting map $f: M' \longrightarrow S^{2n}$ of non-zero degree, where $c^2 < \kappa_0/\alpha$ and α depends upon the dimension of M and data on a fixed Hermitian bundle $E_0 \longrightarrow S^{2n}$ with $c_n(E_0) \neq 0$.   Proposition C3 and computation as in Proposition 3.1 of [GL] imply that

$$
< ch_\tau(\rho^* f^* E_0)\hat{A}_\nu(F), [C_\nu]> = 0.
$$

Since

$$
ch_\tau(\rho^* f^* E_0) = \frac{1}{(n-1)!} \, \rho^* f^* (c_n^\tau(E_0)) + 1
$$

and $\hat{A}_\nu[X] = 0$ by Proposition C1, it follows that

$$
<\rho^* f^* c_n^\tau(E_0), [C_\nu]> = 0.
$$

Since

$$
\int_{S^{2n}} f^* (c_n^\tau(E_0)) \neq 0,
$$

we use the basic computation that $<\rho^* \omega, [C_\nu]> \neq 0$ for foliated bundles, where $\int \omega \neq 0$.   This is a contradiction.   □

Gromov-Lawson show [Cor. A] that any metric of non-negative scalar curvature on the torus T^n is flat.   That suggests the following conjectures.

<u>Conjectures</u> C13.

1)   For foliations over T^n, $\kappa \geq 0$ along the leaves implies that
$\kappa = 0$ along the leaves.

2)   (stronger) If $\kappa \geq 0$ then the leaves are Ricci flat, or even

3)   (still stronger) If $\kappa \geq 0$ then the leaves are flat.

<u>Remark</u> C14.   If M is a manifold with non-negative scalar curvature
and with $\kappa > 0$ at one point of M, then Kazdan and Warner have
shown [KW] that there is a conformal change in the metric of M
such that $\kappa > 0$ everywhere on M.   Suppose that X is a compact
foliated space and suppose that X has positive scalar curvature along
the leaves.   Is it true that the metric on X may be altered so that
$\kappa > 0$ everywhere?   This may be done one leaf at a time; the
difficulty lies in making the change continuous transversely.

REFERENCES

[Ah]   L. Ahlfors. Lectures on quasi-conformal mappings, Van
          Nostrand Math Studies 10, New York, 1966.

[Ar]   W. Arveson. An invitation to C^*-algebras. Graduate
          Textx in Math. Vol 39, Springer, New York, 1976.

[At1]   M.F. Atiyah, K-Theory, Benjamin, Reading, Mass., 1967.

[At2]   M.F. Atiyah, Classical groups and classical differential
          operators on manifolds, Differential Operators on
          Manifolds, C.I.M.E., Rome, Edizioni Cremonesse, 1975.

[At3]   M.F. Atiyah, Elliptic operators, discrete groups, and
          von Neumann algebras, Astérisque 32-33 (1976), 43-72.

[ABP]   M.F. Atiyah, R. Bott, and V.K. Patodi, On the heat
          equation and the index theorem, Invent. Math. 19 (1973),
          279-330. Errata, Invent. Math. 28 (1975), 277-280.

[ABS]   M.F. Atiyah, R. Bott, and A. Shapiro, Clifford modules,
          Topology 3 (1964), 3-38.

[ASI]   M.F. Atiyah and I.M. Singer, The index of elliptic
          operators I, Ann. of Math. 87 (1968), 484-530.

[ASIII]   M.F. Atiyah and I.M. Singer, The index of elliptic
          operators III, Ann. of Math. 87 (1968), 546-604.

[ASIV]   M.F. Atiyah and I.M. Singer, The index of elliptic
          operators IV, Ann. of Math. 93 (1971), 119-138.

[AM]   L. Auslander and C. Moore, Unitary representations of solvable Lie groups. Mem. Amer. Math. Soc. No. 62 (1966). pp. 199.

[BC]   P. Baum and A. Connes, Geometric K-theory for Lie groups and foliations, preprint.

[Bl]   B. Blackadar, K-Theory for Operator Algebras, Math. Sci. Res. Inst. Publ. No. 5, Springer, New York, 1986.

[Bor]   A. Borel, Compact Clifford–Klein forms of symmetric spaces, Topology 2 (1963), 111–122.

[Bo]   R. Bott, Lectures on K(X), W.A. Benjamin, New York, 1969.

[BT]   R. Bott and L. Tu, Differential Forms in Algebraic Topology, Graduate Texts in Math., Springer, New York, 1982.

[Bou]   N. Bourbaki, Topologie Générale, Chap. IX (2e éd.), Act. Sci. Ind. No. 1045, Hermann, Paris, 1958.

[BdM]   L. Boutet de Monvel, A Course on Pseudo Differential Operators and their Applications, Mathematics Department, Duke U., Durham, North Carolina, 1976.

[Bow]   R. Bowen, On Axiom A Diffeomorphisms, C.B.M.S. No. 35, Amer. Math. Soc., Providence, Rhode Island, 1978.

[Bre]   M. Breuer, Fredholm theories in von Neumann algebras I, II. Math. Ann. 178 (1968), 243–254 and 180 (1969), 313–325.

[BGR]   L.G. Brown, P. Green and M.A. Rieffel, Stable isomorphism and strong Morita equivalence of C^*-algebras, Pacific J. Math. 71 (1977), 349–363.

[Bow]   R. Bowen, Anosov foliations are hyperfinite, Ann. of Math. $\underline{106}$ (1977), 549-565.

[CC1]   J. Cantwell and L. Conlon, Leaves with isolated ends in foliated 3-manifolds, Topology $\underline{16}$ (1977), 311-322.

[CC2]   J. Cantwell and L. Conlon, Poincaré-Bendixon theory for leaves of codimension one, Trans. Amer. Math. Soc. $\underline{265}$ (1981), 81-209.

[CGT]   J. Cheeger, M. Gromov and M. Taylor, Finite propagation speed, kernel estimates for functions of the Laplace operator, and the geometry of complete Riemannian manifolds, J. Diff. Geom. $\underline{17}$ (1982), 15-54.

[Ch]   P.R. Chernoff, Essential self-adjointness of powers of generators of hyperbolic equations, J. Functional Anal. $\underline{12}$ (1973), 401-414.

[CMS]   L.A. Coburn, R.D. Moyer and I.M. Singer, C^*-algebras of almost periodic pseudodifferential operators, Acta Math. $\underline{130}$ (1973), 279-307.

[Co1]   A. Connes, Une classification des facteurs de Type III, Ann. Sci. Ecole Norm. Sup. $\underline{6}$ (1973), 133-252.

[Co2]   A. Connes, The von Neumann algebra of a foliation, Proc. Rome Conference, June, 1977, Lecture Notes in Physics, Vol 80, Springer, New York, 1978, pp. 145-151.

[Co3]   A. Connes, Sur la théorie non commutative de l'integration, Algèbres d'Operateurs, Lecture Notes in Math., Vol 725, Springer, New York, 1979, pp. 19-143.

[Co4]   A. Connes, Feuilletages et algèbres d'opérateurs, Sém.

Bourbaki 1979-80, No. 551 (Fevier 1980), 1-17.

[Co5]   A. Connes, C^*-algebres et géométrie différentielle, C. R. Acad. Sc. Paris, t. 290 (1980), 599-604.

[Co6]   A. Connes, An analogue of the Thom isomorphism for crossed products of a C^*-algebra by an action of $\mathbb{R}$, Adv. in Math. 39 (1981), 31-55.

[Co7]   A. Connes, A Survey of foliations and operator algebras, in Operator Algebras and Applications, R. V. Kadison, ed., Proc. Symp. Pure Math., Vol 38, Part I, Amer. Math. Soc., Providence, Rhode Island, 1982, pp. 521-628.

[Co8]   A. Connes, Non commutative differential geometry, Part I, the Chern character in K-homology, Publ. Math. IHES (1986), 257-309.

[Co9]   A. Connes, Non commutative differential geometry, Part II, de Rham homology and non commutative algebra, Publ. Math. IHES (1986), 310-360.

[Co10]   A. Connes, Cyclic cohomology and the transverse fundamental class of a foliation, in Geometric Methods in Operator Algebras, H. Araki and E.G. Effros, ed., Kyoto (1983), Pitman Res. Notes in Math. Vol. 123, Wiley, NY, 1986.

[CM]   A. Connes and H. Moscovici, The L^2 index theorem for homogeneous spaces of Lie groups, Ann. of Math. 115 (1982), 291-330.

[CS1]   A. Connes and G. Skandalis, Théorème de l'indice pour les feuilletages, C. R. Acad. Sci. Paris Ser. I, t. 292 (1981), 871-876.

[CS2]   A. Connes and G. Skandalis. The longitudinal index theorem for foliations. Publ. RIMS. Kyoto Univ. 20 (1984), 1139–1183.

[CT]   A. Connes and M. Takesaki. The flow of weights on factors of type III, Tôhoku Math. J. 29 (1977), 473–575; Errata, ibid. 30 (1978), 653–655.

[Da]   M.W. Davis, Groups generated by reflections and aspherical manifolds not covered by Euclidean space, Ann. of Math. 117 (1983), 293–324.

[Di1]   J. Dixmier, Les Algèbres d'opérateurs dans l'espace Hilbertien. 2nd ed., Gauthier-Villars, Paris, 1969.

[Di2]   J. Dixmier, Les C*-algèbres et leurs représentations, 2nd ed., Gauthier-Villars, Paris, 1969.

[DD]   J. Dixmier and A. Douady, Champs continus d'espaces hilbertiens et de C*-algèbres, Bull. Soc. Math. Fr. 91 (1963), 227–284.

[DM]   M. Duflo and C.C. Moore, On the regular representation of nonunimodular locally compact group, J. Functional Anal. 21 (1976), 209–243.

[DS]   N. Dunford and J. Schwartz, Linear Operators, 3 Vols, Interscience. New York, 1957–1971.

[Du]   J. Dupont, Curvature and Characteristic Classes. Lecture Notes in Math., Vol 640. Springer, New York, 1978.

[Eh]   C. Ehresmann, Structures feuilletées, Proc. 5th Canadian Math. Congress (1961), Univ. of Toronto Press, 1963, 109–172.

[ES]   S. Eilenberg and N. Steenrod, <u>Foundations</u> <u>of</u> <u>Algebraic</u> <u>Topology</u>, Princeton Univ. Press, Princeton, 1952.

[El]   A. El Kacimi-Alaoui, Sur la cohomologie feuilletée, Compositio Math. <u>49</u> (1983), 195–215.

[EMT]   D.B.A. Epstein, K.C. Millett and D. Tischler, Leaves without holonomy, J. London Math. Soc. (2), <u>16</u> (1977), 548–552.

[FS]   T. Fack and G. Skandalis, Sur les representations et indéaux de la C^*-algèbre d'un feuilletage, J. Operator Theory <u>8</u> (1982), 95–129.

[FHM]   J. Feldman, P. Hahn and C.C. Moore, Orbit structure and countable sections for actions of continuous groups, Adv. Math. <u>28</u> (1978), 186–230.

[FM]   J. Feldman and C.C. Moore, Ergodic equivalence relations, cohomology, and von Neumann algebras I, II, Trans. Amer. Math. Soc. <u>234</u> (1977), 289–324 and 325–359.

[Fr]   K. Friedrichs, Symmetric hyperbolic linear differential equations, Comm. Pure Appl. Math. <u>7</u> (1954), 345–392.

[Fu]   H. Furstenberg, Strict ergodicity and transformations on the torus, Amer. J. Math. <u>83</u> (1961), 573–601.

[Gh]   E. Ghys, Strong structural stability for Fuchsian groups, Lecture given at Math. Sci. Res. Inst., Berkeley, California (1984).

[GS]   E. Ghys and V. Sergiescu, Stabilité et conjugation differentiable pour certains feuilletages, Topology <u>19</u> (1980), 179–197.

[Gi1]   P.B. Gilkey, Curvature and eigenvalues for the Laplacian for elliptic complexes, Advances in Math. 10 (1973), 344-382.

[Gi2]   P.B. Gilkey, The Index Theorem and the Heat Equation, Math. Lecture Series No. 4, Publish or Perish, Boston, 1974.

[Gi3]   P.B. Gilkey, Invariant Theory, the Heat Equation, and the Atiyah-Singer Index Theorem, Math. Lecture Series No. 11, Publish or Perish, Boston, 1985.

[Gom]   R. Godement, Topologie Algébrique et Théorie des Faisceaux, Hermann, Paris, 1964.

[Gob]   C. Godbillon, Eléments de Topologie Algébrique, Hermann, Paris, 1971.

[Goo]   S.E. Goodman, Closed leaves in foliations of codimension one, Comment. Math. Helv. 50 (1975), 383-388.

[GP]   S.E. Goodman and J.F. Plante, Holonomy and averaging in foliated sets, J. Differential Geom. 14 (1979), 401-407.

[GL]   M. Gromov and H.B. Lawson, Jr., Spin and scalar curvature in the presence of a fundamental group. I., Ann. of Math. 111 (1980), 209-230.

[Gu]   R. Gunning, Lectures on Riemann surfaces, Math. Notes, Princeton Univ. Press, 1966.

[Haa1]   U. Haagerup, Operator valued weights in von Neumann algebras I, II, J. Funct. Anal. 32 (1979), 175-206; 33 (1979), 339-361.

[Haa2]   U. Haagerup, Connes' bicentralizer problem and uniqueness of the injective factor of type III_1, Acta Math. 158 (1987), 95–148.

[Hae1]   A. Haefliger, Variétés feuilletées, Ann. Scuola Norm. Sup. Pisa (3) 16 (1962), 367–397.

[Hae2]   A. Haefliger, Homotopy and Integrability, Lecture Notes in Math., Vol 197, Springer, New York, (1972), 133–166.

[Hae3]   A. Haefliger, Some remarks on foliations with minimal leaves, J. Differential Geom. 15 (1980), 269–284.

[Hae4]   A. Haefliger, Groupoïdes d'holonomie et classifiants, Structures Transverses des Feuilletages, Toulouse (1982), Asterisque No. 116 (1984), 70–97.

[Hah]   H. Hahn, Reele Funktionen, Akad. Verlag., Leipzig, 1932. Reprinted by Chelsea Publ. Co., New York.

[Hap1]   P. Hahn, Haar measure for measure groupoids, Trans. Amer. Math. Soc. 242 (1978), 1–33.

[Hap2]   P. Hahn, The regular representation of measure groupoids, Trans. Amer. Math. Soc. 242 (1978), 35–72.

[HHi]   G. Hector and U. Hirsch, Introduction to the Geometry of Foliations, Parts A and B, Aspects of Mathematics, Vol E1, E3, Friedr, Vieweg and Sohn, 1983.

[He]   J.L. Heitsch, A cohomology of foliated manifolds, Comment. Math. Helv. 50 (1975), 197–218.

[HHu]   J. Heitsch and S. Hurder, Secondary classes, Weil operators, and the geometry of foliations, J. Differential Geom. 20 (1984), 291–309.

[Her]   M. Herman, Sur la conjugaison différentiable des
          difféomorphisme du cercle à des rotations, Publ. Math.
          IHES No. 49 (1979), 5-233.

[HS]   M. Hilsum and G. Skandalis, Stabilité des C^*-algèbres de
          feuilletages, Ann. Inst. Fourier, Grenoble 33, 3 (1983),
          201-208.

[Hir]   M.W. Hirsch, Differential Topology, Springer, New York,
          1976.

[Hz]   F. Hirzebruch, Automorphe Formen und der Satz von
          Riemann-Roch, Symp. Internacional de Topologia
          Algebraica (Mexico, 1958), 129-144.

[Hu]   D. Husemoller, Fibre Bundles, Springer, New York, 1975.

[KT1]   F.W. Kamber and P. Tondeur, Characteristic invariants of
          foliated bundles, Manuscripta Mathematica 11 (1974),
          51-89.

[KT2]   F.W. Kamber and P. Tondeur, Foliated Bundles and
          Characteristic Classes, Lecture Notes in Math., Vol 493,
          Springer, New York, 1975.

[KM]   J. Kaminker and J.G. Miller, Homotopy invariance of the
          analytical index of certain elliptic operators over
          C^*-algebras, J. Operator Theory 14 (1985), 113-127.

[Kar]   M. Karoubi, K-Theory: An Introduction, Springer, New
          York, 1978.

[Kas1]   G.G. Kasparov, The operator K-functor and extensions of
          C^*-algebras, Izv. Akad. Nauk SSSR, Ser. Mat. 44 (1980),
          571-636 = Math. USSR Izv. 16 (1981), 513-572.

[Kas2]   G.G. Kasparov, Hilbert C*-modules: theorems of Stinespring and Voiculescu, J. Operator Theory 4 (1980), 133–150.

[KW]   J.L. Kazden and F.W. Warner, Scalar curvature and conformal deformation of Riemannian structure, J. Differential Geometry 10 (1975), 113–134.

[KN]   S. Kobayashi and K. Nomizu, Foundations of Differential Geometry, Vols I, II, Wiley, New York, 1963.

[Kr]   W. Krieger, On ergodic flows and the isomorphism of factors, Math. Ann. 223 (1976), 19–70.

[Kum]   A. Kumjian, On C*-diagonals and twisted relations, preprint.

[Kur]   K. Kuratowski, Topology, Vol I, Academic Press, New York, 1966.

[L]   H.B. Lawson, The Quantitative Theory of Foliations, C.B.M.S. No. 27, Amer. Math. Soc., Providence, Rhode Island, 1977.

[Ma1]   G. Mackey, The Laplace transform for locally compact abelian groups, Proc. Nat. Acad. Sci. 34 (1948), 152–162.

[Ma2]   G. Mackey, Point realizations of transformation groups, Illinois J. Math. 6 (1962), 327–335.

[Ma3]   G. Mackey, Ergodic theory, group theory, and differential geometry, Proc. Nat. Acad. Sci. U.S.A. 50 (1963), 1184–1191.

[Ma4]   G. Mackey, Ergodic transformation groups with a pure point spectrum, Illinois J. Math. $\underline{8}$ (1964), 593–600.

[Ma5]   G. Mackey, Ergodic theory and virtual groups, Math. Ann. $\underline{166}$ (1966), 187–207.

[Ma6]   G. Mackey, The Theory of Unitary Group Representations, Chicago Lectures in Mathematics, University of Chicago Press, Chicago, 1976.

[Mac]   S. MacLane, Homology, Springer-Verlag, New York, 1963.

[Mi]   J. Milnor, Construction of universal bundles I, Ann. of Math. (2) $\underline{63}$ (1956), 272–284.

[MS]   J.W. Milnor and J.D. Stasheff, Characteristic Classes, Ann. Math. Studies, Vol 76, Princeton U. Press, Princeton, 1974.

[Mol]   P. Molino, Propriétés cohomologiques et propriétés topologiques des feuilletages à connexion transverse projetable, Topology $\underline{12}$ (1973), 317–325.

[Mr1]   C.C. Moore, Square integrable primary representations, Pacific J. Math. $\underline{70}$ (1977), 413–427.

[Mr2]   C.C. Moore, Ergodic theory and von Neumann algebras, in Operator Algebras and Applications, R.V. Kadison, ed., Proc. Symp. Pure Math., Vol 38, Part 2 (1982), 179–226.

[MRW]   P.S. Muhly, J.N. Renault and D.P. Williams, Equivalence and isomorphism for groupoid C^*-algebras, J. Operator Theory $\underline{17}$ (1987), 3–22.

[N]   T. Natsume, The C^*-algebra of codimension one foliations without holonomy, Math. Scand. $\underline{56}$ (1985), 96–104.

[Pa]   R.S. Palais, editor, Seminar on the Atiyah-Singer Index Theorem, Ann. Math. Studies, Vol 57, Princeton U. Press, Princeton, 1965.

[Ped]   G. Pedersen, C^∞-algebras and their Automorphism Groups, Academic Press, New York, 1979.

[PT]   G. Pedersen and M. Takesaki, The Radon-Nikodym theorem for von Neumann algebras, Acta Math. 130 (1973), 53-87.

[Pen]   M. Penington, K-Theory and C^*-algebras of Lie groups and foliations, D. Phil. Thesis, Oxford, 1983.

[Ph]   J. Phillips, The holonomic imperative and the homotopy groupoid of a foliated manifold, Rocky Mountain J. Math. 17 (1987), 151-165.

[PS]   A. Phillips and D. Sullivan, Geometry of leaves, Topology 20 (1981), 209-218.

[Pl1]   J.F. Plante, Foliations with measure preserving holonomy, Ann. of Math. 102 (1975), 327-361.

[Pl2]   J.F. Plante, Foliations of 3-manifolds with solvable fundamental group, Invent. Math. 51 (1979), 219-230.

[Po]   W.A. Poor, Differential Geometric Structures, McGraw-Hill, New York, 1981.

[Pr]   J. Pradines, Théorie de Lie pour les groupoides differentiables, C.R. Acad. Sc. Paris t. 263 (1966), 909-910.

[Puk]   L. Pukanszky, Unitary representations of solvable Lie groups, Ann. Sci. Ecole Norm. Sup. (4) $\underline{4}$ (1971), 457–608.

[Pu]   A. Purves, Bimeasurable functions, Fund. Math. $\underline{58}$ (1966), 149–157.

[Ra]   A. Ramsay, Topologies on measured groupoids, J. Funct. Anal. $\underline{47}$ (1982), 314–343.

[Ra2]   A. Ramsay, Virtual groups and group actions, Adv. in Math. $\underline{6}$ (1971), 253–322.

[Ree]   G. Reeb, <u>Sur certaines propriétés topologiques des variétés feuilletés</u>, Act. Sci. et Ind., Hermann, Paris, 1952.

[Rei1]   B. Reinhart, Foliated manifolds with bundle–like metrics, Ann. of Math. $\underline{69}$ (1959), 119–132.

[Rei2]   B. Reinhart, <u>Differential Geometry of Foliations</u>, Ergeb. Math. $\underline{99}$, Springer, New York, 1983.

[Ren1]   J. Renault, <u>A Groupoid Approach to C^{*}-algebras</u>, Lecture Notes in Math., Vol 793, Springer, New York, 1980.

[Ren2]   J. Renault, <u>C^{*}-algebras of groupoids and foliations</u>, in Operator Algebras and Applications, R.V. Kadison, ed., Proc. Symp. Pure Math., Vol 38, Part I, Amer. Math. Soc., Providence, Rhode Island, 1982, pp. 339–350.

[Ro0]   J. Roe, Analysis on manifolds, D. Phil. Thesis, Oxford, 1985.

[Ro1]   J. Roe, An index theorem on open manifolds I, J. Diff. Geom., to appear.

[Ro2]   J. Roe, An index theorem on open manifolds II, J. Diff.

Geom., to appear.

[Ro3]   J. Roe, Finite propagation speed and Connes' foliation algebra, preprint.

[Rh]   G. de Rham, Variétiés Differentiables, Herman, Paris, 1955.

[RRW]   H. Rosenberg, R. Roussarie and D. Weil, A classification of closed orientable 3-manifolds of rank two, Ann. of Math. 91 (1970), 449–464.

[RS]   J. Rosenberg and C. Schochet, The classification of extensions of C^*-algebras, Bull. Amer. Math. Soc. (N.S.) 4 (1981), 105–109.

[RuS]   D. Ruelle and D. Sullivan, Currents, flows, and diffeomorphisms, Topology 14 (1975), 319–327.

[Sa1]   K.S. Sakaria, The de Rham cohomology of foliated manifolds, thesis, SUNY, Stony Brook, New York, 1974.

[Sa2]   K.S. Sakaria, A finiteness theorem for foliated manifolds, J. Math. Soc. Japan 30 (1978), 687–696.

[Sh]   P. Shanahan, The Atiyah–Singer Index Theorem, Lecture Notes in Math., Vol 638, Springer, New York, 1978.

[Si]   I.M. Singer, Some remarks on operator theory and index theory, in K-theory and Operator Algebras, Proc. Georgia Conference 1975, Lecture Notes in Math., Vol 575, Springer, New York, 1977.

[St]   N. Steenrod, The Topology of Fibre Bundles, Princeton U. Press, Princeton, 1951.

[Su]   D. Sullivan, Cycles and the dynamical study of foliated manifolds, Invent. Math. 36 (1976), 225–255.

[Ta]   H. Takai, KK-theory of the C*-algebra of Anosov foliations, in Geometric Methods in Operator Algebras, H. Araki and E.G. Effros, ed., Kyoto (1983), Pitman Res. Notes in Math. Vol. 123, Wiley, NY, 1986.

[Tak1]   M. Takesaki, Structure of Factors and Automorphism Groups, C.B.M.S., No. 51, Amer. Math. Soc., Providence, Rhode Island, 1983.

[Tak2]   M. Takesaki, Theory of Operator Algebras I, Springer-Verlag, New York, 1979.

[Tay1]   M.E. Taylor, Pseudodifferential Operators, Princeton U. Press, Princeton, 1981.

[Tay2]   M. Taylor, Fourier integral operators and harmonic analysis on compact manifolds, Proc. Symp. Pure Math. XXXV, 1979, Part 2, 115–136.

[Tho]   R. Thom, Généralisation de la théorie de Morse aux variétés feuilletées, Ann. Inst. Fourier Grenoble 14 (1964), 173–190.

[Th1]   W. Thurston, The theory of foliations in codimension greater than one, Comm. Math. Helv. 49 (1974), 214–231.

[Th2]   W. Thurston, Existence of codimension-one foliations, Ann. of Math. 104 (1976), 249–268.

[To]   A.M. Torpe, K-theory for the leaf space of foliations by Reeb components, J. Functional Anal. 61 (1985), 15–71.

[V]   I. Vaisman, Cohomology and Differential Forms, Dekker,

New York, 1973.

[Var]   V.S. Varadarajan, Groups of automorphisms of Borel
          spaces, Trans. Amer. Math. Soc. 109 (1963), 191–220.

[We]   R.O. Wells, Differential Analysis on Complex Manifolds,
          Prentice-Hall, Englewood Cliffs, New Jersey, 1973.

[Wi]   H.E. Winkelnkemper, The graph of a foliation, Ann.
          Global Analysis and Geometry 1 (1983), No. 3, 53–75.

[Z1]   R.J. Zimmer, Algebraic topology of ergodic Lie group
          actions and measurable foliations, preprint.

[Z2]   R.J. Zimmer, Characteristic classes and smooth
          realizations of measurable foliations, preprint.

[Z3]   R.J. Zimmer, Ergodic actions of semisimple groups and
          product relations, Ann. of Math. 118 (1983), 9–19.

[Z4]   R.J. Zimmer, Ergodic Theory and Semisimple Groups,
          Boston, Birkhauser, 1984.